STANDARD BOILER ROOM
QUESTIONS & ANSWERS

RELATED BOOKS OF INTEREST

Standard Basic Math and Applied Plant Calculations (1978)
Stephen M. Elonka

Standard Plant Operators' Manual (3d ed., 1980)
Stephen M. Elonka

Standard Refrigeration and Air Conditioning Questions and Answers (2d ed., 1973, reprint 1980)
Stephen M. Elonka and Quaid W. Minich

Standard Boiler Operators' Questions and Answers (1969, reprint 1978)
Stephen M. Elonka and Anthony L. Kohan

Standard Industrial Hydraulics Questions and Answers (1967, reprint 1979)
Stephen M. Elonka and Orville H. Johnson

Standard Electronics Questions and Answers (1964)
Stephen M. Elonka and Julian L. Bernstein

Standard Instrumentation Questions and Answers (1962, reprint 1979)
Stephen M. Elonka and Alonzo R. Parsons

Standard Plant Operator's Questions & Answers (2d ed., 1981)
Stephen M. Elonka and Joseph F. Robinson

Electrical Systems and Equipment for Industry (1978)
Arthur H. Moore and Stephen M. Elonka

Marmaduke Surfaceblow's Salty Technical Romances (1979)
Stephen M. Elonka

Plant Energy Systems: Energy Systems Engineering (1967)
The Editors of Power

Mechanical Packing Handbook (1960)
Stephen M. Elonka and Fred Luhrs

STANDARD BOILER ROOM QUESTIONS & ANSWERS

A basic reference on the operation and maintenance of boilers,
complete with license examination questions and answers,
and a math and science refresher chapter.

STEPHEN MICHAEL ELONKA

Contributing Editor, Power *magazine; Licensed Chief Marine Steam Engineer,
Oceans; Unlimited Horsepower; Licensed as Regular Instructor of Vocational High
School, New York State; Member, National Association of Power Engineers (life,
honorary); National Institute for the Uniform Licensing of Power Engineers, Inc.,
Honorary Chief Engineer.*

ALEX HIGGINS

*Late Consulting Engineer; Director, Engineering Extension Courses, Provincial Institute of
Technology and Art, Calgary, Alberta; Member, American Society of Mechanical
Engineers, Engineering Institute of Canada, Alberta Association of Professional
Engineers; Licensed Chief Engineer*

Third Edition

McGRAW-HILL BOOK COMPANY

**New York   St. Louis   San Francisco   Auckland   Bogotá
Hamburg   Johannesburg   London   Madrid   Mexico
Montreal   New Delhi   Panana   Paris   São Paulo
Singapore   Sydney   Tokyo   Toronto**

Library of Congress Cataloging in Publication Data

Elonka, Stephen Michael.
  Standard boiler room questions & answers.

  Revision of: Boiler room questions and answers
Alex Higgins, Stephen Michael Elonka.   2nd ed.
  1. Steam-boilers—Miscellanea.   I. Higgins, Alex.
II. Higgins, Alex.   Boiler room questions and
answers.   III. Title.
TJ289.E398   1982   621.1′6′076   81-3795
                                          AACR2

ISBN 0-07-019301-0

Previous editions published under the title *Boiler Room Questions and Answers.*

The editors for this book were Robert L. Davidson, Frank J. Cerra,
Susan Sexton, and Margaret Lamb, and the production supervisor was Teresa F. Leaden.
It was set in Baskerville by Progressive Typographers.

Dedicated to

the boiler operator, licensing examiner, inspector, and all others involved with boilers who must keep power and process services flowing efficiently, safely, and pollution free.

CONTENTS

Preface   ix

1. MALE AND FEMALE JOB OPPORTUNITIES
   IN THE UNITED STATES AND CANADA   1

2. HOW TO PASS EXAMINATIONS   55

3. MATH AND SCIENCE REFRESHER   61

4. FUELS AND HOW THEY BURN   83

5. BURNING COAL, OIL, AND GAS   97

6. FLUE-GAS ANALYSIS, DRAFT, AND
   DRAFT CONTROL   124

7. HEAT, HOW TO USE STEAM TABLES   151

8. BOILERS (STEAM GENERATORS)   177

9. TEMPERATURE CONTROL OF STEAM,
   FEEDWATER, AND DRAFT   199

10. TRIM FOR STEAM GENERATORS   206

11. FEED SYSTEMS AND BLOWOFF ACCESSORIES   225

12. AUTOMATIC COMBUSTION CONTROL   244

13. STAYS, JOINTS, OPENINGS, AND CALCULATIONS . . . 255

14. CONSTRUCTION MATERIALS, WELDING,
    AND CALCULATIONS 276

15. INJECTORS AND PUMPS 294

16. PUMPING THEORY AND CALCULATIONS 322

17. PIPING, ACCESSORIES, AND CALCULATIONS 336

18. FEEDWATER TREATMENT 364

19. OPERATION, INSPECTION, AND REPAIRS 383

20. BOILER-ROOM MANAGEMENT 412

    APPENDIX 418

    Index 423
    About the Authors 435

PREFACE

Thousands in the United States and Canada and in many other countries have been helped to qualify for higher-paying jobs by the second edition of this popular book, a solid basic review of typical examination questions on boiler room operation, written in clear and concise language that is easily understood by all practical power plant operators.

Though the primary purpose of this third edition is to assist those who are studying for licenses (certificates) as firemen and operating engineers, it also provides an excellent basic reference to help the plant operator generate steam efficiently and solve problems that may come up in a plant.

To this third edition we have added two new chapters: (1) Math and Science Refresher and (2) Temperature Control of Steam, Feedwater, and Draft. Important information on burning fuel oil and coal, automatic combustion control, and also more on instrumentation has been added. Entirely new is "how to calculate steam costs."

The fuel oil crisis today is again making coal-fired boilers a necessity in many localities. Thus we retain the basic information on coal-burning designs. But new material has been included on burning coal, fuel oil, and gas more efficiently. And because of strict air pollution laws today, information on ways to maintain a "clean stack" and on *safe* operating methods are included. To assist the reader in acquiring a well-rounded technical background, suggested reading material has again been added at the end of each chapter.

"Why do boilers explode, although fully protected by safety controls and interlocks?" we asked leading boiler insurance inspectors. The answers in Chapter 12 should not only be known by every owner of boilers, but should be posted in every boiler room.

The first edition of this volume appeared serially in *Power* magazine, where the editors were extremely helpful. Thanks are now due to the many manufacturers of standard engineering equipment who furnished illustrations and data, and also to various machinery insurance firms.

Our deep thanks go to the many license examiners in the United States and Canada for furnishing sample questions and making helpful suggestions to enable the applicant to obtain a passing grade on his examination.

The late   *Stephen Michael Elonka*
Alex Higgins   *Charleston, S.C.*
                                                     1981

Although the masculine pronoun has been used throughout the book to avoid awkwardness, Chapter 1, "Male and Female Job Opportunities in the United States and Canada," makes it clear that no sexism is intended.

STANDARD BOILER ROOM
QUESTIONS & ANSWERS

1

MALE AND FEMALE JOB OPPORTUNITIES IN THE UNITED STATES AND CANADA

Today, of about 20,000 active licenses for operators in the state of Ohio alone, approximately 150 are issued to women. One woman at an electric generating station holds a first class engineer's license, two operate boilers in large steel mills, and another is head custodian for a school district. Women train to take the National Board examination for boiler inspectors, and machinery insurance companies employ women inspectors.

In New Jersey, more than 100 licensed women have jobs from boiler operator to custodian of low-pressure boilers, and right up to steam-turbine tenders in large electric utilities. And each year, more women are being licensed in the United States and in Canada.

Steam today is also generated in nuclear reactors, in which the reactor is the steam generator (boiler), the remainder of the plant being a conventional steam plant. In fact, at least one state (Massachusetts) now requires nuclear plant operators to have a steam license in addition to the Nuclear Regulatory Commission's (formerly Atomic Energy Commission's) license.

At this writing (1980) there are 69 nuclear reactors in the United States, and 1000 are projected to be supplying energy by the year 2000.

Operating steam generating plants offers a bright future for any young person who is mechanically inclined and is ambitious enough to qualify for the job classifications listed in this chapter, in which we show the job requirements in the United States and Canada. And because steam power plants are basically similar, whether ashore or afloat, we also include merchant marine license opportunities in the United States. (Indeed, many of the NRC-licensed operators received their nuclear training aboard nuclear submarines in the United States Navy.)

From Tables 1-1, 1-2, and 1-3 the reader can quickly see what is required to qualify for any job from fireman right up to chief engineer, and also the capacity of the plant each license entitles the holder to operate.

Understanding boilers and their operation is the *first* step for the embryo stationary or marine engineer, who may not have a technical education. Thus this chapter provides basic information guiding the reader to an interesting career that often leads to travel and higher pay.

NOTE: Please keep in mind that this is one of the few fields where a person with only a grammar school education, but *willing* to work and to study, can advance from boiler operator right up to chief engineer.

TABLE 1-1 United States License Requirements for Boiler Operators and Stationary Engineers

States and cities	Class license	Education, experience, and remarks	Plant capacity for class license
Alabama	No state license		
Mobile Board of Engineers Examiners, City Hall, Mobile, 36602	1. Fireman (special permit)	Age 21, 12 months practical experience	One boiler only, up to 50 psi
	2. Wide-open engineer	Same as above	Unlimited
Alaska Dept. of Labor, 650 West International Airport Rd, Anchorage, 99502	1. Fireman apprentice	Depends on examiner's opinion	Under supervision of licensed operator
	2. Boiler operator, third class	Same as above	Up to 3500 lb/hr
	3. Boiler operator, second class	Same as above	Up to 100,000 lb/hr
	4. Boiler operator, first class	Same as above	Unlimited
Arizona	No state license		
Arkansas Boiler Inspection Div., Capitol Hill Bldg., Little Rock, 72203	1. Boiler operator, low-pressure	6 months on-the-job training	Up to 15 psi steam, or 30 psi water for hot-water boiler
	2. Boiler operator, high-pressure	6 months on-the-job training	Unlimited

TABLE 1-1 United States License Requirements for Boiler Operators and Stationary Engineers (Continued)

States and cities	Class license	Education, experience, and remarks	Plant capacity for class license
California	No state license		
Los Angeles: Elevator & Pressure Vessel Div., City Hall, Los Angeles, 90012	1. Auxiliary engine operator	Age 19; 3 months as oiler, wiper; or 90-day training course, 60 days of which is operating, oiling, wiping engines or turbines	Work under licensed operating engineer
	2. Steam engineer	Age 21; 1 year as steam engine operator; or 3 years as oiler, wiper, or auxiliary engine operator	50 hp or more
	3. Steam engineer	Age 21, 3 years as fireman, water tender, or assistant engineer	500 hp and up
	4. Steam engineer, unlimited	Age 21, 3 years as fireman, water tender, or assistant engineer; or 1 year in charge of steam plant	Unlimited
	5. Boiler operator	Age 21, no experience necessary	Up to 35 boiler hp
San Jose: Bureau of Fire Prevention, 476 Park Ave., San Jose, 95110	1. Special boiler operator	Age 18, sufficient experience to operate a specific boiler	Up to 30 boiler hp, but at only one station
	2. Fireman	Age 18, same as above	Work under licensed engineer
	3. Second-grade steam engineer	Age 21, 3 years operating and maintaining boilers, 2 years of which with high-pressure type	Up to 500 boiler hp unless under first-grade engineer
	4. First-grade steam engineer	Age 25, 5 years in charge of boilers, of which 3 years are with high-pressure type	Unlimited steam power plant

Colorado

No state license

Location	License	Requirements	Scope
Denver: Bldg. Inspection Dept., 1445 Cleveland Pl., Denver, 80202	1. Boiler operator, class B	Age 21, 3 years boiler operation or equivalent, 2 years high school	15 to 100 psi and 10 to 100 boiler hp
	2. Boiler operator, class A	Age 21, 3 years boiler operation or equivalent, high school graduate	Over 15 psi and 10 boiler hp for steam boiler, 250°F in entire system for hot-water boiler
	3. Stationary engineer, class A	Age 21, 4 years operating, maintaining boilers, refrigeration machinery; high school graduate	Unlimited steam plant, including refrigeration
Pueblo: City Engineer, Division of Inspection, 211 East D St., Pueblo, 80103	1. Boiler tender, class D	Age 21, experience not specified	Under supervision of licensed engineer
	2. Boiler operator, class C	Age 21, experience not specified	Low-pressure boiler over 300 ft² heating surface
	3. Operating engineer, class B	Age 21, 2 years of operation	Any boiler or steam engine except automatically fired boilers
	4. Chief engineer, class A	Age 21, 5 years operation of boilers and associated machinery	Unlimited steam plant

Connecticut

No state license

Location	License	Requirements	Scope
Bridgeport: Power Engineers Board of Examiners, City Hall, Bridgeport, 06115	1. Low-pressure boiler operator	Age 21; 6 months as boiler or water tender, or assistant to licensed engineer, boiler or water tender; or 6 months at technical school; or 1 year operating low-pressure boilers	Up to 15 psi

TABLE 1-1 United States License Requirements for Boiler Operators and Stationary Engineers (*Continued*)

States and cities	Class license	Education, experience, and remarks	Plant capacity for class license
	2. Boiler operator or water tender	Age 21; 1 year as boiler or water tender, oiler, or assistant to licensed operator or boiler or water tender of plants over 15 psi or 25 boiler hp; or 1 year of technical school and 6 months operating steam boilers	Over 25 boiler hp and 15 psi
	3. Power engineer	Age 21; 4 years as boilermaker or machinist and 1 year in steam boiler plants; or 2 years engineering school and 1 year in steam power plant; or holder of state or U.S. operating engineer license; or 3 years boiler or water tending and work with boilers and steam engines	Unlimited capacity
Delaware	No state license		
Wilmington: Board of Examining Engineers, Public Bldg., Wilmington, 19801	1. Fireman, Red Seal	Age 18, 6 months assisting fireman or engineer with boilers over 15 psi	Boilers of over 15 psi
	2. Engineer, third-class, Red Seal	Age 21, 1 year operating, maintaining steam power plants	Unlimited
	3. Engineer second-class, Blue Seal	2 years experience operating and maintaining steam power plants	Unlimited
	4. Engineer first-class, Gold Seal	5 years experience operating and maintaining steam power plants	Unlimited
	5. Hoisting engineer ID card	Age 21, 1 year operating hoisting engines (steam)	Unlimited

District of Columbia Occupational & Professional Licensing Division, 614 H St., N.W., Washington, 20001	1. Steam and hot water heating boiler engineer, class 6	Age 21, 1 year of low-pressure boiler operating experience other than a miniature boiler	Up to 15 psi steam but not over 200 hp steam or hot water
	2. Steam engineer, class 5, high-pressure boilers	Age 21, 6 months experience in operation of high-pressure steam boilers	Up to 25 hp but not over 125 psi
	3. Steam engineer, class 4, high-pressure boilers	Age 21, 6 months experience in operation of high-pressure steam boilers	Up to 75 hp but not over 125 psi
	4. Steam engineer, class 3, high-pressure and low-pressure steam and hot water heating boilers	Age 21, 2 years high-pressure boiler operating experience, or 4 years of low-pressure boiler operating experience (1 year allowed for mechanical engineer degree)	Up to 400 hp
	5. Steam engineer, class 2	Age 21, in addition to third-class experience, 1 additional year of high-pressure boiler operating experience, or 3 years of high-pressure boiler operating experience if unlicensed	Up to 750 hp
	6. Steam engineer, class 1	Age 21, in addition to third-class experience, 2 additional years of high-pressure boiler operating experience if unlicensed	Unlimited horsepower

TABLE 1-1 United States License Requirements for Boiler Operators and Stationary Engineers (*Continued*)

States and cities	Class license	Education, experience, and remarks	Plant capacity for class license
Florida	No state license		
Tampa: Boiler Bureau, 301 N Florida Ave., Tampa, 33602	1. Fireman, low-pressure	Age 21, no experience required if can pass examination	Take charge, operate up to 15 psi boiler
	2. Fireman, high-pressure	Age 21, 6 months firing or assisting first- or second-class engineer	Take charge, operate 10- to 50-hp plant
	3. Engineer, third-class, hoisting and portable	Age 21, at least 6 months operation of hoisting and portable equipment	Take charge, operate portable boilers with engines and machinery other than boiler feed pump in operation and internal-combustion engines
	4. Engineer, second-class steam	Not specified	Take charge, operate steam plants up to 175 hp
	5. Engineer, second-class steam/refrigeration	Age 21, at least 2 years as oiler, fireman, or assistant under licensed first-class engineer; or graduate of approved technology school with 1 year experience as oiler, fireman, or assistant under licensed first-class engineer	Take charge, operate steam plants up to 175 hp, refrigeration and internal-combustion engines
	6. Engineer, first-class steam	Not specified	Take charge of any steam plant of any capacity or pressure
	7. Engineer, first-class unlimited	Age 21; (*a*) 3 years as oiler or assistant to licensed first-class engineer; or (*b*) 3 years firing or assistant engineer of steamboat or steam locomotive with 1 year as assistant to first-class engineer;	Take charge, operate any type of steam plant, all refrigerating or generating machinery driven by steam, diesel engines, internal combustion, or electricity

Jurisdiction	License	Requirements	Scope
Georgia	No state license	or (c) 2 years machinist or boilermaker plus apprentice time and 2 years as assistant to first-class engineer; or (d) mechanical engineer degree and 1 year under licensed first-class engineer; or (e) 2 years as second-class engineer while holding second-class engineer's license; or (f) holder of first- or second-grade license issued by U.S. government or U.S. Coast Guard	
Guam	No license		
Hawaii	No state license		
Idaho	No state license		
Illinois	No state license		
Chicago: Boiler & Pressure Vessel Inspection, 121 North LaSalle St., Chicago, 60602	1. Boiler tender	Age 21; proof of familiarity with boiler operation and construction	All boilers over 10 psi
	2. Water tender	Same as above	Same as above
	3. Engineer	Age 21; 2 years as machinist or engineer related to boilers, steam engines	Same as above
Decatur: Board of Examiners, Steam Engineers, Decatur, 62521	1. Second-grade engineer	Age 21, 6 months with boilers, steam equipment	Boilers over 15 psi, up to 75 boiler hp
	2. First-grade engineer	Age 21, 2 years in charge of steam boilers, engines	Unlimited

TABLE 1-1 United States License Requirements for Boiler Operators and Stationary Engineers (*Continued*)

States and cities	Class license	Education, experience, and remarks	Plant capacity for class license
East St. Louis: Boiler & Elevator Inspector, City Hall, East St. Louis, 66201	1. Engineer	3 years under licensed engineer in charge of power plant	All boilers over 10 boiler hp and 15 psi
Elgin: City Hall, Elgin, 60120	First-class engineer	Age 21, 1 year with steam-generating equipment	All boilers over 15 psi
Evanston: Building Dept., Municipal Bldg., Evanston, 60201	1. Water tender	Age 21, must prove familiarity with steam-generating equipment	All boilers, pressure vessels above 15 psi
	2. Engineer	Age 21, engineer or machinist with 2 years of managing, operating, or constructing boilers, steam engines, or ice machines	Unlimited
Peoria: City Boiler Inspector, City Hall, Peoria, 61602	1. Boiler tender	Age 21, 3 years in boiler room	All boilers over 15 psi
	2. Second-class engineer	Same as above	Same as above
	3. First-class engineer	Same as above	Same as above
Indiana	No state license		
Hammond: Board of Examiners, City Hall, Hammond, 46320	1. Water tender	Age 21	Any steam boilers
	2. Engineer	Age 21, 2 years in charge of steam boilers and engines—1 year if 2 years as machinist in steam-engine works or 1 year in steam plant if degreed engineer	Same as above

Location	License/Class	Experience	Capacity
Terre Haute: Board of Examiners, City Hall, Terra Haute, 47801	1. Third-class license	Age 21, 2 years firing steam boilers, or 1 year in charge of steam engines, turbines, boilers	Up to 50 boiler hp
	2. Second-class license	Same as above	Up to 200 boiler hp
	3. First-class license	Age 21, 5 years operating steam plants	Unlimited
Iowa			
Des Moines: Dept. Building Inspectors, City Hall, Des Moines, 50307	No state license		
	1. Fireman, second class	Age 18, 1 year helping fireman in boiler plant	Heating plant up to 50 boiler hp
	2. Fireman, first class	Age 18, 2 years as fireman or fireman's helper in boiler plant	Heating plant up to 75 boiler hp
	3. Third-class engineer	Same as above	Steam plant up to 125 boiler hp
	4. Second-class engineer	Age 18, 3 years in steam plant and knowledge of refrigeration, heating, ventilation and electrical apparatus	Steam plant up to 200 boiler hp
	5. First-class engineer	Age 18, 5 years in steam engineering or refrigeration plants, with experience in operating such plants	Unlimited
Sioux City: 6th and Douglas Sts., Sioux City, 51102	1. Fourth-class hydronic heating engineer	1 year operation, construction of steam engines, boilers, internal-combustion engines, turbines, etc., engineering technical school time credited	Low pressure of 51 to 100 hp, high pressure of 0 to 25 hp[a]

[a] Low pressure = not to exceed 15 psi steam, 30 psi water. High pressure = over 15 psi steam, 30 psi water. For operating steam and hot-water boilers (hydronic), horsepower of system = overall aggregate of hp of boiler(s) + all motor(s) or engine(s) connected to boiler(s). One boiler hp = firing rate of: 40 ft³ of natural gas per hr; or 0.285 gal No. 2 fuel oil per hr; or 4 lb of 10,000 Btu coal; or factory input rating on nameplate of boiler; 1 hp of refrigeration = 2546 Btu or 746 W electric output; 1 ton refrigeration = 12,000 Btu or 1 hp electric output; 1 hp on boilers = 33,475 Btu output or 42,000 Btu input; 1 hp on boiler = 13.1 hp refrigeration; 1 hp boiler = 10 kW electric energy.

TABLE 1-1 United States License Requirements for Boiler Operators and Stationary Engineers (*Continued*)

States and cities	Class license	Education, experience, and remarks	Plant capacity for class license
	2. Third-class hydronic heating engineer	2 years operation, construction of steam–internal-combustion engines, boilers, turbines, etc., engineering technical school time credited	Low pressure of 101 to 200 hp, high pressure of 26 to 100 hp[a]
	3. Second-class hydronic heating engineer	3 years operation, construction of steam–internal-combustion engines, boilers, turbines, etc., engineering technical school time credited	Low pressure of 201 to 400 hp, high pressure of 101 to 200 hp[a]
	4. First-class hydronic heating engineer	5 years operation, construction of steam–internal-combustion engines, boilers, turbines, etc., engineering technical school time credited	Low pressure of 401 hp and up, high pressure of 201 hp and up[a]
Kansas	No state license		
Kentucky	No state license		
Covington: Examiner of Engineers, City Hall, Covington, 41011	1. Second-class engineer	Not specified	All boilers over 10 psi
	2. First-class engineer	Not specified	Same as above

[a]Low pressure = not to exceed 15 psi steam, 30 psi water. High pressure = over 15 psi steam, 30 psi water. For operating steam and hot-water boilers (hydronic), horsepower of system = overall aggregate of hp of boiler(s) + all motor(s) or engine(s) connected to boiler(s). One boiler hp = firing rate of: 40 ft³ of natural gas per hr; or 0.285 gal No. 2 fuel oil per hr; or 4 lb of 10,000 Btu coal; or factory input rating on nameplate of boiler; 1 hp of refrigeration = 2546 Btu or 746 W electric output; 1 ton refrigeration = 12,000 Btu or 1 hp electric output; 1 hp on boilers = 33,475 Btu output or 42,000 Btu input; 1 hp on boiler = 13.1 hp refrigeration; 1 hp boiler = 10 kW electric energy.

Louisiana	No state license		
New Orleans: Mechanical Inspection Section, City Hall, New Orleans, 70112	1. Hoisting and portable engineer	Age 18; any engineer passing tests who operates or is in charge of hoisting or portable equipment; boilers, steam or internal-combustion engines, cranes, derricks, hoists, air compressors; power-driven equipment	Unlimited portable equipment
	2. Special operator	Age 18, 2 years operating steam boilers or more than 30 gal water and up to 240 ft² heating surface, of up to 15 psi	Act as assistant under third-class engineer, or in charge of boilers of up to 30 gal of water and up to 120 ft² heating surface, of any pressure; refrigeration systems from 5 to 20 hp
	3. Third-class operating engineer	Age 18; 2 years operating steam engines or boilers from 20 to 50 boiler hp; or at least 2 years apprentice engineer with steam engines or boilers of first-, second-, or third-class hp rating; or hold mechanical engineer degree	Act as assistant under second-class engineer, or in charge of boilers up to 75 hp, steam engines up to 75 hp, internal-combustion engines up to 75 hp, refrigeration and air-conditioning systems totaling 75 hp
	4. Second-class operating engineer	Age 18, 3 years operating steam engines or boilers from 51 to 150 boiler hp, or 6 months in steam plants while holding mechanical engineer degree	Act as assistant under first-class engineer, or in charge of boilers up to 150 hp, steam engines up to 150 hp, internal-combustion engines up to 150 hp, refrigeration and air-conditioning systems totaling up to 150 to 200 hp, depending on refrigerant or system design
	5. First-class operating engineer	Age 18, 4 years as operating engineer with 150 hp boiler or over, or 1 year in steam plants while holding mechanical engineer degree	Unlimited boilers, steam engines, internal-combustion engines, refrigeration and air-conditioning systems

TABLE 1-1 United States License Requirements for Boiler Operators and Stationary Engineers (*Continued*)

States and cities	Class license	Education, experience, and remarks	Plant capacity for class license
Maine: Boiler Rules & Regulations, State Office Bldg. Annex, Augusta, 04330	1. Boiler operator, low-pressure, heating	Any engineer or fireman who has operated in steam plants for 1 year	Heating plant with steam boilers up to 15 psi, or hot-water boilers up to 160 psi on 250°F, or both in schools or municipally owned buildings
	2. Boiler operator	6 months experience in steam power plants	In charge of heating plant up to 20,000 lb/hr, or operate in any plant under engineer licensed for that plant
	3. Fourth-class engineer	High school graduate or equivalent and 1 year operating under licensed engineer in charge of plant	50,000 lb/hr plant, or operate in any plant under engineer licensed for plant
	4. Third-class engineer	1 year operating on fourth-class license	100,000 lb/hr, or operate in plant under engineer licensed for plant
	5. Second-class engineer	2 years operating on third-class license	200,000 lb/hr, or operate in any plant under engineer licensed for plant
	6. First-class engineer	2 years operating on second-class license	Unlimited capacity
Maryland Department of Licensing & Regula- tions, 203 E. Balti- more St., Baltimore, 21202	1. Fourth-grade engineer	Age 21, 2 years with steam boilers and engines	Hoisting or portable boilers
	2. Third-grade engineer	Same as above	Up to 30 boiler hp
	3. Second-grade engineer	Same as above	Up to 500 boiler hp

Massachusetts Board of Boiler Rules, 1010 Commonwealth Ave., Boston, 02215		
4. First-grade engineer	Same as above	Unlimited
1. Second-class fireman	Not specified	Any capacity boiler under first-class fireman or licensed engineer
2. First-class fireman	6 months operating boilers while holding second-class fireman license, or 1 year as engineer or fireman operating boilers	Boilers with safety valves set not over 25 psi, or operate any boilers under licensed engineer or higher grade fireman
3. Portable-class engineer	Not specified	Any portable boilers and engines
4. Special engineer	Not specified	Up to 250 boiler hp and up to 50-hp engine in a specific plant
5. Fourth-class engineer	Not specified	Portable boilers and steam engines
6. Third-class engineer	Operating for 1½ years on steam engine or fireman license	In charge of up to 500 boiler hp and up to 50 hp engines or turbines
7. Second-class engineer	Not specified	In charge of up to 150 boiler hp engines or turbines, or operate in first-class plant under licensed engineer
8. First-class engineer	3 years in charge of plant with one 150-hp engine, or 1½ years in second-class or first-class plant while working on second-class license	Unlimited steam plant

TABLE 1-1 United States License Requirements for Boiler Operators and Stationary Engineers (*Continued*)

States and cities	Class license	Education, experience, and remarks	Plant capacity for class license
	9. Assistant nuclear power plant operator	1 year in steam plant as fireman, water tender, control room assistant to engineer in charge of 8-hr-day shift (Massachusetts is the only state thus far requiring steam license in nuclear plant)	Any nuclear plant; operate auxiliaries, control room and/or related systems under supervision of nuclear power plant operating engineer, or nuclear power plant senior supervising engineer[a]
	10. Nuclear power plant operating engineer	1½ years in steam or nuclear-steam power plant on second-class engineer license or assistant nuclear steam license; 1 year of nuclear plant engineering school equivalent to assistant nuclear operator license, but applicant must complete 6 months operating as assistant nuclear-steam power plant operator, or have B.S. in engineering and 1 year in steam plant	Any nuclear plant; shift supervisor in charge of nuclear power plant, its auxiliaries, prime movers, related control systems
	11. Nuclear power plant senior supervising engineer	2 years in steam plant on Massachusetts first-class engineer or nuclear power plant engineer license; 1 year of nuclear power plant engineering courses equivalent to nuclear power plant operating engineer experience	Any nuclear plant; in full charge of entire nuclear power plant, prime movers, auxiliaries, control systems as the designated shift supervisor
Michigan	No state license		
Dearborn: Bldg. & Safety Division, 4500 Maple, Dearborn, 48126	1. Low-pressure boiler operator	1 year experience	Low-pressure boilers, unlimited. All boilers of over 300 ft² heating surface and over 15 psi must have licensed operator on duty

[a] This license does not entitle holder to operate plant using fossil fuel as major source of heat energy.

License class	Experience	Scope
2. High-pressure boiler operator	2 years experience	Unlimited hp or high-pressure boiler
3. Fourth-class engineer	2 years experience	Portable equipment operator
4. Third-class engineer	3 years experience[a]	Limited to aggregate 1000 ft² boiler heating surface and prime movers to 100 hp
5. Second-class engineer	4 years operating experience or M.E. degree and 1 year on high-pressure boilers or prime movers[a]	High-pressure boiler up to 2500 hp (aggregate) and 250-hp prime movers
6. First-class engineer	5 years experience as stipulated in by-laws[a]	Unlimited boilers capacity and/or prime movers
7. Turbine and reciprocal engine operator	4 years experience with turbines and engines	Unlimited horsepower of prime movers
Detroit: Safety Engineering, Examination Div., City-County Bldg., Detroit, 48226		
1. Miniature boiler operator	Age 18, 1 month operating high-pressure boilers	Up to 16-in. inside-diameter boiler shell, 5 ft³ gross volume of furnace and 100 psi gage
2. Low-pressure boiler operator	Age 18, 1 year with high-pressure boilers and steam prime movers	Low-pressure plants up to 5000 ft² boiler heating surface
3. High-pressure boiler operator	Age 18, 2 years with low- or high-pressure boilers, or 1 year on low-pressure license	Boiler plants of all pressures, but not over 4000 ft² heating surface, and engine turbine up to 10 hp
4. Portable steam equipment operator	Age 19, 1 year with high-pressure boilers or steam prime movers	Steam locomotives and portable boilers to 2000 ft² heating surface: portable steam equipment to 150 hp

[a]Varying credits given to mechanical engineers or registered engineers, or for approved steam apprentice programs. Experience may be waived on documentary evidence of equivalent education.

TABLE 1-1 United States License Requirements for Boiler Operators and Stationary Engineers (*Continued*)

States and cities	Class license	Education, experience, and remarks	Plant capacity for class license
	5. Third-class stationary engineer	Age 20; 1 year on high-pressure boiler operator license; or 1 year on low-pressure boiler license and 1 year with high-pressure boiler; or 1 year in high-pressure boiler plant of over 4000 ft² heating surface on high-pressure license; or 3 years in high-pressure boiler plant of over 4000 ft² heating surface; or 1 year in high-pressure boiler plant of over 4000 ft² plus 3 years with steam prime movers of over 10 hp; or 3 years in high-pressure plant up to 4000 ft² heating surface	Boilers up to 7500 ft² heating surface and up to 100 steam engine-turbine hp
	6. Second-class stationary engineer	Age 21; 1 year on third-class stationary engineer license; or M.E. or E.E. degree and 1 year in steam-electric power plant; or 4 years in high-pressure boiler plant of 7500 ft² heating surface; or 1 year in high-pressure boiler plant of over 7500 ft² plus 4 years with steam prime movers over 100 hp; or 1 year in high-pressure boiler plant of over 7500 ft² plus 4 years in high-pressure boiler plant of over 4000 ft² heating surface	Boiler plants up to 20,000 ft² heating surface and steam engine-turbine up to 200 hp
	7. First-class stationary engineer	Age 22, 6 years in high-pressure boiler plant of over 20,000 ft² heating surface plus 6 years with prime movers of over	Unlimited capacity

		200 hp, or 2 years with high-pressure boilers of over 20,000 ft² plus 6 years with high-pressure boilers of over 7500 ft²	All boilers of 10 boiler hp and 15 psi must have licensed operator[a]
Grand Rapids: Inspection Services, City-County Bldg., Grand Rapids, 49502	1. Boiler operator	1 year with high-pressure boiler	All boilers of 10 boiler hp and 15 psi must have licensed operator[a]
	2. Boiler engineer	3 years as boiler operator under licensed boiler engineer	Same as above
Saginaw: Board of Examiners, Stationary Engineers, City Hall, Saginaw, 48602	1. Fireman, limited-horsepower	Age 21, grammar school education and 2 years firing boilers	Same as above
	2. Fireman, unlimited-horsepower	Same as above	Same as above
	3. Engineer, low-pressure	Same as above	Same as above
	4. Engineer, high-pressure	Same as above	Same as above
Minnesota Division of Boiler Inspection, 444 Lafayette Rd., St. Paul, 55101	1. Fourth-class engineer[b]	Age 18, must satisfy inspector that applicant can safely operate low-pressure boiler of up to 30 boiler hp	All boilers require licensed operators except those in private residence or apartment houses under five families
	2. Third-grade engineer	Age 18, 6 months with steam boilers up to 30 boiler hp	Up to 30 boiler hp
	3. Second-class engineer, grade C	Age 19, 1 year with low-pressure boilers of up to 100 boiler hp	Up to 100 boiler hp

[a] No plant capacity specified.
[b] Shift engineers can operate a shift up to one grade license higher than required for engineer in charge of plant.

TABLE 1-1 United States License Requirements for Boiler Operators and Stationary Engineers (*Continued*)

States and cities	Class license	Education, experience, and remarks	Plant capacity for class license
	4. Second-class engineer, grade B	Age 20, 1 year with steam boilers of up to 100 boiler hp	Up to 100 boiler hp
	5. Second-class engineer, grade A	Age 21, 1 year with all classes steam boilers, steam engines, or turbines	Up to 100 hp steam boiler, engine, or turbine
	6. First-class engineer, grade C	Age 21, 3 years with all classes low-pressure steam plants	Up to 300 boiler hp, low-pressure
	7. First-class engineer, grade B	Age 21, 3 years with all classes steam boiler plants	Up to 300-boiler-hp plants
	8. First-class engineer, grade A	Age 21, 3 years with all classes steam boilers, engines or turbines	Up to 300-boiler-hp plants, engines, turbines
	9. Chief engineer, grade C	Age 21, 5 years with all classes low-pressure steam plants	All classes low-pressure boilers
	10. Chief engineer, grade B	Age 21, 5 years with all classes steam boilers	All classes steam boilers
	11. Chief engineer, grade A	Age 21, 5 years with all classes steam boilers, engines or turbines	Unlimited steam plant
Mississippi	No state license		
Missouri	No state license		
Kansas City: Codes Administration Division, City Hall, Kansas City, 64106	1. Plant fireman	Age 21, Letter from employer stating ability to operate boiler plant equipment safely	Limited to designated plant or system

License type	Requirements	Scope
2. Fireman	Age 21, 1 year experience with boilers, must satisfy examiner of safe operation of boiler and fuel burning equipment	Boilers over 15 psi and up to 126 psi
3. Plant operating engineer	Age 21, 3 years with fuel-burning equipment, or 2 years with fuel-burning equipment while holding M.E. degree from school accredited by the state of Missouri	In charge of plant specified on license
4. Operating engineer	Age 21, 3 years with all kinds of power plant equipment, know ASME codes for power plant equipment, or M.E. degree from state of Missouri school and 1 year in power plants	Unlimited steam plants
St. Joseph: Dept. of Public Works, City Hall, St. Joseph, 64501		
1. Special permit	Prove to examiner competence and experience to have charge and operate power, heating, and refrigeration equipment	Only specified equipment under first-class licensed operator
2. Heating engineer	Same as above	All boilers of 15 psi or less for heating purposes only
3. Second-class engineer	1 year practical experience in boiler room or power plant or equipment	All boilers of 15 psi or more
4. First-class engineer	3 years practical experience in boiler room or power plant, or equivalent	Unlimited steam boilers, pressure vessels, refrigeration and air-conditioning units
St. Louis: Dept. of Public Safety, City Hall, St Louis, 63103		
1. Boiler operator	Not specified	Boilers generating saturated steam of 15 psig minimum to 125 psig maximum, each boiler of over 100 ft² rated heating surface

States and cities	Class license	Education, experience, and remarks	Plant capacity for class license
	2. Class 2 stationary engineer	Age 19, 1 year operating boilers under Class 1 or Class 2 engineer, or 1 year in maintenance work on steam boilers and/or engines or turbines or on direct-fired natural gas or manufactured gas engines of over 200 brake hp, and/or ammonia compressors of over 50 tons, or registered M.E. or engineer in training	Boilers up to 1500 ft² heating surface from 15 to 300 psig, or hot water or other liquid defined in code. Also operate associated compressor, ammonia compressors, pump and feed-water heaters, electric generators, etc.
	3. Class 1 stationary engineer	Age 21, 2 years on Class 2 license, or registered M.E. or engineer in training, and employed in engineering or research division of power plant for 12 months	Unlimited capacity
Montana Dept. of Labor & Industry, 815 Front St., Helena, 59601	1. Traction engineer	Age 18, 6 months as assisting traction engineer	Any steam traction unit
	2. Low-pressure engineer	Age 18, 3 months operating low-pressure boiler	Up to 15 psi steam, 50 psi hot-water boiler
	3. Third-class engineer	Age 18, 6 months with boilers in this classification	Up to 100 psi steam, 150 psi hot-water boiler
	4. Second-class engineer	Age 18, 2 years with boilers and steam-driven machinery in this classification under second- or first-class engineer, or hold third-class license and 1 year with steam units under second- or first-class engineer	Up to 250 psi steam, 375 psi hot-water boiler

5. First-class engineer	Age 18, 3 years with steam boilers and units under first-class engineer; or hold second-class license and 1 year with steam boilers under first-class engineer; or hold third-class license and 2 years under first-class engineer	Unlimited steam power plants

Nebraska No state license

Lincoln:
Board of Examiners, City-County Bldg., Lincoln, 68508

1. Fireman	Age 19, 1 year operating steam boilers	All boilers of 15 psi and over require licensed operator, but class license not specified
2. Third-grade engineer	Age 19, 1 year operating steam boilers	
3. Second-grade engineer	Age 19, 3 years operating steam boilers	
4. First-grade engineer	Age 19, 5 years operating steam boilers	

Omaha:
Dept. of Public Safety, City Hall, Omaha, 68102

1. Fireman	Age 21, 1 year with steam boilers under the licensed operator	Up to 15 psi and heating surface up to 750 ft^2
2. Limited stationary engineer	Age 21, 1 year with steam boilers under the licensed operator of like or higher grade	Up to 100-hp steam prime movers
3. Third-grade engineer	Same as above	Any steam boiler plant but no prime mover
4. Second-grade engineer	Age 21, 3 years with steam boilers under the licensed engineer of like or higher grade	Any steam power plant up to 100 boiler hp

States and cities	Class license	Education, experience, and remarks	Plant capacity for class license
	5. First-grade engineer	Age 21, 5 years with steam power and/or heating plants	Any steam-power, refrigeration, or compressor plant
Nevada	No state license		
New Hampshire	No state license		
New Jersey Trenton: Mechanical Inspection Bureau, Trenton, 08625	1. Boiler fireman, low-pressure	Age 17, 3 months fireman helper or coal passer	(a) Steam or hot-water boilers over 15 psi and 6 boiler hp, (b) steam or hot-water heating plant above 499-ft² heating surface, 1000 kW, or 4,000,000 Btu; (c) prime movers over 6 hp; (d) 6-ton system with flammable or toxic refrigerant
	2. Boiler fireman-in-charge, low pressure	Age 17, 3 months in boiler room under licensed operator	Act in any capacity in low-pressure heating plant
	3. Boiler fireman-in-charge, high pressure	Age 17, 6 weeks training by licensed operator; 6 months as low-pressure fireman reduces 6 weeks to 30 days	Act as chief of 500 boiler hp plant or assume shift under chief engineer of up to 1000 boiler hp plant
	4. Third-grade engineer (blue seal)	Age 18, 6 months while holding fireman-in-charge license in third-grade engineer plant, or as assistant in such plant	Act as chief engineer of up to 1,000 boiler hp, 100 engine hp, or 65 tons refrigeration, or operate in any plant under chief engineer
	5. Second-grade engineer (red seal)	Age 18, 1 year on third-grade license, or equivalent experience	Act as chief engineer of up to 3000 boiler hp, 500 engine hp, or 300 tons refrigeration, or operate in any plant under chief engineer

Jurisdiction	License	Requirements	Capacity
	6. First-grade engineer (gold seal)	Age 18, 1 year on second-grade license as supervising or chief engineer, or 2 years as operating engineer under first-grade engineer, or equivalent experience	Act as chief engineer of any steam generating capacity
	7. Nuclear engineer (gold, red, or blue seal)[a]	Hold certificate of U.S. Nuclear Regulatory Commission qualifying for operation of nuclear power plant equipment	
New Mexico	No state license		
New York	No state license		
Buffalo: Division Fuel Devices, City Hall, Buffalo, 14202	1. Second-class engineer	Age 21, 2 years as fireman, oiler, or helper on repairs of boilers and engines	30 to 100 boiler hp
	2. First-class engineer	Age 22, 1 year operating on second-class license	100 to 150 boiler hp
	3. Chief engineer, unlimited	Age 24, 3 years operating on first-class license	Unlimited plant capacity
Mt. Vernon: Boiler Inspector, City Hall, Mt. Vernon, 10500	1. Fireman	Age 21	Steam boiler or battery of boilers of 75 boiler hp (combined), or combined 825 ft² heating surface, 15 psi and evaporating 2586 lb of water/hr from 212°F
	2. Second-class engineer	Age 21	Same as above, except boiler hp of up to 150 instead of 75

[a] Gold seal designates a first-grade license; red seal a second-grade license; and blue seal a third-grade license. Black seal designates a boiler fireman. If "in charge," the license is so stamped on the face.

TABLE 1-1 United States License Requirements for Boiler Operators and Stationary Engineers (Continued)

States and cities	Class license	Education, experience, and remarks	Plant capacity for class license
	3. First-class engineer, unlimited	Age 21	Unlimited plant capacity
New York City: Bureau of Examinations, Civil Service Comm., Municipal Bldg., New York, 10013[a]	1. Fireman, with or without oil-burner endorsement	Age 21, 2 years as fireman, water tender, oiler, or assistant stationary or marine engineer	All boilers over 15 psi and 100 ft² heating surface require licensed operator, but class plant for license not specified
	2. Portable engineer	Age 21; 5 years as fireman, oiler, or assistant to licensed operating engineer in N.Y.C. within last 7 years; or 5 years as boilermaker or machinist engaged in prime mover repair or manufacturer, 1 year being under licensed engineer in N.Y.C. in last 3 years; or at least 1 year in stationary plant under N.Y.C. engineer while holding M.E. degree; or 1 year in stationary plant under recognized N.Y.C. engineer while holding any engineer's license issued by U.S. government, state, or jurisdiction	
	3. Third-grade engineer (with or without oil-burner endorsement)	Same as above	
	4. Second-grade engineer	Age 21, 2 years continuous work on third-grade N.Y.C. license	

[a] Administered by the Department of Buildings, License Division, Room 1521, 120 Wall Street, New York, NY 10005. New York City has a variety of requirements, and the Department of Buildings should be consulted for details.

Jurisdiction	License	Requirements	Scope
	5. First-grade engineer	Age 21, 1 year continuous work on second-grade N.Y.C. license	All steam boilers
Niagara Falls: Board of Examiners, Stationary Engineers, City Hall, Niagara Falls, 24302	1. First-class fireman	Age 19, 6 months as helper in boiler room	Not stipulated
	2. First-class engineer	Age 20, 1 year on fireman license	Not stipulated
	3. Chief engineer	Age 21, 3 years on first-class engineer's license	Any boiler or pressure vessel up to 100 boiler hp
Rochester: City Public Safety Bldg., Rochester, 14614	1. Third-class engineer	Age 21, 1 year in steam plant	Up to 500 boiler hp
	2. Second-class engineer	Age 21, 2 years in steam stationary plant	Up to 1500 boiler hp
	3. First-class engineer	Age 21, 3 years in stationary steam plant	Unlimited steam plant
	4. Chief engineer	Age 21, first-class license and 5 years in power steam plants	
Tonawanda: Smoke Abatement Office, City Hall, Tonawanda, 14150	1. Special engineer	Age 21, must prove capability of operating a specific power plant	Limited to specific plant, up to 30 boiler hp and or engine hp, up to 100 psi
	2. Second-class engineer	Age 21; 3 years firing boilers or operating engines, etc., or building power units; or technical school graduate	Up to 100 boiler hp
	3. First-class engineer	Age 21, 2 years on second-class license	Up to 225 boiler hp
	4. Chief engineer	Age 21, 3 years on first-class license	Power plant of any capacity

TABLE 1-1 United States License Requirements for Boiler Operators and Stationary Engineers (*Continued*)

States and cities	Class license	Education, experience, and remarks	Plant capacity for class license
White Plains: Fire Dept., Municipal Bldg., White Plains, 10601	1. Second-class engineer	Age 21, 2 years under second-class engineer, or equivalent	High-pressure plant of 30 to 100 hp
	2. First-class engineer	Age 21, 3 years under first-class engineer, 1 year holding second-class license	High-pressure steam plant of 100 to 500 hp
	3. Chief engineer	Age 21, 4 years under chief engineer in high-pressure plant, and 1 year holding first-class license	Any high-pressure steam plant
Yonkers: Board of Examiners, City Hall, Yonkers 10700	1. Fireman	Age 21	All boilers require licensed operators, but class plant for license not stipulated
	2. Portable engineer	Age 21	
	3. Second-class engineer	Age 21	Up to 500 boiler hp
	4. First-class engineer	Age 21	Unlimited
North Carolina	No state license		
North Dakota	No state license		
Ohio Division of Steam Engineers 2323 West Fifth Ave., Columbus, 43216	1. Low-pressure boiler operator	Age 18, 600 hr as boiler room attendant, operating, maintaining, or erecting boilers	All boilers over 30 boiler hp, but under 15 psi must have licensed operators; class plant for license stipulated

	License	Requirements	Scope
	2. High-pressure boiler operator	Age 18, 1200 hr as boiler room attendant, operating, maintaining, or erecting boilers	Same as above
	3. Stationary steam engineer, third class	Age 18, 1800 hr as engineer, oiler, fireman, or water tender of steam boilers	Same as above
	4. Stationary steam engineer second-class	Age 18, 3600 hr practical experience as operating stationary steam engineer	Same as above
	5. Stationary steam engineer first-class	Age 18, 5400 hr practical experience as operating steam engineer	Unlimited capacity
Oklahoma	No state license		
Oklahoma City: Boiler Inspector, City Hall, Oklahoma City, 73102	First-class engineer	Age 21, 2 years in steam plants under practical engineer	Unlimited capacity
Tulsa: Board of Examiners, 200 Civic Center, Tulsa, 74103	1. Fireman	1 year in steam plants	All heating boilers above 15 psi and 1,000,000 Btu, steam boilers above 15 psi and 100 ft² heating surface require licensed operator, but class plant on license not stipulated
	2. Third-class engineer	1 year in steam and refrigeration plants	
	3. Second-class engineer	3 years in steam and refrigeration plants	Up to 15-psi boilers and up to 75-ton refrigeration and air-conditioning plants
	4. First-class engineer	5 years in steam and refrigeration plants	Unlimited
Oregon	No state license		

States and cities	Class license	Education, experience, and remarks	Plant capacity for class license
Panama Canal Zone Cristobal, Canal Zone[a]	Yes, but not stipulated		
Pennsylvania	No state license		
Erie: Bureau of Licenses, Municipal Bldg., Erie, 16501	1. Class 3 water tender	Age 21, 1 year with steam-generating equipment, high school grad; tech school degree in lieu 1 year experience	Up to 100-hp steam-driven machinery, 2000 ft² heating surface boiler; any size power equipment under licensed engineer
	2. Class 2 stationary engineer	Age 21, 2 years holding water tender license	Up to 200-hp steam-driven machinery, 4000 ft² heating-surface boiler; any size power equipment under licensed chief engineer
	3. Class 1 chief stationary engineer	Age 21, 2 years holding class 2 license	Unlimited
Philadelphia: Bureau of Licenses & Inspections, Municipal Services Bldg., Philadelphia, 19107	1. Grade D fireman	Age 21, 2 years in boiler room	Over 15 psi, 30-boiler-hp steam prime movers; 25-ton refrigeration equipment
	2. Grade C portable and stationary engineer	Age 21, 2 years assistant engineer or helper	Not specified
	3. Grade A stationary engineer	Age 21, 2 years assistant engineer or helper	Unlimited

[a]The republic of Panama may have jurisdiction over licenses in the Canal Zone.

	License	Requirements	Scope
Pittsburgh: Bureau Bldg. & Inspections 100 Grant St., Pittsburgh, 15219	1. Fireman	Age 21, 2 years assistant engineer or helper	All boilers and pressure vessels require licensed operators, but license class for plant not specified
	2. Portable engineer	Age 21, 2 years as engineer, oiler, fireman of steam boilers	Not specified
	3. Stationary engineer	Not specified	Unlimited
Rhode Island	No state license		
Providence: Dept. Public Service, City Hall, Providence, 02903	1. Boiler operation, fireman, and water tender	Age 18, 6 months apprentice under licensed operator while holding permit	All boilers over 30 boiler hp but not over 100 hp refrigeration over 15 tons require licensed operator, but class license for plant not specified
	2. Stationary engine operator	Age 21; 5 years boiler operator; or 3 years assistant to licensed engineer; or 1 year in steam plants while holding permit if degreed engineer	Unlimited
Woonsocket: Licensing Steam Engineer, City Hall, Woonsocket, 02895	Boiler operator	Age 21, 6 months apprentice under licensed operator while holding permit	All boilers up to and including 150 boiler hp
South Carolina	No state license		
South Dakota	No state license		

TABLE 1-1 United States License Requirements for Boiler Operators and Stationary Engineers (*Continued*)

States and cities	Class license	Education, experience, and remarks	Plant capacity for class license
Tennessee	No state license		
Memphis: Safety Engineer Division, 125 N. Main St., Memphis, 38103	1. Third-grade engineer	Age 21, 3 years in steam plants	Up to 50 boiler hp, 50-hp compressor, 40-ton refrigeration, or assist second- or first-grade engineer
	2. Second-grade engineer	Same as above	Up to 100 boiler hp, 100-hp internal-combustion engine, 100-hp compressor, 75-ton refrigeration, or assist first-grade engineer
	3. First-grade engineer	Age 21, hold third- and second-grade licenses	Unlimited
Texas	No state license		
Houston: Public Works Department Bldg., Inspection Division, Air Condition & Boiler Section, Houston, 77001	1. Boiler operator	No prerequisite	Low-pressure hot-water-heating boilers, high-pressure steam boilers up to 1,004,000 Btu
	2. Third-grade engineer	2 years licensed engineer, oil, water tender, boiler repairman, fireman, or 6 months of above if degreed engineer	Up to 1,674,000 Btu boiler, or act as shift engineer of up to 6,696,000 Btu under second-grade engineer
	3. Second-grade engineer	3 years stationary engineer, oiler, water tender, fireman, if 1 of 3 years as licensed engineer, or of degreed engineer and 1 year of above	Up to 6,696,000 Btu boiler, or unlimited capacity under licensed operator

4. First-grade engineer	5 years stationary engineer, oiler, water tender, fireman if 3 of 5 years as licensed engineer, or if 3-year apprentice program completed need only 2 years as licensed engineer	Unlimited
Utah	No state license	
Salt Lake City: Power and Heating Div., City & County Bldg., Salt Lake City, 84111		
1. Second-class fireman	Age 18, 1 year in steam plants	Low-pressure boilers up to 50 boiler hp
2. First-class fireman	Age 18, 2 years in steam plants	Low- or high-pressure boilers, unlimited capacity
3. Second-class engineer	Age 20, 3 years in steam plants	Low- or high-pressure plant up to 100 hp
4. First-class engineer	Age 20, 4 years in steam plants	Any capacity above 100 hp
Vermont	No state license	
Virginia	No state license	
Washington	No state license	
Seattle: Dept. of Bldgs., 503 Seattle Municipal Bldg., Seattle, 98104		
1. Small power-boiler fireman	Same as fifth-grade fireman	Under 100 psi and 350,000 Btu/hr boilers
2. Grade 5, boiler fireman	1 year in care of boilers; or completion of in-service training course	Up to 15 psi and up to 500 ft² heating surface, 5,000,000 Btu/hr input boilers
3. Grade 4 boiler fireman	3 years of operation in steam power plants	Boilers of 20,000,000 Btu/hr input

TABLE 1-1 United States License Requirements for Boiler Operators and Stationary Engineers (*Continued*)

States and cities	Class license	Education, experience, and remarks	Plant capacity for class license
	4. Grade 3 steam engineer	3 years in steam plants; or 1 year in steam plant after serving 3 years in steam engine works; or 1 year in steam plants if degree from school of technology	Boiler capacity up to 50,000,000 Btu input
	5. Grade 2 steam engineer	Same as above	Boiler capacity up to 300,000,000 Btu/hr input and up to 1500 boiler hp
	6. Grade 1 steam engineer	Same as above	Unlimited
Spokane: Dept. of Buildings, City Hall, Spokane, 99201	1. Fireman	Age 18	Steam boiler up to 300 ft² heating surface or heating boiler up to 15 psi
	2. Third-class engineer	Age 21, 1 year as steam engineer	Up to 100 boiler hp, or shift engineer under second-class engineer
	3. Second-class engineer	Age 21, 1 year as steam engineer	Up to 200 boiler hp, or shift engineer under first-class engineer
	4. First-class engineer	Age 21, 2 years on steam license, or 1 year steam engineer while holding M.E. degree	Unlimited
Tacoma: Boiler Inspector, 930 Tacoma Ave. South, Tacoma, 98402	1. Provisional license (restricted)	License indicates applicant is not qualified, can be reexamined after 6 months	All restrictions noted on face of license

Class	Requirements	Scope
2. Class 5 small power boiler fireman	Produce satisfactory evidence of qualification	Boiler plant up to 440,000 Btu/hr input
3. Class 4 low-pressure boiler fireman	Same as above	Up to 5,000,000 Btu/hr input, up to 15 psi, or hot water boiler up to 160 psi of 240°F water temperature
4. Class 3 high-pressure boiler fireman	1 year as low- or high-pressure plant helper, trainee or apprentice for 2 years, or engineering/technology graduate and 3 months in boiler room	Unlimited under licensed engineer, or up to 20,000,000 Btu/hr input
5. Class 2 operating engineer	3 years operating power boilers, or engineering/technical graduate, or engineering apprentice with 2 years of power boiler operation	Unlimited under chief engineer, or up to 300,000,000 Btu/hr input
6. Class 1 chief operating engineer	5 years operating high-pressure boiler plants, or engineering graduate, or apprentice in training program with 3 years high-pressure boiler experience	Unlimited

West Virginia No state license

Wisconsin No state license

Kenosha:
Board of Examiners, City Hall, Kenosha, 53140

Class	Requirements	Scope
1. Third-class engineer	Age 21, 2 years in steam plants	Up to 50 boiler hp
2. Second-class engineer	Age 21, 2 years in steam plants	Up to 150 boiler hp
3. First-class engineer	Age 21, 2 years in steam plants	Unlimited

TABLE 1-1   United States License Requirements for Boiler Operators and Stationary Engineers (*Continued*)

States and cities	Class license	Education, experience, and remarks	Plant capacity for class license
Milwaukee: Dept. Bldg. Inspection & Safety Engineering, Municipal Bldg., Milwaukee, 53202	1. Low-pressure fireman	Age 21, 1 year in steam boiler plants	Operate up to 3 low-pressure plants (within 3-block area) of up to 10 boiler hp
	2. High-pressure fireman	Age 21, 1 year in steam boiler plants	Up to 75 boiler hp but no steam engines except boiler auxiliaries
	3. Third-class engineer	Age 21, 2 years in high-pressure steam plants	Up to 75 boiler hp and steam engines
	4. Second-class engineer	Age 21, 3 years in high-pressure steam plants	Up to 300 boiler hp
	5. First-class engineer	Age 21, 4 years in high-pressure steam plants	Unlimited
Racine: Stationary Engineer Examiner, City Hall, Racine, 53203	1. Third-class engineer	Age 19, course of instruction in steam boiler operation	Up to 75 boiler hp, or act as assistant to second-class engineer
	2. Second-class engineer	Age 21, 2 years in steam boiler plants	Up to 300 boiler hp, or act as assistant to first-class engineer
	3. First-class engineer	Age 21, 2 years in steam boiler plants	Unlimited steam-generating plant
Wyoming	No state license		

Source: Stephen M. Elonka survey of licensing agencies (1980).

TABLE 1-2   U.S. Merchant Marine QMED Ratings and Marine Engineer's Steam License Requirements

U.S. Coast Guard	Class rating or license	Education, experience for 4000-hp vessels and over[a]
Apply to: Merchant Marine Licensing, U.S. Coast Guard, any U.S. port	1. QMED (Qualified Member of Engine Department) *a.* Fireman *b.* Oiler *c.* Water tender *d.* Machinist *e.* Refrigerating engineer *f.* Electrician *g.* Deck engineer *h.* Junior engineer *i.* Boilermaker *j.* deck and engine mechanic *k.* Pumpman *l.* Engineman	QMED is any person below rating of licensed officer (engineer) and above rating of coal passer or wiper who holds certificate of service as QMED issued by Coast Guard. Must (*a*) speak English language, (*b*) present certificate of health from U.S. Public Health Service attesting that eyesight, hearing, and physical condition are adequate for performing duties required by a QMED. Medical examination is same as for an original license as an engineer. Applicant for QMED must have: (1) 6 months service at sea in a rating at least equal to that of coal passer or wiper in engine department of vessels having such certificated persons, or in engine department of tugs or towboats operating on high seas or Great Lakes, or bays or sounds directly connected with the seas; or (2) graduate from a schoolship approved by Coast Guard, or (3) courses of training approved by Coast Guard and service aboard a training vessel; or (4) graduate from U.S. Naval Academy or U.S. Coast Guard Academy. QMED applicants as *boilermaker* or *pumpman* must prove by examination they are qualified for rating. QMED applicants for deck engine mechanic or engineman must also prove by examination they are qualified for rating. An *engineman* must hold endorsements as fireman/water tender and oiler, or junior engineer—have 6 months service as junior engineer, fireman/water tender, or oiler on steam vessels of 4000 hp, or over, or proof from "partially automated" 4000-hp vessel, or over, of 2 weeks completion of training in

U.S. Coast Guard	Class rating or license	Education, experience for 4000-hp vessels and over[a]
		engine department. Any holder of a merchant mariner's document endorsed for any unlicensed rating in the engine department, QMED—any rating, or deck engine mechanic, is qualified as an engineman. Examination subjects required for QMED ratings are listed in Table 21-3
Apply to: U.S. Coast Guard, Merchant Marine Licensing, any U.S. port	1. Third assistant engineer	(1) 3 years in engine department of steam or motor ships, one-third may be on motor vessels; 2½ years must be as qualified member of engine department, 1 year 6 months must be as fireman, oiler, water tender, or junior engineer on steam ships; or (2) 3 years as machinist apprentice in building, repairing of marine, locomotive, or stationary engines, together with 1 year in engine department of steam ships as oiler, water tender, or junior engineer, one-third may be on motor vessels; or (3) engineering graduate from (*a*) U.S. Merchant Marine Academy; (*b*) nautical schoolship; (*c*) U.S. Naval Academy; or (*d*) U.S. Coast Guard Academy; or (4) graduate of U.S. government training schools acceptable as 4 months sea service if additional service obtained prior to enrollment; or (5) graduate from marine engineering or technology school with 3 months service in engine department of steam ships, one-third acceptable from motor vessels; or (6) graduate of mechanical or electrical engineering from technology school with 6 months in engine department of steam vessels; or (7) 1 year as oiler, water tender, or junior engineer on steam ships while holding license as third assistant engineer of motor vessels; or (8) completion of an approved 3-year apprentice engineering training program

U.S. Coast Guard	Class rating or license	Education, experience for 4000-hp vessels and over[a]
	2. Second assistant engineer	(1) 1 year as engineer in charge of watch, while holding license as third assistant engineer of steam vessels; or (2) 2 years service as assistant engineer to engineer in charge of watch, while holding license as third assistant engineer of steam vessels; or (3) 5 years in engine department of steam or motor vessels, 1 year of which may be on motor ships; 4 years and 6 months as qualified member of engine department, 2 years and 6 months as fireman, oiler, water tender, or junior engineer on steam vessels; or (4) while holding license as second assistant engineer of motor vessels, either (*a*) 6 months as third assistant engineer of steam ships; (*b*) 6 months as observer second assistant engineer on steam vessels; or (*c*) 1 year as oiler, water tender, or junior engineer of steam vessels
	3. First assistant engineer	(1) 1 year as second assistant engineer of steam vessels; or (2) 2 years as third assistant or junior second assistant engineer in charge of watch on steam vessels while holding license as second assistant engineer of steam ships; or (3) while holding license as first assistant engineer of motor vessels, either: (*a*) 6 months as second assistant engineer of steam vessels; (*b*) 6 months as observer first assistant engineer on steam vessels; or (*c*) 1 year as oiler, water tender, or junior engineer of steam vessels; or (4) 3 years as oiler, water tender, or fireman on steam ships for license as first assistant engineer of steam towing or ferry vessels of up to 2000 hp; or (5) while holding license as third assistant engineer of steam ships of any horsepower, 3 months as third assistant engineer or observer first assistant engineer on steam vessels for license as first assistant engineer of steam

U.S. Coast Guard	Class rating or license	Education, experience for 4000-hp vessels and over[a]
		towing or ferry vessels of not over 2000 hp; or (6) 3 years as oiler, water tender, or fireman on steam vessels for license as first assistant engineer of steam vessels of up to 1000 hp
	4. Chief engineer	(1) 1 year as first assistant engineer of steam vessels; or (2) 2 years as second assistant or junior first assistant engineer in charge of watch on steam ships while holding license as first assistant engineer of steam vessels; or (3) while holding license as chief engineer of motor vessels, either (*a*) 6 months as first assistant engineer of steam vessels; (*b*) 6 months as observer chief engineer on steam vessels; or (*c*) 1 year as oiler, water tender, or junior engineer of steam vessels

Source: U.S. Coast Guard.

[a] Limitation of horsepower on license is based on applicant's qualifying experience, but in no case is an applicant's license limited to a lower horsepower than the highest horsepower on which 25 percent or more of experience was obtained. An applicant whose qualifying service has all been on Coast Guard inspected vessels of 4000 hp or over is considered eligible for an engineer's license of *unlimited* horsepower.

Applicant for license as engineer officer of either steam or motor (diesel–internal-combustion engine) vessels shall pass written examination listed in Table 21-4.

TABLE 1-3 Examination Subjects Required for QMED Ratings

Subjects	Machinist	Refrigerating engineer	Fireman	Water tender	Oiler	Electrician	Junior engineer	Deck engineer
1. Application, maintenance, and use of hand tools and measuring instruments	X	X	X	X	X	X	X	X
2. Uses of babbitt, copper, brass, steel, and other metals	X	X		X	X	X	X	X
3. Methods of measuring pipe, pipe fittings, sheet metal, machine bolts and nuts, packing, etc.	X	X		X	X	X	X	X
4. Operation and maintenance of mechanical remote control equipment	X				X	X	X	X
5. Precautions to be taken for the prevention of fire and the proper use of firefighting equipment	X	X	X	X	X	X	X	X
6. Principles of mechanical refrigeration; and functions, operator, and maintenance of various machines and parts of the systems		X					X	
7. Knowledge of piping systems as used in ammonia, Freon, and CO_2, including testing for leaks, operation of bypasses, and making up of joints		X					X	
8. Safety precautions to be observed in the operation of various refrigerating systems, including storage of refrigerants, and the use of gas masks and firefighting equipment	X	X	X	X	X	X	X	X
9. Combustion of fuels, proper temperature, pressures, and atomization		X	X	X	X		X	
10. Operation of the fuel oil system on oil burning boilers, including the transfer and storage of fuel oil			X	X	X		X	X
11. Hazards involved and the precautions taken against accumulation of oil in furnaces, bilges, floorplates, and tank tops; flarebacks, leaks in fuel oil heaters, clogged strainers and burner tips	X	X	X	X	X	X	X	X
12. Precautions necessary when filling empty boilers, starting up the fuel oil burning system, and raising steam from a cold boiler			X	X	X		X	
13. The function, operation, and maintenance of the various engineroom auxiliaries	X	X		X	X	X	X	
14. Proper operation of the various types of lubricating systems	X	X		X	X	X	X	
15. Safety precautions to be observed in connection with the operation of engineroom auxiliaries, electrical machinery, and switchboard equipment	X	X		X	X	X	X	X
16. The function, operation, and maintenance of the bilge, ballast, fire, freshwater, sanitary, and lubricating systems	X	X		X	X	X	X	X
17. Proper care of spare machine parts and idle equipment	X	X		X	X		X	X
18. The procedure in preparing a turbine, reciprocating, or diesel engine for standby; also the procedure in securing						X	X	X
19. Operation and maintenance of the equipment necessary for the supply of water to boilers, the dangers of high and low water and remedial action			X	X	X		X	
20. Operation, location, and maintenance of the various boiler fittings and accessories	X		X	X	X		X	
21. The practical application and solution of basic electrical calculations (Ohm's law, power formula, etc.)						X	X	X
22. Electrical wiring circuits of the various two-wire and three-wire dc systems and the various single-phase and polyphase ac systems						X	X	X
23. Application and characteristics of parallel and series circuits						X	X	X
24. Application and maintenance of electrical meters and instruments						X	X	X

TABLE 1-3 Examination Subjects Required for QMED Ratings (*Continued*)

Subjects	Machinist	Refrigerating engineer	Fireman	Water tender	Oiler	Electrician	Junior engineer	Deck engineer
25. The maintenance and installation of lighting and power wiring involving testing for, locating and correcting grounds, short circuits and open circuits, and making splices						X	X	X
26. The operation and maintenance of the various types of generators and motors, both ac and dc						X	X	X
27. Operation, installation, and maintenance of the various types of electrical controls and safety devices						X	X	X
28. Testing and maintenance of special electrical equipment, such as telegraphs, telephones, alarm systems, fire-detecting systems, and rudder angle indicators						X	X	
29. Rules and regulations and requirements for installation, repair, and maintenance of electrical wiring and equipment installed aboard ships						X	X	X
29a. Pollution laws and regulations, procedures for discharge containment and cleanup, and methods for disposal of sludge and waste from cargo and fueling operations	X	X	X	X	X	X	X	
30. Such further examination of a nonmathematical character as the Officer in Charge, Marine Inspection, may consider necessary to establish the applicant's proficiency	X	X	X	X	X	X	X	

Source: U.S. Coast Guard.

TABLE 1-4 Subjects for Engineer Officers' Licenses of Steam or Motor Vessels

Subjects	Steam				Motor			
	Chief engineer		Assistant engineer		Chief engineer		Assistant engineer	
	Over 2000 hp	2000 hp and less	Over 2000 hp	2000 hp and less	Over 2000 hp	2000 hp and less	Over 2000 hp	2000 hp and less
General								
1. Pumps and compressors	X	X	X	X	X	X	X	X
2. Heat exchangers	X	X	X	X	X	X	X	X
3. Propellers and shafting	X	X	X	X	X	X	X	X
4. Steering and miscellaneous machinery	X	X	X	X	X	X	X	X
5. Valves—reducing, control, etc.	X	X	X	X	X	X	X	X
6. Condensers, air-ejectors and vacuum	X	X	X	X				
7. Engineering definitions and principles	X	X	X	X	X	X	X	X
8. Instruments	X	X	X	X	X	X	X	X
9. Lubrication	X	X	X	X	X	X	X	X
10. Inspection	X	X	X	X	X	X	X	X
11. Mathematics	X	X	X	X	X	X	X	X
12. Sketch		X	X	X		X	X	X
13. Three-dimensional drawing	X				X			
14. Ship construction and repair	X	X	X	X	X	X	X	X
Steam engines								
21. Reciprocating—construction, operation, maintenance	X	X	X	X				
22. Turbine—construction, operation, maintenance	X	X	X	X				
23. Reduction gear and miscellaneous	X	X	X	X				
24. Steam governors	X	X	X	X				
Motor								
31. Construction, operation, maintenance					X	X	X	X
32. Operating principles					X	X	X	X
33. Fuel injection					X	X	X	X
34. Air compressors					X	X	X	X
35. Operation and maintenance of auxiliary diesel engines	X	X	X	X				
36. Air—starting, combustion					X	X	X	X
37. Governors					X	X	X	X
Boilers								
41. Watertube—construction, operation, maintenance	X	X	X	X	X		X	
42. Firetube—construction, operation, maintenance	X	X	X	X	X		X	
43. General—construction, operation, maintenance	X	X	X	X	X		X	
44. Safety valves	X	X	X	X	X		X	
45. Corrosion and feed water	X	X	X	X	X		X	
46. Fuels and combustion	X	X	X	X	X		X	

Subjects	Steam				Motor			
	Chief engineer		Assistant engineer		Chief engineer		Assistant engineer	
	Over 2000 hp	2000 hp and less	Over 2000 hp	2000 hp and less	Over 2000 hp	2000 hp and less	Over 2000 hp	2000 hp and less
Electricity								
51. Direct current - - - -	X	X	X	X	X	X	X	X
52. Alternating current - - - - - - - - -	X	X	X	X	X	X	X	X
53. General—switchboards, controls, wiring- - - - - - - -	X	X	X	X	X	X	X	X
54. Storage batteries - - - - - - - - - - - -	X	X	X	X	X	X	X	X
55. Electric drive - - - - -	X	X	X	X	X	X	X	X
56. Problems - - - - - - - -	X	X	X	X	X	X	X	X

Source: U.S. Coast Guard.

TABLE 1-5 Canadian Stationary Engineer's License Requirements

Province	Class of certificate	Education, experience, and remarks	Plant capacity for class of certificate
Alberta Labour, General Safety Services Division, 10339–124 Street, Edmonton, T5N 3W1	1. Fireman	6 months operating any boiler	Chief of power plant up to 50 boiler hp
	2. Fourth-class engineer	12 months assisting in operation of power plant exceeding 25 boiler hp; or 12 months as operator in a pressure plant; or holder of degree in mechanical engineering or equivalent; or completion of vocational course if satisfactory to chief inspector or as one-half above with completion of an approved course	Shift engineer of plant up to 100 boiler hp Chief of power plant up to 100 boiler hp Chief of power plant up to 500 boiler hp of pressure not exceeding 20 psi Chief of power plant of coil-type drumless boilers up to 500 boiler hp when used for underground thermal flooding in oil fields Shift engineer of power plant up to 500 hp Shift engineer of power plant of coil-type boilers up to 1000 boiler hp used for underground thermal flooding Assistant engineer in power plant up to 1000 boiler hp
	3. Third-class engineer	12 months as chief in plant exceeding 50 boiler hp while holding fourth-class certificate; or 12 months as chief in plant of coil-type boilers exceeding 100 boiler hp while holding fourth-class certificate; or 12 months	Chief of power plant up to 500 boiler hp Chief of any power plant not exceeding 20 psi Chief of any power plant of coil-type drumless boilers up to 1000 boiler

45

TABLE 1-5 Canadian Stationary Engineer's License Requirements (*Continued*)

Province	Class of certificate	Education, experience, and remarks	Plant capacity for class of certificate
		shift engineer in plant exceeding 100 boiler hp while holding fourth-class certificate; or 12 months shift engineer in power plant of coil-type boiler exceeding 500 boiler hp while holding fourth-class certificate; or 12 months as assistant engineer in power plant exceeding 500 boiler hp while holding fourth-class certificate; or one-half above and holder of degree in mechanical engineering or equivalent; or one-half above and 12 months maintenance; or one-half above and completion of approved course in power engineering	hp when used for underground thermal flooding in oil fields Shift engineer of power plant up to 1000 boiler hp Shift engineer of power plant of coil-type boiler up to 1500 boiler hp Assistant engineer in any power plant
4. Second-class engineer		Hold third-class certificate and 24 months chief in power plant exceeding 100 boiler hp; or 24 months chief in power plant of coil-type boilers exceeding 500 boiler hp; or 24 months shift engineer in power plant exceeding 500 boiler hp; or 24 months shift engineer in power plant of coil-type boilers exceeding 1000 boiler hp; or 24 months assistant engineer in power plant exceeding 1000 boiler hp; or 36 months shift engineer in power plant exceeding 100 boiler hp; or one-half above and	Chief of power plant up to 1000 boiler hp Chief of power plant of coil-type boilers up to 1500 boiler hp Shift engineer in any power plant

Certificate	Requirements	Capacity
	holder of degree in mechanical engineering or equivalent. Credit of 9 months in lieu of experience for completion of approved course in power engineering	Chief of any power plant Shift engineer of any power plant
5. First-class engineer	Hold second-class certificate and 30 months chief of power plant exceeding 500 boiler hp; or 30 months chief of power plant of coil-type boilers exceeding 1000 boiler hp; or 30 months shift engineer in power plant exceeding 1000 boiler hp; or 30 months shift engineer in power plant of coil-type boilers exceeding 1500 boiler hp; or 45 months as assistant shift engineer in power plant exceeding 1000 boiler hp; or one-half above and holder of degree in mechanical engineering or equivalent; or 30 months as inspector of boilers and pressure vessels under Canadian Shipping Act; credit of 12 months in lieu of experience for completion of approved course in power engineering	Chief of any power plant Shift engineer of any power plant
6. Special oil well	6 months in power plant on oil rig	Chief on power plant up to 100 boiler hp on oil rig
7. Temporary certificate	Hold certificate one grade less than certificate required	Any capacity

TABLE 1-5 Canadian Stationary Engineer's License Requirements (Continued)

Province	Class of certificate	Education, experience, and remarks	Plant capacity for class of certificate
	8. Provisional stationary engineer, fourth, third, second, or first class[a]		
British Columbia Safety Engineering Services Division, 501 West 12th Ave., Vancouver V5Z 1M6		All provinces and the two territories, except Ontario and Quebec, have an *interprovincial agreement*; thus the service times and examination for power engineers in those provinces and territories are virtually the *same* as in Alberta. All plant ratings, qualifications for examinations, examinations, and marking procedures are very similar for the first-, second-, third-, and fourth-class engineers. Anything *below* fourth is a provincial item and may vary from province/territory to province	
Manitoba Mechanical & Engineering Div., 611 Norquay Bldg., Winnipeg, R3C 0P8		Same as Alberta above, except that the old fireman class has been eliminated and replaced with a fifth class. All age limitations have been dropped	
New Brunswick Dept. of Labor, P.O. Box 6000, Fredericton, E3B 5H1		Same as Alberta (also see British Columbia), except that horsepower ratings have been changed to therm hour ratings. Therm hour of a boiler other than electric boiler is horsepower of boiler shown on certificate multiplied by 2 and divided by 3. Therm hour rating of electric boiler is horsepower of boiler shown on certificate divided by 3	
Newfoundland and Labrador Engineering & Technical Services Div., Confederation Bldg., St. John's A1C 5T7		Same as Alberta (also see British Columbia)	

[a] A provisional certificate is one grade lower than the certificate of qualification that corresponds to the certificate issued by the other provinces.

Northwest Territories Yellowknife, X0E IHQ		Same as Alberta (also see British Columbia)	
Nova Scotia Board of Examiners, P.O. Box 697, Halifax, B3J 2T8		Same as Alberta (also see British Columbia)	
Ontario Operating Engineers Branch, 400 University Ave., Toronto, M7A 2J9	1. Stationary engineer, fourth-class	1 year operation in Ontario plants on provisional fourth-class stationary engineer certificate; or 3 months in stationary plant or low-pressure stationary power plant; or a third-class certificate of Canadian Shipping Act, or 12 months on boilers, engines, and auxiliaries of naval or merchant steam ships	(1) Act as chief operating engineer (*a*) in charge of any stationary power plant up to 50 therm hours[a] where rating of refrigeration compressors is up to 2.544, including total of not more than 5.088; or (*b*) any low-pressure stationary plant up to 134 therm hours; or (*c*) any steam-powered plant up to 7.632 therm hours, or (*d*) any compressor plant up to 10.176, or (*e*) any plant in *a* or *b* with total rating up to 3.816 therm hours, including refrigeration compressors up to 7.632 therm hours (2) Shift engineer in (*a*) any stationary power plant up to 134 therm hours where therm hour rating of refrigeration compressors is up to 5.088 and that of compressors including refrigeration compressors is up to 10.176; or (*b*) any low-pressure stationary plant up to 400 therm hours; or (*c*) any steam-powered plant; or (*d*) any refrigeration plant up to 20.352 therm hours; or (*e*) any compressor

[a] Therm hour = 100,000 Btu/hr or 39.3082 brake hp.

TABLE 1-5 Canadian Stationary Engineer's License Requirements (*Continued*)

Province	Class of certificate	Education, experience, and remarks	Plant capacity for class of certificate
			plant; or (*f*) any plant in *b* or *c* above, with total therm hour rating of refrigeration compressors up to 15,264 and rating of compressors including refrigeration up to 30,528 therm hours (3) Assistant shift engineer in (*a*) any stationary power plant up to 400 therm hours; or (*b*) any low-pressure stationary plant, steam-powered plant, refrigeration plant, or compressor plant
	2. Stationary engineer, third-class	1 year operation in Ontario plants on provisional certificate as a third-class stationary engineer; or hold fourth-class certificate with 1 year in stationary low-pressure power plant	(1) Act as (*a*) chief operating engineer in charge of any power plant up to 134 therm hours with refrigeration compressors up to 5,088 and compressors including refrigeration up to 10,176 therm hours; or (*b*) any low-pressure stationary plant up to 400 therm hours; or (*c*) any steam-powered plant; or (*d*) any refrigeration plant up to 20,352 therm hours; or (*e*) any compressor plant; or (*f*) any plant of *a* or *b* where total therm hour rating includes refrigeration compressors up to 15,264 or compressors, including refrigeration compressors up to 30,528 (2) Act as shift engineer in (*a*) any sta-

tionary plant up to 400 therm hours including rating of refrigeration compressors up to 15.264, with combined rating up to 30.528; or (*b*) any low-pressure stationary plant, steam-powered, with compressor including refrigeration

(3) Act as assistant shift engineer in any plant

3. Stationary engineer, second-class	1 year operation in Ontario on provisional certificate as second-class stationary engineer; or (1) hold certificate as stationary engineer (third-class) with experience on certificate and 18 months in stationary power plant up to 134 therm hour rating; or (2) hold certificate as first-class steam engineer (Merchant Shipping Act or Canadian Shipping Act) with 1 year in plant up to 134 therm hour rating; or (3) approved degreed engineer with 24 months experience in plant up to 12,000 therm hours in Ontario hydro plants	Act as chief operating engineer in (*a*) stationary plant up to 400 therm hours including that of refrigeration compressors of not more than 15.264, or that of combined compressors up to 30.528 therm hours; or (*b*) chief operating engineer in any low-pressure stationary plant, steam-powered, compressor, or refrigeration plant; or (*c*) act as shift engineer in any plant
4. Stationary engineer, first-class	Hold second-class certificate, with experience as second-class stationary engineer and 30 months in stationary power plant: (1) have spent not less than 6 months of the 30 in plant up to 134 therm hour rating; and (2) 24 of the 30 months in therm	Chief operating engineer in charge of any plant

51

TABLE 1-5 Canadian Stationary Engineer's License Requirements (*Continued*)

Province	Class of certificate	Education, experience, and remarks	Plant capacity for class of certificate
		hour rating above 300; or (3) hold approved engineering degree and 36 months operating experience in 12,000 therm hour plant	
	5. Provisional stationary engineer, fourth-, third-, second-, or first-class[a]	Not specified	
Prince Edward Island Boiler Inspection Branch, Box 2000, Charlottetown, C1A 7N8		Same as Alberta (also see British Columbia)	
Quebec Edifice La Laurenti-enne 425 St-Amable, 3ieme étage Québec, G1R 4Z1	1. Fifth-class, heating–steam engines	4 months in heating or steam engine plant, or fabrication, installation, or repair of steam engines and boilers	Any heating or steam plant up to 100 hp, or fourth- or lower-class plant for one day, or any electric boilers up to 200 boiler hp
	2. Fourth-class, heating–steam engines	Hold fifth-class license and serve with such for 2 years in fifth-class plant, or 9 months in fourth-class or higher plant, or 6 months in fifth-class plant; or hold fourth- or higher-class license in marine, heating, or steam engines for past 1 year; or complete technical courses; or work under	Any heating or steam plant up to 300 hp, or third-class plant for one day, or electric boiler plant up to 400 boiler hp

[a] A provisional certificate is one grade lower than the certificate of qualification that corresponds to the certificate issued by the other provinces.

Class of license	Experience and qualifications required	Qualifies to operate
	fourth-class or higher engineer for: (a) 1 year as helper in fourth-class plant, (b) 10 months as helper in fourth-class plant plus completed technical courses, (c) 8 months as helper in fourth-class plant and completed technical courses	
3. Third-class, heating–steam engines	Hold fourth-class license and serve under third-class operator for one of the following periods: (a) 1 year third-class plant, (b) 10 months in third-class plant plus completing technical courses, (c) 8 months in third-class plant plus completing technical courses, (c) 8 months on license in marine heating or steam engines for past 1 year; or complete 3-year course in technical school	Any heating or steam plant up to 600 hp, or second-class plant for one day, or electric boiler plant up to 600 boiler hp
4. Second-class, heating–steam engines	Hold third-class license and serve under second-class operator in heating and steam engine plant for one of the following periods: (a) 18 months in second-class plant, (b) 16 months in second-class plant. plus completed technical courses, (c) 14 months in second-class plant plus completed technical courses; or hold second-class license for marine, heating, or steam engines for 1 year	Any heating or steam engine plant up to 1000 hp, or any electric boiler plant
5. First-class, heating–steam engines	Hold second-class license in heating and steam engines plus serving un-	Unlimited capacity

TABLE 1-5 Canadian Stationary Engineer's License Requirements *(Continued)*

Province	Class of certificate	Education, experience, and remarks	Plant capacity for class of certificate
		der first-class operator for: (*a*) 2 years in first-class plant of which 1 year in charge, (*b*) 22 months in first-class plant of which 1 year in charge plus completed technical courses, (*c*) 20 months in first-class plant of which 1 year in charge plus completed technical courses; or hold first-class license in heating and marine steam engines for 1 year	
Saskatchewan Boiler & Pressure Vessel Unit, 1150 Rose St., Regina, S4P 2YR		Same as Alberta (also see British Columbia)	
Yukon Territory Box 2703, Whitehorse, YIA 2C6		Same as Alberta (also see British Columbia)	

Source: Stephen M. Elonka survey of licensing agencies (1980).

SUGGESTED READING

Refrigeration License Data

Elonka, Stephen M., and Quaid Minich: *Standard Refrigeration and Air Conditioning Questions and Answers*, 2nd ed., McGraw-Hill Book Company, New York, 1973 (contains chapter on license requirements for refrigeration engineers, which are tied in closely with steam licenses where refrigeration compressors are steam-driven).

2

HOW TO PASS
EXAMINATIONS

Chapter 1 shows that for a fireman's or operating engineer's certificate (license) on land, river, lake, or sea, you must meet requirements in age, experience, and character. Details can usually be obtained in printed form from the examining boards and U.S. Coast Guard listed in the previous chapter.

In addition, the applicant must be familiar with laws and regulations covering operation of power plants, those defining qualifications, duties, and responsibilities of the engineer in charge of a plant, and the penalties that may be incurred for neglect. Also today's strict pollution standards must be considered. In this chapter, we briefly cover the important items that will help you not only pass your examinations but become a better operator, protect you and the plant owner from fines, and help you to advance more quickly to higher pay and responsibility.

TYPICAL QUESTIONS AND HOW TO ANSWER THEM

Q   What are the duties of an engineer in charge of a power plant?
A   The engineer sees that his plant is installed so as to comply in every way with regulations governing these matters and that it is operated with due regard to safety and efficiency. He satisfies himself that his assistants are competent to perform their duties and are faithfully following his instructions and orders. He draws up rules, and sees that they are carried out, for general machinery adjustment and repair and for all routine operations,

such as testing water gages, soot blowing, and blowing down and washing out boilers. He draws up a logbook form (see Chap. 20) suitable for his plant and insists that it be an orderly record of all routine operations, repairs, and happenings worthy of note in and around the plant. He personally inspects all important parts, such as interiors of boilers, when they are shut down for cleaning; sees that all repairs and replacements are carried out as soon as reasonably possible; and reports immediately any serious defect to his employer or to the boiler inspector. As he is ultimately held responsible for the plant's safe operation, he uses his own judgment, in the event of any unusual or dangerous situation, as to whether to keep the plant operating or shut down.

Q   What penalties may be incurred by an engineer who is guilty of gross neglect of duty, or serious infraction of the laws and rules governing operation of power plants?

A   The engineer's certificate may be suspended or canceled. He may also be fined or even imprisoned, in extreme cases, when gross neglect of duty has resulted in grave bodily injury to other persons.

Q   What experience have you had as an operating engineer?

NOTE: This question may be answered in either of two ways: (*a*) by listing jobs in order held, or (*b*) by an account of experience in narrative form. Following are examples of typical good answers to (*a*) and (*b*), using fictitious names.

A   (a) Listing of Jobs in Order   May 1970 to August 1980. Shift engineer in plant of the Newbury Food Products Co., Newbury (state or province).

Boiler plant: two 200-boiler-hp Cleaver Brooks fire-tube packaged-type fire-tube boilers, Detroit Stoker coal-fired; two Worthington $5 \times 4 \times 6$ in. duplex-type boiler feed pumps, Koerting closed-type feedwater heater; two Coffin condensate-return pumps for heating-system returns.

Turbine plant: one 75-hp General Electric steam turbine, direct-connected to 50-kW alternator; one Coppus 50-hp noncondensing steam turbine (exhaust used for process heating), driving screw-type Sullair air compressor.

August 1980 to present date. Chief engineer in plant of the Standard Mfg. Co., Hillsboro (state or province).

Boiler plant: two 250-boiler-hp York-Shipley fire-tube boilers, Bunker C oil-fired, one electric-motor-driven Gould centrifugal split-casing feed pump, one $6 \times 4 \times 10$ in. Warren simplex steam pump for factory water supply, one Bell & Gossett closed feedwater heater.

Turbine plant: One 100-hp DeLaval steam turbine and one 200-hp Elliott steam turbine, direct-connected to 150-kW alternator for supplying current for plant motors and lighting load. One Onan diesel-driven 30-kW emergency electric generator set.

(b) Experience in Narrative Form   I sat for and obtained a second-class engineer's certificate in April 1969, started in May 1970 as shift engineer in the plant of the Newbury Food Products Co., Newbury (state or province), and continued in this position until August 1980. The boiler plant consisted of [NOTE: Write out list of major equipment as in (a)].

In August 1980 I secured my present position as chief engineer with Standard Mfg. Co., Hillsboro (state or province). [NOTE: Again, detail all the major pieces of equipment as in (a).]

In answers like either (a) or (b) give names of makers of plant equipment, such as boilers, engines, and pumps, where possible so that the examiner knows *exactly* what kinds of equipment you operated.

Q   Should applicants know how to draw and use drawing instruments?
A   Yes. Ability to make a freehand sketch and a scale drawing with instruments is a valuable asset to any engineer in his daily work. It is especially useful in an examination when a sketch or drawing may answer a question better than several pages of writing. Mechanical drawing is the *universal* language of the engineer, understood the world over. Very few people have a knack for really good freehand work, but neat sketches, not necessarily to scale, may be made with a few drawing instruments, such as a com-

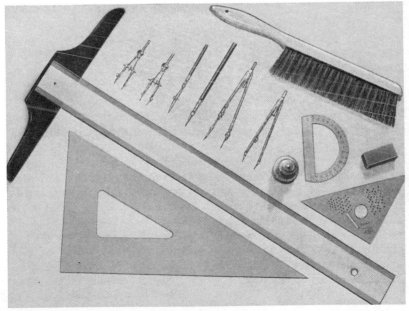

FIG. 2-1.   These drawing instruments may be needed for some license examinations.

pass, 60 and 45° triangles, and a 12-in. rule. Every progressive engineer should own a complete drafting outfit (Fig. 2-1).

Q What basic education and background must the applicant have?

A Three qualities are essential to proper understanding and answering of examination questions: (1) sufficient command of English to express ideas clearly after ascertaining the exact meaning of the question, (2) a good working knowledge of elementary mathematical operations, and (3) ability to sketch neatly and to draw in proportion.

The first quality is needed to answer questions for any grade of certificate, the second is needed more particularly for the higher grades, and the third is essential for higher grades but is useful for all. For higher grades of certificates (especially) one must have considerable knowledge of elementary mathematics and mechanical drawing. Mathematics required for lower grades seldom covers more than simple arithmetic (see Chap. 3), but accuracy in solving even the simplest arithmetical problems is always important.

> REMEMBER: Correctness of all mathematical solutions depends on accuracy in addition, subtraction, multiplication, and division.

RULES FOR BOILER CONSTRUCTION

Q What codes must you be familiar with to pass examinations?

A The various codes drawn up by the American Society of Mechanical Engineers (ASME) in connection with the design, construction, and installation of boilers and other pressure vessels, because they have been adopted over most of the North American continent. Common agreement on a uniform and reliable set of standards is a big help to designers, manufacturers, and operators.

> NOTE: Sometimes a candidate is allowed to take copies of the ASME code and of local regulations into the examination room and to consult them freely. Whether their use is permitted or not, study the codes and regulations beforehand, especially the sections dealing with use and location of boiler fittings. Use of logarithmic tables, steam tables, slide rules, and calculators may be permitted by examiners.

In addition to these ASME codes, some authorities draw up supplementary regulations covering inspection fees, examination requirements, and other items that may have a purely local application. Good examples are today's EPA (Environmental Protection Agency) and OSHA (Occupational Safety and Health Administration) requirements.

One day let us hope we will have greater uniformity in examination requirements and candidates' qualifications for engineer's certificates. The best example is the marine license, which qualifies the holder for a similar berth aboard any ship of similar horsepower and machinery in any United States port.

Q Does the ASME cover all the required boiler standards?

A No. Matters that do not come within the scope of the codes and other published standards of the ASME are usually covered in the published standards of such other organizations as the American Society for Testing and Materials (ASTM) and the American National Standards Institute (ANSI) in the United States, and the Canadian Engineering Standards Association in Canada.

Q What boiler acts and rules govern examinations for license?

A Examiners sometimes ask about laws and rules governing power-plant operation and qualifications and duties of the engineer. Such questions can be answered by quoting the particular section or sections of the act or regulation mentioned in the question. No particular instruction or advice is needed on the method of answering this type of question other than to repeat that a candidate should get copies of laws and regulations governing these matters and study them carefully, paying special attention to sections dealing with *qualifications* of the engineer, *duties* of the engineer, and *penalties* for neglect of duty or infraction of laws or rules.

An engineer is expected to be familiar with laws, rules, and regulations governing the conduct of the profession.

Q What kind of performance does the examiner expect from you during the examination?

A Arrange your work neatly. Number each page and leave margins around your writing. Number each answer to correspond with its question and leave spaces between answers. Always put down enough steps in working mathematical problems to explain how you arrived at the result. Write your answers plainly, and state units in which answers are given, that is, feet (ft), inches (in.), or pounds per square inch (psi), as the case may be.

NOTE: If more than one question is given at one time, answer the ones you are sure of and come back to the ones for which you must give further thought, especially if a limited time is required for completing the examinations.

STUDY FOR ADVANCEMENT

Q What publications and books are necessary?

A All stationary and marine engineers should subscribe to one or more engineering magazines and have a few good textbooks dealing with the various branches of their profession. The "Standard" series of books listed as reference at the end of chapters in this volume provides the student with a complete home-study course. This volume might be considered a *primer* on boiler-room operation. Those wishing to advance will find that another book in this series, *Standard Boiler Operators' Questions and Answers,* goes into the subject in greater depth.

Standard Plant Operators' Manual, 3rd ed., has over 2000 illustrations, showing step-by-step procedures for such operations as rolling in boiler tubes, lining up a pump, and timing a diesel engine, thus providing an "instructor" at the reader's elbow. With many types of instruments installed in modern plants, *Standard Instrumentation Questions and Answers* covers instruments the operator needs to understand. Because electronic instruments (solid-state) are common in today's power plants, the two-volume set, *Standard Electronics Questions and Answers,* provides many needed answers.

Standard Plant Operator's Questions and Answers, 2nd ed. (two volumes) covers every type of power-plant equipment from boilers and pumps to gas turbines and nuclear reactors. *Standard Industrial Hydraulics Questions and Answers* introduces the operator to the many hydraulic devices used. *Standard Refrigeration and Air Conditioning Questions and Answers* has basic information to pass examinations and operate equipment indicated.

The newest volume, *Standard Basic Math and Applied Plant Calculations,* ties in with all the subjects covered in this series. But to put the "icing on the cake," *Marmaduke Surfaceblow's Salty Technical Romances* should make technical book publishing history because it is (1) the first technical book written as fiction and (2) the first book of fiction with complete technical index; so the reader can quickly find one of the 121 solutions the legendary engineer used to solve a similar perplexing problem. This volume is a trouble shooter's manual. To date, these books have been translated and published by 10 publishers in the United States and in foreign countries, including Japan, Poland, Taiwan, India, the Philippine Islands, and Brazil.

SUGGESTED READING

Magazines

Power, 1221 Ave. of the Americas, New York, NY 10020.
Steam & Heating Engineer, 35 Red Lion Square, London WC1, England.

3

MATH AND SCIENCE
REFRESHER

If this is one of your first technical books, you probably are rusty on mathematics and basic physics. Or, for upgrading a license, you also may need a brief refresher. Most examiners and technical school instructors do not accept answers punched out on a pocket calculator but want all problems worked out on paper, showing all the steps in proper sequence. In this chapter we'll quickly run through basic steps which are needed to work calculations in succeeding chapters, also to help you understand the operation of energy systems more clearly and, we hope, to prepare you for engineering studies and eventually for an engineer's degree.

ADDITION

Below is an addition of two numbers as commonly set down:

$$
\begin{array}{r}
26.13 \\
415.38 \\
\hline
441.51 \quad \textit{Ans.}
\end{array}
$$

Here 3 plus 8 is 11. Put down 1 and carry 1. Then, 1 plus 1 plus 3 is 5, etc. This is the usual procedure, and we can't suggest any improvement.

In contrast, try adding a long column of figures. First, the customary way:

```
  26.13
 415.38
 260.41
  22.85
 191.06
 205.43
  88.55
 739.11
 263.05
2211.97      Ans.
```

Following the usual procedure, we added the digits in the last column to get 37. We set down the 7 in the last place of the sum and added the 3 to the sum of the digits in the next to the last column, and so on.

This is faster:

```
  26.13
 415.38
 260.41
  22.85
 191.06
 205.43
  88.55
 739.11
 263.05
    .37
   2.6
  39.
  37
  18
2211.97      Ans.
```

NOTE: When adding decimal numbers as in these three examples, be sure to keep the decimal point in line.

SUBTRACTION

```
  833,610
 −541,881
  291,729      Ans.
```

Here 1 cannot be taken from zero; so we borrow 1 from the next column above, which gives us 10. Now 1 from 10 = 9. Then 8 cannot be taken from zero above it (zero because we borrowed 1); so we borrow 1 from 6.

Now 8 from 10 = 2. Eight from 15 (we borrowed 1) = 7, and so on.

Because subtraction is the opposite of addition, always check your answer with addition, thus:

$$\begin{array}{r} 541,881 \\ +291,729 \\ \hline 833,610 \end{array} \quad \textit{Ans.}$$

DIVISION

Now we divide 70,000 by 16, thus:

$$\begin{array}{r} 4,375 \quad \textit{Ans.} \\ 16\overline{)70,000} \\ \underline{64} \\ 60 \\ \underline{48} \\ 120 \\ \underline{112} \\ 80 \\ 80 \end{array}$$

Here 16 goes into 70 four times. We multiply and find that $4 \times 16 = 64$. Place the 64 under 70 and subtract, which leaves 6, and so we bring down the next number, which is zero. Now 16 into 60 goes 3 times, and $3 \times 16 = 48$; this we subtract from 60, leaving 12. Now 16 into 120 (we brought down a zero) goes 7 times, and $7 \times 16 = 112$. Subtracting, we have 8 left over. We bring down the remaining zero and find that 16 goes into 80 exactly 5 times with nothing left over.

MULTIPLICATION

Handy Multiplication Table

1	2	3	4	5	6	7	8	9	10	11	12
2	4	6	8	10	12	14	16	18	20	22	24
3	6	9	12	15	18	21	24	27	30	33	36
4	8	12	16	20	24	28	32	36	40	44	48
5	10	15	20	25	30	35	40	45	50	55	60
6	12	18	24	30	36	42	48	54	60	66	72
7	14	21	28	35	42	49	56	63	70	77	84
8	16	24	32	40	48	56	64	72	80	88	96
9	18	27	36	45	54	63	72	81	90	99	108
10	20	30	40	50	60	70	80	90	100	110	120
11	22	33	44	55	66	77	88	99	110	121	132
12	24	36	48	60	72	84	96	108	120	132	144

How to Use

To multiply 9 × 9, for example, just read down and over to the center of the table above, and find the answer 81.

RULE: First multiply the two numbers as whole numbers. Then the given product has as many decimal places as are in the *two* numbers combined.

EXAMPLE: 24.3 × 0.0613. First, 243 × 613 = 148,959. Decimal places in the two numbers multiplied total five; so the correct product is 1.48959.

To check a multiplication, don't repeat the original operation. It is much safer to reverse the numbers and remultiply, like this:

```
462 × 893 = ?
      462          Reversing:      893
      893                          462
     1386                         1786
     4158                         5358
     3696                         3572
    412,566    Ans.             412,566      Ans.
```

FRACTIONS

To change a proper fraction to a decimal fraction, divide the numerator by the denominator and carry out the quotient (answer) to the desired number of decimal places.

EXAMPLE: Solve for decimal fraction equivalent of $7/16$

SOLUTION:

```
       0.4375      Ans.
   16)7.0000
      6 4
        60
        48
       120
       112
         80
         80
```

EXAMPLE: Solve for decimal fraction equivalent of $5/13$, correct to the third decimal place.

SOLUTION:

$$
\begin{array}{r}
0.384 \qquad Ans. \\
13\overline{)5.000} \\
\underline{3\ 9} \\
1\ 10 \\
\underline{1\ 04} \\
60 \\
\underline{52} \\
8
\end{array}
$$

To change a decimal fraction to a proper fraction, multiply the decimal fraction by the denominator of the desired proper fraction and place the product obtained over the denominator.

EXAMPLE: Change 0.032 to 125ths.

SOLUTION: $0.032 \times 125 = 4$
Thus, $0.032 = {}^4/_{125}$ *Ans.*

To change a decimal fraction to the nearest indicated proper fraction, multiply the decimal fraction by the denominator of the desired proper fraction and then round off the product obtained to the nearest whole number and place it over the denominator.

EXAMPLE: Change 0.109 to the nearest 64th.

SOLUTION: $0.109 \times 64 = 6.976 = 7$ *Ans.* (nearest whole number)
Thus, $0.019 = {}^7/_{64}$ *Ans.* (nearest 64th)

To add mixed numbers (whole numbers and fractions), first convert all fractions to least common denominator.

EXAMPLE: Add $25 + {}^7/_8 + 5\,{}^3/_4 + {}^6/_8 + 6\,{}^5/_8$

SOLUTION: 25

$$
\begin{array}{r}
{}^7/_8 = {}^{21}/_{24} \\
5\,{}^3/_4 = {}^{18}/_{24} \\
{}^6/_8 = {}^{21}/_{24} \\
+6\,{}^5/_8 = {}^{20}/_{24} \\
\hline
36 \qquad = {}^{77}/_{24}
\end{array}
$$
$= 77$ divided by $24 = 3\,{}^5/_{24}$, which we add to the whole number:

$$
\begin{array}{r}
36 \\
+3\,{}^5/_{24} \\
\hline
39\,{}^5/_{24} \qquad Ans.
\end{array}
$$

NOTE: It is not necessary to find the least common denominator when multiplying two or more fractions. You can use cancellation where possible, to shorten the work, which is dividing both the numerator and denominator of the multiplied fractions by the same whole number.

EXAMPLE: Multiply $^6/_7 \times \, ^5/_8 \times \, ^7/_{25} = \dfrac{6 \times 5 \times 7}{7 \times 8 \times 25}$

$$= \dfrac{\overset{3}{\cancel{6}} \times \overset{1}{\cancel{5}} \times \overset{1}{\cancel{7}}}{\underset{1}{\cancel{7}} \times \underset{4}{\cancel{8}} \times \underset{5}{\cancel{25}}} = \dfrac{3}{20} \qquad Ans.$$

NOTE: Cancellation is permitted only when fractions are multiplied. In the above, 7 goes into 7 one time; so these two numbers cancel. Then 5 goes into 25 five times. Next, 2 goes into 6 three times, and also into 8 four times. Thus we have $3 \times 1 \times 1 = 3$, over $1 \times 4 \times 5 = 20$, or $^3/_{20}$ *Ans.*

Subtracting Fractions

First we find the least common denominator.

EXAMPLE: Subtract $^{11}/_{15}$ from $^5/_6$

SOLUTION: Least common denominator is 30. Thus:

$^{11}/_{15} = \, ^{22}/_{30}$ and $^5/_6 = \, ^{25}/_{30}$

Then $^{25}/_{30}$

$\dfrac{- \, ^{22}/_{30}}{}$

$^3/_{30} = \, ^1/_{10}$ *Ans.*

Dividing Fractions

Invert the divisor and proceed as in the multiplication of fractions.

EXAMPLE: Divide $^8/_9$ by $^4/_{15}$

SOLUTION: $\dfrac{^8/_9}{^4/_{15}} = \dfrac{8 \times 15}{9 \times 4} = \dfrac{\overset{2}{\cancel{8}} \times \overset{5}{\cancel{15}}}{\underset{3}{\cancel{9}} \times \underset{1}{\cancel{4}}} = \dfrac{10}{3} = 3^1/_3$ *Ans.*

POWERS AND ROOTS

A power is the product obtained by using a base number a certain number of times as a factor. Thus 3 to the fourth power is written 3^4 and is equal to $3 \times 3 \times 3 \times 3 = 81$ *Ans.*

EXAMPLE: Find the square of 25 (which is written as 25^2).

SOLUTION: $25 \times 25 = 625$ *Ans.*

A root is the opposite of a power. It is one of the equal factors of a number.

EXAMPLE: Find the square root of 64, which is written as $\sqrt{64}$.

SOLUTION: $8 \times 8 = 64$; thus the answer is 8.   *Ans.*

EXAMPLE: Find the square root of 535.713 correct to three decimal places.

SOLUTION: 5'35.71'30 Here we first separate the number into groups of two figures each, starting with the *decimal* point, and going from that point both ways. Next determine the largest number whose square is contained in the left-hand-most group. Thus,

$$
\begin{array}{r}
2 \\
\sqrt{5'35.71'30} \\
4 \\
\hline
1'35
\end{array}
$$

Now subtract the square of the root ($2 \times 2 = 4$) from the first group and annex the second group (35) to the difference (1'35). A trial divisor is formed by doubling the number (2) contained in the root thus far ($2 \times 2 = 4$). Now determine the number of times this trial divisor is contained in the partial remainder when the right-hand digit of the partial remainder is omitted (1'35, omitting the 5 gives us 13). Thus 13 divided by $4 = 3$. This same number 3 is also annexed to the trial divisor 4, making it 43. Now multiply the complete divisor by the last number that appears in the root (3), thus $3 \times 43 = 129$. Repeat the last step until the desired number of places have been obtained in the root, thus:

$$
\begin{array}{r}
2\ \ 3.\ 1\ \ 4\ \ 5 \\
\sqrt{5'35.71'30'00} \\
4 \\
\hline
43\quad\ \ 1\ 35 \\
1\ 29 \\
\hline
461\qquad\ \ 6\ 71 \\
4\ 61 \\
\hline
4624\qquad\ 2\ 10\ 30 \\
1\ 84\ 96 \\
\hline
46285\qquad 25\ 34\ 00 \\
23\ 14\ 25 \\
\hline
2\ 19\ 75
\end{array}
$$

Carrying to three decimals gives us 23.145.   *Ans.*

NOTE: This has been a brief refresher course, but complete enough to enable the reader to work the problems in this volume. The reader, if deficient, should pursue this subject in a good book on mathematics, such as *Standard Basic Math and Applied Plant Calculations*.

BASIC PHYSICS

Efficiency

If a machine could give out as much as went into it, the efficiency would be 100 percent. That never happens, although big generators and hydro turbines get close to this perfect mark.

A machine that delivers half its input has an efficiency of 50 percent. One that delivers three quarters has an efficiency of 75 percent, and so on.

The input and output being considered may be work, heat, mechanical power, or electrical power. The important thing is to figure both the input and output in the same kind of units, whether they be horsepower, kilowatthours, Btu, or foot-pounds.

EXAMPLES: Input and output are in like units in Fig. 3-1a and b. In Fig. 3-1 both input and output are measured in foot-pounds. The man has to pull 50 lb to lift 90 lb and he pulls 2 ft for every 1 ft the load rises. Then:

Input = 50 × 2 = 100 ft · lb
Output = 90 × 1 = 90 ft · lb
Efficiency = 90 ÷ 100 = 0.90 = 90 percent   *Ans.*

In the motor generator set in Fig. 3-1b both input and output are kilowatts. By feeding 103 kW to the motor, you get 79 kW out of the generator. Then:

Efficiency = 79 ÷ 103 = 0.767 = 76.7 percent   *Ans.*

In Fig. 3-1c the two units might appear to be different, since input is *indicated* horsepower and output is *brake* horsepower. Yet both are mechanical horsepower. The indicated horsepower is the mechanical power delivered by the steam to the piston and the brake horsepower is the mechanical power delivered by the engine to the load. So the efficiency of the engine in converting cylinder power into useful power (called the mechanical efficiency) is

39 ÷ 42 = 0.929 = 92.9 percent   *Ans.*

The case of the diesel generating set (Fig. 3-1d) is somewhat different. Here input is heat and output is electrical. For every kilowatthour delivered this engine burns 0.53 lb of 19,300-Btu oil. Then the input per kilowatthour is 0.53 × 19,300 = 10,230 Btu.

Table 3-1 shows that one kilowatthour equals 3413 Btu; so:

Input = 10,230 Btu
Output = 3413 Btu
Efficiency = 3413 ÷ 10,230 = 0.334 = 33.4 percent   *Ans.*

For a final problem, assume that the simple steam plant shown in Fig. 3-1e blows all exhaust to waste and that, when delivering 82 hp, the fuel burned is 310 lb/hr of 13,500-Btu coal.

This figures 310 ÷ 82 = 3.78 lb coal/bhp · hr; so:

Input = 3.78 × 13,500 = 51,030 Btu
Output = 2544 Btu (see Table 3-1)
Efficiency = 2544 ÷ 51,030 = 0.498 = 4.98 percent   *Ans.*

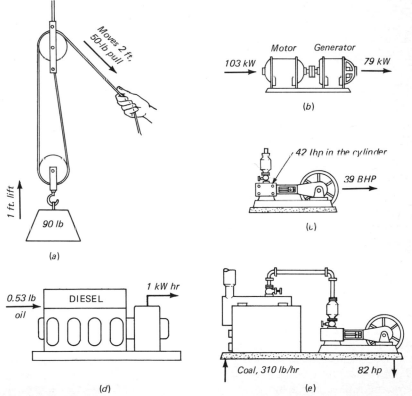

FIG. 3-1. Efficiency of a machine or device is always output divided by input.

TABLE 3-1 **Conversion Factors**

1 kilowatt = 1.3415 horsepower	1 horsepower = 424 Btu/min
= 738 ft · lb/sec	= 2544 Btu/hr
= 44,268 ft · lb/min	
= 2,656,100 ft · lb/hr	
= 56.9 Btu/min	1 Btu = 778.3 ft · lb
= 3413 Btu/hr	
	1 kilowatthour = 3413 Btu
	= 1.342 hp · hr
1 horsepower = 0.7455 kW	
= 550 ft · lb/sec	
= 33,000 ft · lb/min	1 horsepower-hour = 2544 Btu
= 1,980,000 ft · lb/hr	= 0.7455 kWh

FORCE MULTIPLIERS

Figure 3-2 shows some common ways to multiply force. Under each sketch you find the multiplication factor, neglecting friction.

For example, the force-multiplication factor for a lever (Fig. 3-2a) is the length of the long (input) arm divided by the length of the short (output) arm. If you pull 3 ft from the fulcrum, and the load is 1 ft from the fulcrum, you can lift three times the amount you pull.

In the usual case of a lever and certain other devices pictured, friction would waste only a small percentage of the applied energy. At the other extreme, friction might use up more than half the power in the case of a wedge (Fig. 3-2g) or a screw jack (Fig. 3-2f). Of course, the actual percentage wasted will depend on lubrication and many other factors.

It's easy to figure the force-multiplication factor for any special case if you remember that (neglecting friction) *the work you get out of any such machine equals the work you put in.*

That's another way of saying *what you lose in distance you can make up in force.* In every device that multiplies force the applied force must move a greater distance than the delivered load. The force-multiplication factor is merely the movement of the applied force divided by the movement of the load.

In short, if you have to push the handle 4 in. to lift the load 1 in., the machine will then multiply force by 4.

Where the motion is continuous and rotary (Fig. 3-2b, c, and d), you can substitute torque for force and r/min for speed. Thus, in a pulley or gear drive, if the driving shaft rotates four times as fast as the driven shaft, the delivered torque will be four times the input torque.

In the case of the worm and gear wheel, count how many times the worm must rotate to cause one complete rotation of the wheel. That number is the multiplication factor.

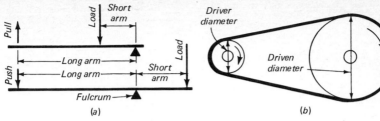

Lever: With the pull on long arm, and the load on short arm, multiplication factor is the long arm divided by the short arm.

Pulley: Factor is the diameter of driven pulley divided by the diameter of the driver.

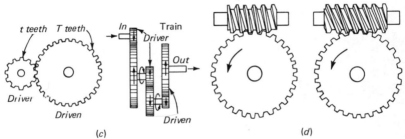

Spur gear: Factor is number of teeth on driven gear divided by number of teeth on driver. For gear *train* apply this rule to each meshing pair. Factor for whole train (right) is product factors of all the pairs.

Worm and gear: Factor is the number of teeth on gear in case of single threaded worm. For double thread divide this factor by 2. For triple thread divide it by 3.

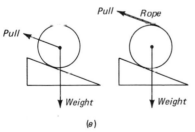

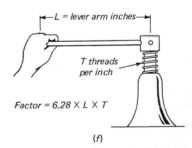

Inclined plane: See how many inches the object moves in the direction of applied pull *for each inch of rise*. Former is the multiplication factor. Usually pull is applied (as pictured) in the slant direction—the most effective. Scheme pictured at right (often used in handling barrels and heavy pipe) gives a factor double that for direct push or pull.

Screw jack: Factor is distance hand moves to raise the screw 1 in. This is 2 X 3.14 = 6.28 times the lever arm multiplied by the number of threads per inch. But beware of very high friction (see the text).

FIG. 3-2. Force multipliers can build up tremendous forces but may demand lots of time to do so, especially if the input is slight and the output great.

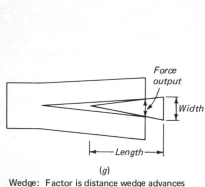

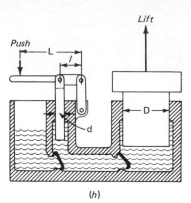

(g)

Wedge: Factor is distance wedge advances
for every inch of additional separation of the
parts being wedged. This equals the length
of wedge, divided by width.

(h)

Hydraulic jack: Here, multiplication is
proportional to number of times cross-sectional
area of small plunger goes into that of big
plunger.

FIG. 3-2. (*Continued*)

In Fig. 3-2*h* divide *D* by *d*, then square. Thus, if *D* is 5 and *d* is 1 in., $D \div d = 5$. This squared is 25; so lift is 25 times the force on the small plunger. In the sketch, force is further increased by the leverage of the handle, $L \div l$. In the problem just worked, if *L* were 3 times *l*, the total multiplication factor would be $3 \times 25 = 75$.

TORQUE AND SPEED

Power is usually delivered through rotating parts rather than in a straight-line push. In either case horsepower is the force in pounds times the distance through which it acts in 1 min, divided by 33,000, because one horsepower is 33,000 ft · lb/min.

Figure 3-3(*a*) shows a hoisting drum turning with a rim speed of 330 ft/min, winding up a weight of 100 lb at the same speed. Work done is $330 \times 100 = 33,000$ ft · lb/min, or 1 hp.

Now look at Fig. 3-3*b*. The two spring balances pull on the leather-strap brake.

The net drag on the drum surface is $140 - 40 = 100$ lb. The surface moves 330 ft in one minute; so the work done is $330 \times 100 = 33,000$ ft · lb/min. Again this is one horsepower, but the power is delivered as heat rather than as "useful work."

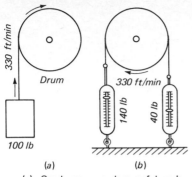

(a)

(b)

(a) One horsepower does useful work

(b) One horsepower delivered as heat

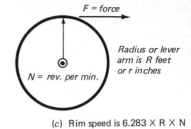

(c) Rim speed is 6.283 × R × N

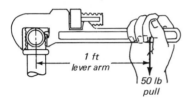

(d) Here torque is 50 ft·lb or 600 in.·lb

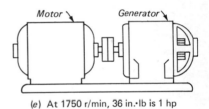

(e) At 1750 r/min, 36 in.·lb is 1 hp

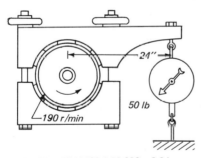

(f) 24 × 50 × 190 ÷ 63,000 = 3.6 hp

FIG. 3-3. Torque and speed are present in most things that move.

Torque

Now let's consider torque and the relation of torque and rotative speed to power delivered. Torque is nothing but twisting effort measured as the product of the pull by the length of the lever arm of the pull (Fig. 3-3c).

The lever arm may be measured in feet, in which case the torque is in foot-pounds. Often it is more convenient to measure the lever arm in inches, in which case the product of lever arm in inches by the force in pounds gives the torque in inch-pounds.

For example, in Fig. 3-3d, if the man pulls 50 lb at a point 12 in. from the center of the fitting, the torque is $1 \times 50 = 50$ ft · lb, also $12 \times 50 = 600$ in. · lb.

Torque and Power

Suppose (Fig. 3-3c) that we have the radius of a wheel and the rotative speed in r/min. In one revolution every point on the rim moves $2 \times 3.1416 \times R$ feet, where R is the radius in feet. This is $6.283 \times R$ *feet per revolution*.

If N is the rotative speed in r/min, the distance traveled by the rim in one minute will be $6.28 \times R \times N$ feet. If the force at the rim is called F, the horsepower will be $6.28 \times R \times N \times F \div 33,000$.

Note that this boils down to $R \times N \times F \div 5250$. Now $R \times N$ is nothing but the torque in foot-pounds. Call this T. Then horsepower is $T \times N \div 5250$.

If the radius is measured in inches, the product of radius and force is the torque in inch-pounds (call it t) and horsepower will be $t \times N \div 63,000$. For convenient reference the torque-speed-power formulas are repeated in Table 3-2. Let's apply these to the 1-hp motor generator set pictured in Fig. 3-3e. The set runs at 1750 r/min. Find the torque. Then $t = 63,000 \div 1750 = 36$ in. · lb.

Now Fig. 3-3f shows a prony brake measuring the power of a small

TABLE 3-2 Formulas for Rotary Power

R = radius of torque arm, feet
r = radius of torque arm, inches
F = force at end of torque arm, pounds
$T = R \times F$ = torque, foot-pounds
$t = r \times F$ = torque, inch-pounds
N is rotative speed, r/min
P is horsepower

Then:

$$P = T \times N \div 5250$$
$$P = t \times N \div 63,000$$

engine. Here torque $t = 24 \times 50 = 1200$ in. · lb. At 190 r/min the brake power is $190 \times 1200 \div 63,000 = 3.6$ hp.

TENSION, COMPRESSION, AND SHEAR

Tensile Stress

When a tensile test specimen (Fig. 3-4a) is stretched to the breaking point, the material tends to neck down. *Tensile strength* of the material in pounds

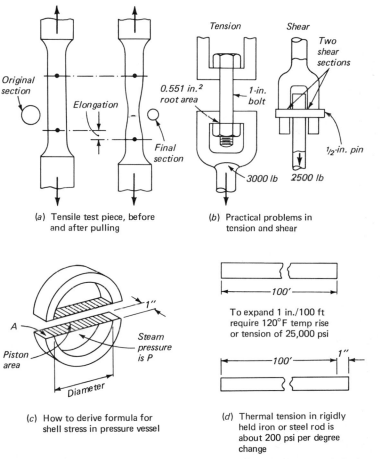

(a) Tensile test piece, before and after pulling

(b) Practical problems in tension and shear

(c) How to derive formula for shell stress in pressure vessel

(d) Thermal tension in rigidly held iron or steel rod is about 200 psi per degree change

To expand 1 in./100 ft require 120°F temp rise or tension of 25,000 psi

FIG. 3-4. Tension, compression, and shear must be kept within safe limits, thus must be known.

per square inch (psi) is the total breaking tension divided by the *original* cross section.

At any tension short of breaking, the *unit stress* is the actual total tension divided by this same original cross section.

Up to a certain unit stress, called the *elastic limit,* the metal stretches in exact proportion to the applied load and the unit stress. Beyond that the metal starts to give a little and stretches more and more for a given increase in unit stress.

Compression

The *compressive strength* of a material is tested by crushing a short block. Total crushing load, divided by original cross section, is the compressive strength in pounds per square inch.

Shear

In the same way, if the section is sheared across (as in a power shear), the *shearing strength* is the total force required divided by the cross-sectional area sheared.

Factor of Safety

Ultimate strength divided by actual stress.

Bolts and Pins

Two applications are shown in Fig. 3-4*b*. What is the unit tensile stress at the thread root of the bolt at the left? The cross-sectional area at the thread root of a 1-in. (U.S. Standard) bolt is 0.551 in.^{2}. Then unit stress is 3000 ÷ 0.551 = 5440 psi. For Class B bolt steel of 60,000 psi tensile strength, the factor of safety is 60,000 ÷ 5440 = 11.0.

The $^1/_2$-in. pin at the right in Fig. 3-4*b* has a cross-sectional area 0.5 × 0.5 × 0.785 = 0.196 in.^{2}. Since the pin is in *double shear,* the total sheared area is 2 × 0.196 = 0.392; so the shearing unit stress is 2500 ÷ 0.392 = 6380 psi.

Stress in Shell

In a seamless or welded tube or drum under internal pressure, the tendency to split lengthwise is double the tendency to split around. Thus, unless the circular seams are very weak, such shells or drums *always* fail by splitting lengthwise (longitudinally).

First calculate the shell strength without seams; then correct for the efficiency of the longitudinal joint. Consider a half slice of shell 1 in. long (Fig. 3-4c). Fluid pressure tends to split this into two halves, both of which are shown. If the internal diameter (inches) is D and the fluid pressure (psi) is P, the force on the shaded piston area is $P \times D$. This force is resisted and balanced by the pull of the two metal sections A. This resisting force is $2 \times t \times s$, where t is shell thickness and s is unit stress. From this relation we derive the shell formula $P = 2 \times t \times s \div D$.

If the diameter is 30 in., the thickness 0.5 in. and the safe unit stress 15,000 psi, the safe pressure for a seamless shell is $2 \times 0.5 \times 15{,}000 \div 30 = 500$ psi. If the shell has a longitudinal riveted seam of 80 percent efficiency, the allowable pressure for 15,000 psi stress is $0.80 \times 500 = 400$ psi.

Heat Stresses

Almost any form of iron or steel (except cast iron) has a *modulus of elasticity* around 30,000,000; that is, a stress of 1 psi will stretch it 1/30,000,000 part of its own length. This relation holds right up to the elastic limit.

To put it another way, stretching iron or steel 1 in./100 ft (1200 in.) produces a tensile stress of $^1/_{1200} \times 30{,}000{,}000 = 25{,}000$ psi. The same increase in length can be caused by a temperature rise of 120°F (say, from 80 to 200°F).

It follows that if a piece of steel rod or pipe of any length is heated to 200°F, then rigidly held at both ends and cooled to 80°F, the resulting tensile stress will be 25,000 psi (Fig. 3-4d). This is about 200 psi for each degree of change.

TANK CAPACITIES

Tank dimensions are generally measured in feet and inches, but these compound numbers are inconvenient to multiply; so use the measurements in inches straight or in feet and decimals. If you turn inches into decimal parts of a foot, one decimal place is close enough for most tank work. To show the relative convenience of the two methods, we will work some of the following calculations in feet, some in inches:

Before starting, note the simple geometrical and other facts given in Fig. 3-5.

Q (Fig. 3-5a): What is the capacity of a rectangular tank 4 ft 8 in. $\times$ 3 ft 6 in. $\times$ 8 ft 4 in.?
A Dimensions in feet and decimals are $4.7 \times 3.5 \times 8.3$. $4.7 \times 3.5 = 16.4$, discarding second decimal place. $16.4 \times 8.3 = 136$ ft³, discarding both decimal places. Volume in gallons is $136 \times 7.5 = 1020$ gal. Weight of water contained $= 62.4 \times 136 = 8486.4$ lb, call it 8490 lb.

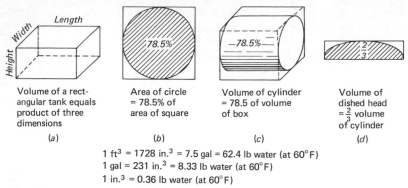

Volume of a rect-
angular tank equals
product of three
dimensions

(a)

Area of circle
= 78.5% of
area of square

(b)

Volume of cylinder
= 78.5 of volume
of box

(c)

Volume of
dished head
= $\frac{2}{3}$ volume
of cylinder

(d)

1 ft³ = 1728 in.³ = 7.5 gal = 62.4 lb water (at 60°F)
1 gal = 231 in.³ = 8.33 lb water (at 60°F)
1 in.³ = 0.36 lb water (at 60°F)

FIG. 3-5.   Volume of tanks is easy to calculate by sketching out areas.

In the last operation above, we discarded 3.6 lb to round out the figure to 8490 lb. In the following calculations the results will be rounded out in the same way. Note that in each case we carry the actual figures for three or four digits, not more than four, and use ciphers after that.

Q (Fig. 3-5c): What is the capacity of a flat-ended, round, vertical tank 72 in. in diameter and 94 in. high?
A Note that this tank is a cylinder and that the volume of a cylinder is 78.5 percent of the volume of the rectangular box it fits into (Fig. 3-5b). First figure volume of the imaginary box. Volume of box (in cubic inches) = 72 × 72 × 94. 72 × 72 = 5184, say 5180. 5180 × 94 = 94 × 486,920, say 487,000 in.³.

Volume of tank = 78.5 percent of 487,000 = 382,295, say 382,000 in.³. Volume in cubic feet = 382,000 ÷ 1728 = 221 ft.³ Capacity in gallons = 382,000 ÷ 231 = 1650 gal. Capacity in pounds = 382,000 × 0.36 = 137,-000 lb.

Q How about the volume of a tank with dished or bumped ends?
A The capacity of a bumped or dished head is two-thirds that of a cylinder of the same length and diameter. Thus to get the volume of such a tank, assume it to be a flat-ended cylindrical tank of a length equal to the length of the cylindrical part plus two-thirds the length of each bumped head minus two-thirds the length of each dished head (Fig. 3-5d).

Q A tank 72 in. in diameter has a full-diameter length of 110 in. and two bumped heads, each 12 in. deep. What is the equivalent flat-headed tank?
A Each 12-in. bumped head is equivalent to ²/₃ × 12 = 8 in. of straight tank; so the equivalent flat-headed tank is 72 in. in diameter and has a length of 110 + 8 + 8 = 126 in. This figures 300 ft.³

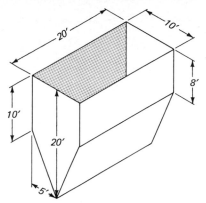

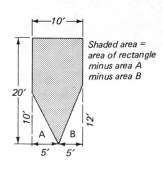

FIG. 3-6. Each specific shape of a coal bunker can be calculated as a separate area; then all of them are added together.

If the original tank had two dished heads, its equivalent length would be $110 - 8 - 8 = 94$ in. If it had one bumped head and one dished head, its equivalent length would be $110 + 8 - 8$, same as the original tank. In this case the bumped head would cancel the dished head.

NOTE: In all computations of tank capacity use the inside tank dimensions, not the outside.

COAL BUNKER

Q The bunker in Fig. 3-6 is level full of coal running 50 ft³/ton. How much coal does it contain?

NOTE: Often, in problems of this sort, engineers set down a number like 2,365,430. Now this doesn't make sense because none of these problems can be correct to the thousandth part of one percent. Here the 4 and the 3 are useless. Use zeros like this: 2,365,000.

A First draw the end view of the bunker and fill in all indicated dimensions by common sense. Note that the end is a 10×20 ft rectangle minus the two bottom triangles.

Triangle $A = 5 \times 10 \times \frac{1}{2} = 25$ ft²
Triangle $B = 5 \times 12 \times \frac{1}{2} = 30$ ft²
$\overline{}$ Sum 55 ft²
Net area is $(10 \times 20) - 55 = 145$ ft²
Volume $145 \times 20 = 2900$ ft³
Weight coal $= 2900 \div 50 = 58$ tons $= 116,000$ lb.   *Ans.*

WATER PRESSURE

At 60°F, water weighs 62.4 lb/ft³, and its weight does not depart from this figure by as much as 1 percent between 32 and 100°F.

A column of water 1 in. square and 2.3 ft high weighs 1 lb. Therefore, 2.3 ft of head will create a pressure of 1 psi—convenient to remember.

Q What head of water will create a pressure of 50 psi?
A 50 × 2.3 = 115 ft.

TERMINOLOGY USED IN BASIC PHYSICS

ABSOLUTE HUMIDITY: The weight of water vapor in a unit volume, commonly expressed as grains per cubic foot (7000 grains = 1 lb).

ABSOLUTE PRESSURE: The sum of the gage reading and the atmospheric pressure. At sea level, the atmospheric pressure used is 14.7 psi. This atmospheric value decreases with the elevation; for example, at 2000 ft altitude the pressure is 13,664 psi, at 10,000 ft only 10.1 psi, and at 16,000 ft 7.7 psi, which affects all power-plant equipment using air.

ABSOLUTE TEMPERATURE: The observed temperature plus 459.67°F, or on the Celsius scale, the observed temperature plus 273.15°C.

ABSOLUTE ZERO: A temperature of −459.67°F or −273.15°C. This is the base point from which all absolute temperatures are calculated. At this point there is no molecular action and therefore no heat.

ALKALINITY: A condition of water expressing the amount of carbonates, bicarbonates, hydroxides, silicates, or phosphates in solution, and measured in parts per million (ppm) or grains per gallon as calcium carbonate.

AMBIENT TEMPERATURE: The temperature of air in the area of study or consideration. Example: the ambient air temperature in the boiler room is 93°F.

BAROMETRIC PRESSURE: The atmospheric pressure as indicated by a barometer. Barometers are calibrated in inches of mercury. At sea level, the normal barometric pressure is accepted as 30 in., although the exact pressure under standard conditions of air temperature of 70°F and 60 percent relative humidity is 29.921 in. of mercury.

CALORIE: The calorie represents one-hundredth of the heat required to lift the temperature of 1 gram of water from 0 to 100°C. Thus it is approximately equal to the amount of heat required to heat 1 gram of water 1°C at a constant atmospheric pressure. One calorie is the equivalent of 0.003968 Btu.

CHANGE OF STATE: The change of a material from solid to gaseous, from gaseous to solid, from solid to liquid, etc. The opposite of steady state.

DEGREE DAY: For each calendar day, the number of heating degree days is the difference in degrees between the mean temperature for that day and 65°F. For example, a mean for the day of 50°F produces 65 − 50 = 15 degree days. The figure of 65°F is the dividing temperature between heat needed and not needed.

DEW POINT: The temperature of the air at which any reduction of the ambient temperature would produce some condensation of the air's water-vapor content. Thus the dew point is the temperature at which the air's relative humidity becomes 100 percent.

ENTHALPY: Term used instead of longer *total heat* or *heat content*. It is expressed in Btu per pound.

FUSION: A term used to describe the melting of ash or the melting of ice. The temperature of fusion is the temperature at which the change from solid to liquid takes place.

HEAT, SPECIFIC: The heat absorbed or given up by a unit of a substance in changing its temperature 1°F. The common unit is the pound, and the unit expressing heat is the British thermal unit (Btu) for liquids and solids. The specific heat of water is 1; the specific heat of air is 0.24, also taken on the basis of the pound. To convert cubic feet of air to pounds, refer to the standard tables of the American Society of Heating, Refrigerating and Air Conditioning Engineers (ASHRAE) appearing in its Guide.

ISOBARIC: The word applies to a change carried through a constant pressure. Isobars are lines on a map that connect points of equal barometric pressure. Superheating steam is an isobaric procedure.

LOAD FACTOR: The ratio of the average load during a given period and the normal maximum, or full capacity, load rating of the unit under consideration. Applicable also to the complete system.

MERCURY COLUMN: A measure of pressure. One inch of mercury at 62°F equals 0.4897 psi and is the equivalent of a water column 13.57 in. high. At 32°F, the inch of mercury (the chemical symbol Hg is found in many books but is seldom explained) exerts a pressure of 0.49112 psi.

MICROMETER: A unit of length (formerly called a micron) equal to $1/25{,}400$ in. and to $1/1{,}000{,}000$ m. This unit is highly useful in expressing the diameter of dust particles and in stating the capacity of a filter.

NATURAL CIRCULATION: Another term for the movement of fluids in a system resulting from differences of density rather than from the actions of a pump or blower. See Convection in Chap. 7, How to Use Steam Tables.

NITROGEN: This is an inert gas making up about 79 percent by volume of the atmosphere and used extensively as a purging medium for refrigeration systems and for combustible-gas systems. It is stored commercially in tanks and cylinders.

NORMAL BAROMETER: The atmospheric-pressure reading at sea level. This is 29.9 in. of mercury, or 14.7 psi, approximately. In practical calculations, a barometer value of 30 in. is usually close enough, since the pressure may vary almost from minute to minute because of weather fluctuations.

OXYGEN: An elemental gas that constitutes about 21 percent of the atmosphere by volume, the remainder being nitrogen. Traces of other elements, such as argon and helium, are present in the atmosphere, but they exert minimal effect on reactions between the air and other material and thus are generally ignored.

OZONE: An unstable compound of oxygen produced by the electric field immediately surrounding an electrode operating at high voltage. The presence of ozone suggests to a plant operator that a high-voltage discharge has taken place, such as a lightning stroke, and thus may have importance in correcting a problem.

PEAK LOAD: The maximum load carried by a system or a unit of equipment over a designated period of time.

PLENUM: An enclosed space that functions as a reservoir for air or gas and from which the air or gas is bled off through one or more ducts, similar to piping manifolds.

PRESSURE VESSEL: A closed vessel of any shape that confines some kind of fluid, either liquid or gas, at some pressure above atmospheric. When the vessel is not subjected to direct heating, such as an oxygen flask, it is termed an "unfired pressure vessel." This distinguishes it from a fired boiler or heater.

RADIATION, EQUIVALENT DIRECT (EDR): Heat delivery in the amount of 240 Btu, the amount of heat assumed to be delivered by 1 square foot of direct radiation under standard conditions of 215°F steam temperature and 70°F ambient.

REACTION: The process by which the bringing together of two or more materials produces a new substance or a release of energy. Reactions occur most often in chemistry.

RELATIVE HUMIDITY: The weight of water vapor present in air (or a gas) expressed as a percentage of the maximum weight of water vapor possible in the considered air (air is a gas) at the given temperature and pressure. Relative humidity appears as curved lines on the psychrometric chart. See Absolute Humidity.

SATURATION: A term used to describe a condition of air or gas in which the maximum possible amount of water vapor is present in the mixture for the existing temperature and pressure. The saturation line on the psychrometric chart is useful in air-conditioning design for the determination of the dew point. Saturation also describes a condition of 100 percent relative humidity. See Dew Point.

SPECIFIC VOLUME: The reciprocal of density. Whereas density refers to the weight per cubic foot, specific volume is based on the number of cubic feet in 1 lb. This number of cubic feet, compared with the number of cubic feet for 1 lb of water where liquids are involved and with air for gases and vapors, indicates the specific volume of the subject liquid or gas.

SPECTROSCOPIC ANALYSIS: The identification of materials by a common microscope is not always satisfactory, and where possible, the use of a spectroscope will eliminate doubts. In this instrument, the color of each element may be determined by light rays passing through or by reflection from the material's surface.

STANDARD AIR: This is air at 70°F, at sea level, and at 29.92 in. barometric pressure. Under these conditions, it weighs 0.078 lb/ft^3 and 12.82 ft^3 is needed to weigh 1 lb.

TEMPERATURE, CRITICAL: The temperature at which the properties of a liquid and its vapor phase are identical. For example, the critical temperature of water and steam is 705.4°F. At this temperature, the pressure is 3206.2 psi (absolute) and the volume of 1 lb of water is the same volume as that of 1 lb of steam. Enthalpy, or total heat content, of the water is 902.7 Btu/lb, and that of the steam which it becomes is the same 902.7 Btu. No latent heat is needed to turn the water into steam; the change of state results wholly from the combination of temperature and pressure.

TEMPERATURE, DRY-BULB: The temperature read on the conventional thermometer.

SUGGESTED READING

Elonka, Stephen M.: *Standard Basic Math and Applied Plant Calculations*, McGraw-Hill Book Company, New York, 1978 (has full chapter on metric conversion).

4

FUELS AND HOW THEY BURN

The reaction by which fuel and oxygen combine and give off heat is still the single most important energy-producing process in the world. Heat is the operating engineer's stock in trade, and combustion is the process the engineer lives with daily. As fuels get more expensive, and as industry demands more power and more steam, the stationary engineer faces the need to understand better how fuels burn and how to get the most out of them without polluting the atmosphere. This chapter gives the basics on combustion chemistry that all persons working with fuel should know.

CHEMISTRY OF MATTER

Q What is *matter?*
A *Matter* is the general name for all the material substances, gaseous, liquid, or solid, forming the earth and its surrounding atmosphere.

Q What is the composition of matter?
A All matter is composed of simple substances called *elements,* or combinations of these elements. For example, by combining the two simple substances iron and carbon we get steel. Water is a combination of two gases, hydrogen and oxygen. There are 92 elements, ranging from hydrogen, the lightest element, to uranium, the heaviest.

Elements seldom occur in their pure state but are usually combined with other elements in varying proportions to form the infinite variety of material things in our universe.

Q What is an *atom?*
A An *atom* is the smallest particle of matter that can take part in a chemical change. It is composed of yet smaller particles called *electrons,* and the number of electrons in the atom determines its weight. Hydrogen, the lightest element, has one electron rotating around a *proton,* whereas uranium has 92 rotating electrons.

Q What is meant by *atomic weight?*

A This term refers to the *comparative* weight of the atom. For convenience, oxygen is usually taken as 16 on the atomic-weight scale, and the weights of the other atoms are given as compared with oxygen. This makes the atomic weight of hydrogen 1.008, or slightly more than 1.

Q What is a molecule?
A A molecule is the smallest particle of matter that can exist alone. It is composed of two or more atoms of the same or different elements. For example, a molecule of oxygen is composed of two atoms of oxygen, and a molecule of carbon dioxide is composed of one atom of carbon and two of oxygen.

Q What is meant by molecular weight?
A Molecular weight, or weight of a molecule, is calculated by adding the atomic weights of the atoms contained. Thus, the atomic weight of oxygen is 16, and the oxygen molecule contains two atoms; so the molecular weight of oxygen is $16 \times 2 = 32$.

Q What is a chemical combination?
A It is the combination of atoms of two or more different elements to form another substance, very often having entirely different physical properties. For example, two atoms of hydrogen gas combine with one atom of oxygen gas to form one molecule of water.

In a given chemical compound the atoms always combine in the same proportions. Their combining is a chemical change, and a chemical process is required to break up the combination into its component atoms.

Q What is a mechanical mixture?
A A mechanical mixture is a physical mixing together of two substances and does not entail any chemical change. A mechanical mixture may be again split up into its component parts by some mechanical or physical process, such as screening or washing. For example, we could mix sugar and salt and then separate them by washing out the sugar. This would be a purely mechanical process.

Q How do you usually denote the composition of a complex substance?
A Writing out the names of the elements in full every time they are used would be rather a laborious process; so we use a system of symbols instead. Usually the symbol is the first letter, or letters, of the name, or its Latin equivalent. Thus, for carbon we write C, for oxygen O, and for iron Fe, from the Latin word *ferrum.* If more than one atom of an element is to be denoted, a small figure, called a *subscript,* is placed after and below the letter; thus O_2 means two atoms of oxygen, O_3 means three atoms, and so on.

Any combination of elements can be shown very simply and plainly by these symbols and subscripts. For example, hydrogen sulfide is a chemical combination of hydrogen and sulfur, consisting of two atoms of hydrogen and one atom of sulfur in each molecule. All this information can be conveyed briefly by writing H_2S. If more than one molecule is to be denoted, a large figure is placed in front of the symbol. Thus, $2H_2S$ means two molecules of hydrogen sulfide.

Q Name the most common elements in fuels and give their chemical symbols and atomic weights.

A

Element	Chemical symbol	Atomic weight
Hydrogen	H	1.008
Carbon	C	12.005
Nitrogen	N	14.01
Oxygen	O	16.00
Sulfur	S	32.06

These are the atomic weights, taking oxygen as 16. In calculations where atomic weights are used, the fractions are often neglected, as they are too small to make any practical difference in the answers.

Q What is air?
A By volume, the air we breathe is 21 percent oxygen and 79 percent nitrogen. The nitrogen is inactive, slow to react chemically with other substances. Not so the oxygen. The oxygen of the air is always busy tarnishing silver, coating copper with green, rusting iron and steel. All these processes are oxidation, which is the combining of substances with oxygen. All give off heat, but too slowly to start a fire.

COMBUSTION

Q What is combustion?
A Combustion is high-speed oxidation—so speedy that the heat of reaction keeps relighting the unburned part of the fuel, keeping the flame or burning continuous. Our eyes seem to tell us that wood, coal, and gasoline burn. Yet, strictly speaking, nothing truly burns unless it is a gas. When we burn coal, wood, or gasoline, we really burn gas produced from those solids or liquids.

A candle proves the point. The solid wax won't burn (Fig. 4-1). Even the liquid wax in the wick won't burn. But the vapor formed from this liquid as it rises up the wick to meet the flame does burn and generates further heat to melt and vaporize the next layer of wax. So the candle keeps burning as

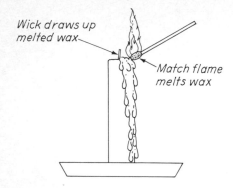

Wick draws up
melted wax

Match flame
melts wax

FIG. 4-1. Wick draws up melted wax by capillary action.

the wick sucks up new liquid to be vaporized and ignited by the heat of what went before.

Q How does gas burn?

A The gas must be combustible. It must be mixed with air in proper proportions. The mixture must be raised to the ignition or firing temperature and held there.

The simplest combustible gas is the element hydrogen. Chemists tell us that two atoms of hydrogen combine with one atom of oxygen to produce H_2O, which is ordinary water. When generated by combustion at high temperature, this water is first an invisible vapor that may later condense as liquid water. For every pound of hydrogen burned to water, 62,000 Btu of heat energy is given off.

Burning 1 lb of hydrogen uses up 8 lb of oxygen to produce 9 lb of water. The large amount of nitrogen carried along with the oxygen takes a free ride and does not enter into the chemical reaction.

Natural gas is mostly methane, CH_4. This chemical formula means that one molecule of natural gas contains one atom of carbon and four of hydrogen. When this natural-gas molecule burns, the hydrogen goes to H_2O (water) as before and the carbon to CO_2 (carbon dioxide, two atoms of oxygen and one of carbon). See Table 4-1.

Q How does hydrocarbon burn?

A There are many hydrocarbon fuels consisting of carbon and hydrogen in various proportions. Whatever the proportions, the hydrogen finally burns to H_2O. The carbon normally burns to CO_2, but sometimes burns incompletely to CO (carbon monoxide).

Even if a cold combustible gas is mixed with exactly the right amount of oxygen for combustion, nothing happens until the temperature of some portion of the mixture is raised to the ignition temperature. This starts the burning, and the heat of the burning gas ignites the next portion of mix-

TABLE 4-1   Characteristics of Gases Used for Power Generation

Property	Propane	Butane
Boiling temp., °F at atmospheric pressure	−51	15
Weight per gal at 60°F, liq form, lb	4.24	4.84
Vapor pressure at 100°F, psig	195	65
Volume of gas, ft³ per liq gal	36	32
Heating value, Btu:		
Per ft³	2550	3200
Per gal	91,800	102,400
Per lb	21,650	21,500
Specific gravity:		
Liquid	0.509	0.576
Gas	1.52	2.00
API at 60°F	147	111

Kind of gas and values	Specific gravity*	Heating value Btu/ft³
Carbureted water gas	0.63	550
Coke-oven gas:		
Max	0.44	645
Avg	0.40	559
Min		466
Natural gas:		
Max	1.21	1548
Avg	0.72	1090
Min	0.57	977

* Specific gravity is referred to air.

ture, so that fire spreads rapidly. An example is the cylinder spark in a gas engine which starts the combustion in a small area around the spark gap. Soon the whole mass is aflame.

If a mixture contains too much air, or too little air, it is harder to ignite. It is possible to get the proportions so far from a perfect mixture in either direction that a flame cannot be maintained. Every car driver knows he can be stalled by too rich a mixture as well as by one too lean (a too-lean mixture has a lot of air that can't be used). In either case the dead mass absorbs heat and lowers the temperature produced by combustion. When combustion can't create a temperature high enough for ignition, flame cannot spread in a mixture.

A continuous flame requires a continuous supply of fuel, as in the old-style fishtail gas burner (Fig. 4-2) and in the bunsen burner (Fig. 4-3). The fishtail gives a yellow flame and the bunsen a blue flame.

Q   What do yellow and blue flames indicate?

A   If pure hydrogen were burned, both flames would be pale blue. Hydrogen cannot produce a yellow flame because it contains no carbon. However, city gas and natural gas do contain substantial quantities of carbon atoms in the molecules of hydrocarbons—compounds of hydrogen and carbon. Here, if the hydrogen can be burned a small fraction of a second ahead of the carbon, the carbon particles are released as a cloud of individual molecules, which can glow brightly for an instant before they are burned. Millions of these incandescent particles constitute a yellow flame.

The fishtail burner soots this gas out, without any contained air, in a thin flat stream through the slotted opening in the tip. The dark area just above the tip is the cold gas in the process of picking up and mixing slightly with the surrounding air. Suddenly, at a certain level, the mixture reaches cor-

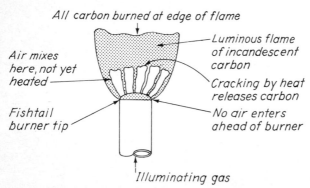

FIG. 4-2.   Fishtail gas burner shows yellow flame.

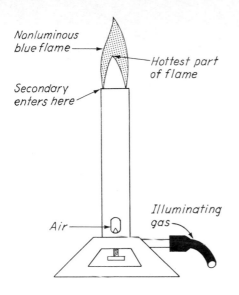

FIG. 4-3. Bunsen burner operates with blue flame.

rect proportion and starts to burn rapidly. Note that the heat of the flame at any instant is igniting the cooler mixture on the way up.

Q Explain the term *cracking*.

A The temperature splits the hydrocarbon molecule into carbon particles and hydrogen gas. The hydrogen gas burns first with a pale flame, but the flame is colored bright yellow by the millions of incandescent carbon particles. The level at which these particles are completely burned to CO_2 is the top edge of the flame (Fig. 4-2). No further heat is generated above this point but, of course, the heat previously released in the flame below carries up in the invisible combustion products.

If a cold spoon is held well above the fishtail flame, no soot deposits because all the carbon has been burned. The same spoon stuck into the flame (or into a candle flame) is quickly coated with lampblack (soot). Here the spoon cools the mixture below the ignition point before the carbon is all consumed, so that the unburned carbon deposits on the spoon.

Q How are soot and smoke formed?

A There is an important point in the previous question for boiler operators. Wherever a yellow flame is allowed to strike relatively cold surfaces (even those as hot as a boiler tube), it will be cooled below the ignition temperature before the carbon is all burned. The unburned carbon deposits as soot on the heating surfaces or goes up the stack as smoke. In either case, there are waste and pollution.

To prevent smoke and soot, make sure combustion is complete before yellow flames reach the tubes in the boiler. The main ways to do this are to increase furnace height, volume, or temperature, or to ensure more perfect mixing so that the combustibles burn up sooner.

Q Explain how flame behaves in a boiler's furnace.

A In one area of the furnace, the mixture may be too rich to burn. In other areas there may be almost all air and hardly any combustion gas. It may be so lean a mixture that it burns badly or not at all. Thus, even if the average amount of air is right, the combustion may be bad because some parts of the mixture are too rich and some are too lean.

These explanations may make burning seem simpler than it really is. Actually, the hydrocarbons from coal, oil, and gas may go through a whole series of reactions. Yet the operating engineer must remember that, however it starts, the hydrogen finally ends up in H_2O and the carbon (if it burns fully) in CO_2. But if the carbon burns incompletely, it may produce CO gas, which is capable of further burning, and which is a big waste if allowed to pass up the stack unburned.

FUELS AVAILABLE

Q What are commercial fuels burned in boilers?

A All commercial fuels considered in this book consist of one of the following, or of a mixture of two or more of them: (1) gaseous hydrocarbon, (2) solid carbon, or (3) a mixture of carbon monoxide and hydrogen.

Coke and charcoal have little combustible material beyond the solid carbon. Bituminous (soft) coal includes in addition quite a bit of distillate hydrocarbons. Natural gas consists almost entirely of gaseous hydrocarbons. Ordinary carbureted water gas is largely carbon monoxide and hydrogen with a fair amount of hydrocarbons.

Fuel oil vaporizes into gaseous hydrocarbons before it actually burns. The oil does not burn as oil because the oil is cracked in the flame, thereby producing solid carbon (soot) and hydrogen. Coal is another fuel that decomposes into elementary fuels before actual burning starts. When coal is first exposed to the fire's heat, gaseous hydrocarbons, carbon monoxide, and hydrogen are distilled off, leaving solid carbon behind.

Q Explain the actual firing operation in a furnace.

A Broadly speaking, fuel may be burned (1) in suspension or (2) on a fuel bed. The simplest example of suspension burning is gas firing, in which the burner delivers the right proportions of air and gas.

An oil burner first has to convert the oil into a gas. The coal burner must be arranged to distill off the volatile matter soon after the coal enters the furnace. In both cases the furnace heat does the job, but the firing equipment puts the fuel into a condition to make the best use of this heat. For suspension firing of both coal and oil, the condition is the same: the fuel must be broken up into many small particles to expose as much surface as possible.

With oil, this means good atomization. With coal, it means fine grinding. In addition, turbulence is needed to strip away the protective coating of dead gas from the particles.

Q What are the functions of the burner compared with those of the furnace?

A The furnace takes over where the burner leaves off. Within the combustion zone, fuel must be vaporized or distilled, mixed with air, ignited, and the resulting reaction between fuel and oxygen carried to a finish. Mixing fuel and air is chiefly the job of the furnace. Heating and igniting are chiefly the burner's job. We know how hard it is to keep one log burning in a fireplace. With two or more logs the fire keeps going. In the same way, a furnace must maintain a heat supply to prepare and ignite incoming fuel.

Q How are soot and smoke prevented?

A To prevent soot and smoke, the combustion steps (mixing, ignition, turbulence, burning carbon, etc.) must be finished in the short time it takes for the particles to travel from the burner to the furnace outlet. Turbulence is the whirling and eddying of air and combustion gas. For boiler furnaces this is much better than streamlined flow. Turbulence (1) increases the time available for combustion, (2) mixes fuel and air better, and (3) helps the air wash away the dead combustion-product gases and expose fresh surfaces for quick combustion. Without this turbulence, a burning particle soon surrounds itself with a protective layer of unburnable combustion products, mainly CO_2 and nitrogen.

Just remember that good combustion depends upon the three T's: Temperature, Time, and Turbulence. Ash is the part of the coal that can't burn, and hence might not seem to enter this picture. Just the same, it can make trouble. Ash particles, out of control, melt, stick on furnace walls and tubes as slag, wash down refractory walls, and tend to clog the gas passages on their way up to the stack.

Q What are the principal constituents of the fuels used by the power engineer?

A The principal constituents of all fuels used for boiler firing purposes are carbon, hydrogen, oxygen, nitrogen, sulfur, and, for coal, the incombustible elements that form ash.

Q What is the most accurate method of finding the heating value of a fuel?

A   The heating value of a fuel can be found most accurately by burning a measured sample with pure oxygen in an apparatus called a *calorimeter,* where the heat of combustion is absorbed in water and the heating value is determined by noting the rise in the temperature of the water. The heating value of a solid or liquid fuel is usually given as so many Btu per pound, and the heating value of a gaseous fuel as so many Btu per cubic foot, measured at some standard temperature and pressure.

Q   What is a Btu?

A   Btu is the abbreviation for *British thermal unit.* It is the amount of heat required to raise the temperature of one pound of water one degree Fahrenheit. At different temperatures the amount of heat required to raise the temperature of one pound of water one degree may be slightly more or less, but the variation is so small that we usually take 1 Btu as the heat required to raise the temperature of one pound of water through one degree at any temperature.

Q   What are the principal substances used as fuels by the engineer?

A   The principal ones are solids, such as wood and coal; liquids, such as fuel oils derived from petroleum; and gases, such as natural gas, producer gas, blast-furnace gas, and coal gas.

COAL, FUEL OIL, AND GAS

Q   Into what classes is coal divided?

A   Numerous methods of classifying coal have been used and suggested, based variously on carbon content, ratio between fixed carbon and volatile matter, coking or noncoking qualities, and on other physical characteristics. Nevertheless, the great differences in the composition and physical appearance of coals from the various coal fields, and even from different sections of the same coal seam, make it hard to find any entirely suitable and satisfactory classification system for all coal.

In general there are three main coal classes: anthracite, bituminous, and lignite, but no clear-cut line exists between them, and we have other coals classed as semianthracite, semibituminous, and subbituminous. Anthracite is the oldest coal, geologically speaking. It is a hard coal composed mainly of carbon with little volatile content and practically no moisture. Coming up the scale toward the youngest coal (lignite), carbon content decreases and volatile matter and moisture increase.

Fixed carbon refers to carbon in its free state, not combined with other elements. *Volatile* matter refers to those combustible constituents of coal that vaporize when coal is heated.

Q   What are the approximate heat values of the various classes of coal?

A   Because of the wide variation in composition it is impossible to give definite figures for each class. Generally speaking, the heat value varies from as low as 7000 to 8000 Btu/lb for lignite to around 15,000 Btu/lb for high-grade bituminous.

Q   What is meant by pulverized coal?
A   This is coal ground to a fine powder before being fed into the furnace. It is blown into the furnace by a strong blast of air and burns in much the same manner as gas.

Q   What are the average composition and heat value of wood?
A   Heat value of wood depends largely upon its moisture content. This may run as high as 50 percent in green wood. Even seasoned wood contains 15 to 25 percent moisture—about 8 percent for kiln-dried.

The heat value of perfectly dry wood is about 8500 Btu/lb. The same wood with 25 percent moisture would run about 6400 Btu/lb. Sawdust has much the same heat value as the wood from which it is cut.

Q   What are the average composition and heat value of fuel oil?
A   Fuel oil may be crude oil direct from the well but is commonly the residue left after the lighter oils, such as gasoline and naphtha, have been distilled from the crude. It is composed mainly of hydrocarbons (compounds of hydrogen and carbon) with small quantities of moisture, sulfur, oxygen, and nitrogen. The following is a typical analysis:

	Percent
Carbon	84
Hydrogen	13
Sulfur	1
Oxygen	1
Nitrogen	1

Heat value is approximately 19,000 Btu/lb.

Q   What are the composition and heat value of natural gas?
A   Natural gas is composed mainly of hydrocarbons, such as methane (CH_4) and ethane (C_2H_6), with small amounts of carbon dioxide (CO_2), oxygen (O_2), nitrogen (N_2), and sometimes hydrogen sulfide (H_2S). The following is a typical analysis:

	Percent
Methane	80.2
Ethane	17.3
Nitrogen	2.3
Oxygen	0.2

The average high heat value is around 1000 Btu/ft³, measured at 14.7 psi and 60°F.

Q What are the combustible elements in fuels?
A The combustible elements are carbon, hydrogen, and sulfur.

Q What are the noncombustible elements in fuels?
A The noncombustible elements are nitrogen and those elements that make up the compounds forming the moisture and the ash.

Q What element in fuels is harmful in metals?
A Sulfur is harmful because it will combine with condensate to form sulfurous and sulfuric acids (H_2SO_3 and H_2SO_4), which have a corrosive action on iron and steel.

Q What are the products of the complete combustion of carbon, hydrogen, and sulfur?
A If burned with a sufficient supply of oxygen, carbon will burn to carbon dioxide (CO_2), hydrogen will burn to water vapor (H_2O), and sulfur will burn to sulfur dioxide gas (SO_2).

Q What are the products of the incomplete combustion of carbon, hydrogen, and sulfur?
A If the supply of oxygen is insufficient for complete combustion, only part of the hydrogen and sulfur will be burned to H_2O and SO_2; the rest will pass off unchanged. If burned with an insufficient supply of oxygen, carbon will burn to carbon monoxide (CO), instead of CO_2. This carbon monoxide is a combustible gas and is also highly poisonous.

Q What is meant by *excess air,* and why is it necessary?
A The amount of air *theoretically* necessary to burn a fuel completely can be calculated from the ultimate analysis of the fuel, but in practice we must introduce more air into the furnace than the theoretical amount to make sure that all the combustible elements of the fuel receive enough oxygen for complete combustion.

Q What are the usual percentages of excess air admitted to boiler furnaces with (1) coal—hand-fired? (2) coal—stoker-fired? (3) oil, gas, or powdered fuel?
A 1. With hand firing: (*a*) Under good conditions—50 percent; (*b*) Under poor conditions—100 percent and over.
   2. With mechanical stokers—20 to 50 percent
   3. With oil, gas, or powdered fuel—10 to 30 percent

Q What are the disadvantages of using large amounts of excess air?
A This excess air does not take part in combustion but absorbs and carries off heat from the furnace to the chimney, thus lowering the efficiency of the steam generating unit.

COAL ANALYSIS

Q What methods are used to analyze coal?

A There are two methods: the *ultimate analysis* splits up the fuel into all its component elements, solid or gaseous; and the *proximate analysis* determines only the fixed carbon, volatile matter, moisture, and ash percentages. The first must be carried out in a properly equipped laboratory by a skilled chemist, but the second can be made with fairly simple apparatus. Note that *proximate* has no connection with *"approximate."*

Q Explain how a proximate analysis is made.

A **To Determine Moisture** Crush a sample of raw coal until it will pass through a 20-mesh screen (20 meshes per linear inch), weigh out a definite amount, place it in a covered crucible, and dry in an oven at about 225°F for 1 hr. Then cool the sample to room temperature and weigh again. The loss in weight represents moisture.

To Determine Volatile Matter Weigh out a fresh sample of crushed coal, place in a covered crucible, and heat over a large bunsen burner until all the volatile gases are driven off. Cool the sample and weigh. Loss of weight represents moisture and volatile matter. The remainder is coke (fixed carbon and ash).

To Determine Carbon and Ash Remove the cover from the crucible used in the last test, and heat the crucible over the bunsen burner until all the carbon is burned. Cool and weigh the residue, which is the incombustible ash. The difference in weight from the previous weighing is the fixed carbon.

Q Of what value is a coal analysis to the engineer?

A It enables the engineer to compare one fuel with another on the basis of moisture, ash, and combustible matter.

TERMINOLOGY USED IN FIRING FUELS

AIR-FUEL RATIO: The amount of air, in cubic feet or pounds, being furnished for combustion per cubic foot or per pound of the fuel. For example, 12 lb of air per pound of coal is a 12:1 ratio and gives almost perfect combustion.

FLASH POINT: The temperature at which sufficient vapor is emitted from fuel oil to produce a momentary flame when subjected to ignition, this flame immediately expiring.

HEAT BALANCE: The name given to a procedure by which the efficiency of a combustion process is determined, all heat losses being added together and the total being subtracted from 100 percent. The remaining figure is the efficiency. The losses are heat loss in dry gas, percent; loss in moisture in the fuel, percent; moisture in the air, percent; loss by radiation from hot surfaces, percent; other losses not accounted for, 1.5 percent; also, loss due to the burning of hydrogen in the fuel, percent; loss in unburned carbon in the ash (coal fuel), percent; and finally, the loss from the incomplete combustion of the carbon, percent.

OVERFIRE AIR: Secondary air introduced into a furnace above the grate for the purpose of completing combustion and producing turbulence in the furnace. This turbulence increases the efficiency of the combustion.

PREHEATED AIR: Air that is warmed above the atmospheric temperature for the purpose, in combustion, of promoting faster ignition of the fuel and more complete burning.

PRIMARY AIR: Combustion air introduced into a furnace in such a way as to make initial contact with the fuel, usually being mixed with the fuel either immediately before it ignites or simultaneously with the ignition. In pulverized-coal systems, the primary air enters the pulverizer mill, where it is mixed with the coal and proceeds to the burners as a mixture.

RADIATION LOSS: *See* HEAT BALANCE.

RESIDUAL FUEL: Crude petroleum after the removal of volatile hydrocarbons and water.

SECONDARY COMBUSTION: The burning of fuel in areas or spaces beyond the outlet of a furnace and where combustion is not intended by the design.

TERTIARY AIR: The injection of air into a furnace in addition to the primary and secondary air. The purpose of the tertiary injection is generally the establishment of greater furnace turbulence and the reduction of smoke. This is not a universal practice, but with today's strict EPA restrictions, it may be a solution.

WASTE FUEL: Any combustible substance that is a by-product of a process or product. For example, bagasse is an otherwise waste by-product of a sugar mill.

WASTE HEAT: Heat created or released as a by-product of some process or of some manufacturing or generating activity; it is usually wasted if no useful application is found. As an example, the heat released by the operation of a diesel engine in its exhaust is wasted to the atmosphere unless some application is found. Common uses are the heating of water and the passing of the exhaust gases through the heat-exchange surfaces of an unfired steam boiler, thus generating steam for heating. Less common is the use of the waste heat in an absorption refrigeration or air-conditioning system.

SUGGESTED READING

Books

Elonka, Stephen M., and Anthony L. Kohan: *Standard Boiler Operators' Questions and Answers,* McGraw-Hill Book Company, New York, 1969.

Elonka, Stephen M., and Joseph F. Robinson: *Standard Plant Operator's Questions & Answers,* vols. I and II, McGraw-Hill Book Company, New York, 1981.

5

BURNING COAL, OIL, AND GAS

After the stoppage of fuel oil shipments from the Mideast to North America in 1973 and the rapidly rising prices after shipments resumed, many plants that had earlier converted to oil or gas turned back to burning coal. North America has abundant supplies of coal, which is stoker-fired for most industrial boilers. But only petroleum products can be used in internal-combustion engines, for which they must be conserved.

Here we cover the methods and equipment needed for burning coal, oil, and gas, from which fuels the operator must squeeze the last possible Btu because of spiraling fuel prices and the restrictions on air pollution. And in case you ever have to hand-fire coal to at least supply heat for your plant, this is also covered.

HAND-FIRING COAL

Q   What tools are used in hand-firing steam boilers?
A   They are:

1.   A *shovel* spreads fresh fuel on the fire and loads coal or ashes if they are handled by wheelbarrow.

2.   A *slice bar*, before the fire is cleaned, breaks up clinkers that adhere to the grate. With the *alternate method* of cleaning fires, the bar sweeps fire from one side of the grate to the other.

3.   A *hoe* pulls out ashes from the top of the grate when the fire is being cleaned, and cleans out the ashpit.

4. A *rake* cleans fires and spreads fuel evenly over the grate.

5. A *lazy bar* is a short bar whose ends are hooked over the door hinge and latch of the fire door and ashpit door. It supports hoe and rake when they clean the fire and ashpit.

Q What hand-firing systems are used?

A 1. Spreading Method: Coal is spread in an even layer over the entire grate at each firing, usually commencing at the back of the grate and working out toward the fire door. Coal must be fired frequently in small quantities.

2. Alternate method: Each side of the grate is fired alternately so that the volatile gases distilled from fresh fuel will be ignited by the bright fire on the other side of the grate.

3. Coking method: Coal is fired at the front of the grate and allowed to coke there. Afterward it is pushed back and spread over the grates, and more coal is fired at the front. Thus the hot gases distilled from fresh fuel at the front of the grate ignite and burn as they pass over the glowing fire to the rear of the grate.

The alternate and coking methods are preferable to the spreading method when coal is high in volatile matter, since they do not lower the furnace temperature by blanketing the entire fire surface with fresh fuel at any one time. Also, there is less heat loss from volatile gases passing to the chimney unburned.

The spreading method, though probably most generally used, is efficient only when firing is light and frequent so as not to reduce the furnace temperature below the ignition point of the gases distilled from the fuel. Covering the entire fire surface at infrequent intervals with a thick layer of fresh fuel is most wasteful and inefficient.

Q Describe two methods of cleaning fires when hand firing.

A Two methods are:

1. Cleaning alternate sides: The top layer of coked fuel on one side of the grate is swept over to the other side by the hoe and slice bar. The clinker and ash thus laid bare are pulled out with the hoe. The coked fuel is swept back over the clean part of the grate and fresh fuel fired. When the fire again burns brightly, the other side of the grate is cleaned in the same manner.

2. Front to rear: The coked fuel on the grate front is pushed to the rear after ashes and clinkers have been worked up to the front of the grate with the hoe. Ashes are then pulled out through the fire door and coked fuel is pulled forward and spread evenly over the grate. The alternate is better than the front-to-rear method, but sometimes the latter is the only one that can be used where grates are small and the fire door is narrow.

Q   Sketch and describe several grate bars for stationary grates.

A   A *plain single-grate* bar is shown in Fig. 5-1. This is a straight cast-iron bar with side projections at ends and center to keep the bars forming the grate far enough apart to leave an air space.

Figure 5-1 also shows a *plain double-grate* bar and a *herringbone-*, or *Tupper-grate* bar. The latter is wider and has less tendency to warp than the plain grate bar. For a grate bar for burning sawdust or other fine material, a large number of fine holes permit air for combustion to pass up through the grate, and yet allow very little unburned fuel to drop into the ashpit.

Q   Sketch and describe a *pyramid-type* grate bar.

A   This bar (Fig. 5-1) has a layer of flat-topped hollow pyramids with slits in the sides through which air passes. The great number of small air openings gives very even air distribution through the fire bed, with no direct upward current and no large openings through which fuel falls into the ashpit. This grate is made in both stationary and shaking types.

Q   How are shaking grates constructed, and what are their advantages over stationary grates?

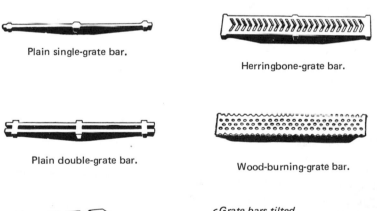

Plain single-grate bar.

Herringbone-grate bar.

Plain double-grate bar.

Wood-burning-grate bar.

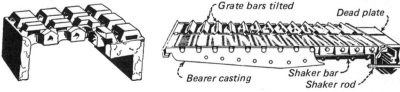

Pyramid grate.

Shaking grate.

FIG. 5-1.   Six types of grate bars used today for stationary grates.

A Shaking-grate bars are supported at both ends by pins resting on a notched side plate, and have projecting lugs attached to a long flat bar running the entire length of the grate, at the side. This bar extends through the boiler front or connects to an extension rod (Fig. 5-1). Moving this rod to and fro by a hand lever or a power cylinder attached to the boiler front rocks or tilts all the grate bars. If the grate is made in two sections, two levers or rocking cylinders are used. Figure 5-1 shows a shaking grate with the bars on one side tilted, as they would be when fires are being cleaned, and the bars on the other side flat, as they would be in the usual firing position.

With stationary grates, fire doors must be kept open when the fires are being cleaned, but with shaking grates the fires can usually be cleaned by rocking the grate bars and working the ashes and clinkers down through the grate into the ashpit, without having to keep the fire doors open for long periods, or temporarily baring part of the grate and thus allowing excessive quantities of cold air to enter and cool the furnace.

Q What regulates the size of air spaces in a coal-burning grate?

A The main factors regulating their size are the kind of coal being burned and the nature of the draft. Much coal is lost if a wide air space is used in grates burning slack or fine coal. Air openings can also be smaller if mechanical draft is used. In general, air spaces run from $1/4$ to $3/8$ in. for small coals and from $1/2$ to $3/4$ in. for larger sizes. The quantity of ash in the fuel and the coking or noncoking properties of the coal also have a bearing on the proper size of air space to use.

MECHANICAL STOKERS

Q Why are mechanical stokers preferable to hand firing?

A Less labor and attendance are needed to operate mechanical stokers. Cheaper grades of fuel are burned successfully. More fuel can be burned per square foot of grate surface. Much higher efficiencies can be attained. Better furnace conditions can be maintained. Production of smoke can be eliminated. Firing conditions can be controlled more exactly to meet varying loads, and overloads can be carried that would be impossible with hand firing.

Q Into what general classes may mechanical stokers be divided?

A Most stokers in common use fall into one or the other of these classifications: (1) sprinkler stoker, (2) traveling or chain-grate stoker, (3) overfeed stoker, (4) underfeed stoker.

1. Sprinkler stoker: May be used in conjunction with stationary grates, shaking grates, or even chain grates. In the last-mentioned case, there is no

need to open fire doors or clean fires; good results are claimed for this combination. The sprinkler stoker spreads coal over the grate much the same as an expert hand fireman does with his shovel. It may be mounted on the boiler front above the fire doors, leaving them clear for cleaning fires if the grates are stationary or for hand firing if the stoker should break down or be cut out for repairs.

2. Traveling or chain-grate stoker: Coal is fed into the front of a revolving grate, and burns as it travels toward the rear of the grate where the ashes are dumped over the back.

3. Overfeed stoker: Coal is fed onto a stepped reciprocating grate, either at the front or the side, and is gradually consumed as it travels toward the rear or the bottom of the grate.

4. Underfeed stoker: In one form, coal is fed into one or more deep retorts by power-driven rams and overflows onto side grates or dead plates which slope downward to dump plates that can be tilted to dump ashes when necessary. In the other type, grates slope from front to rear, and coal is also fed by power-driven rams into narrow retorts between the grate bars, whence it flows over onto the grates. The grates have a reciprocating motion that moves the coal downward to dump plates at the bottom that discharge the ashes to an ashpit or hopper.

Chain-grate and overfeed stokers in small or medium-sized power plants may operate on natural, induced, or forced draft. Traveling or chain-grate stokers in large plants are usually constructed for forced-draft operation. Because of the thick fuel bed, the underfeed stoker is essentially a forced-draft stoker.

Q Is it practical to install stokers for burning coal when modernizing very small boiler plants?
A Yes, Fig. 5-2 shows one such installation for a cast-iron boiler plant. Here a screw conveyor is direct-fed from the bin. Similar systems are used in plants up to 100 boiler hp for such services as supplying heating steam

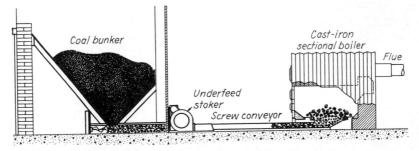

FIG. 5-2. Bin-fed stoker for hand-fired boiler being modernized.

for small apartment and office buildings, hotels, or for process steam on a year-round basis in smaller industrial plants, dairies, etc.

Q Are chain-grate stokers practical for smaller boiler plants?

A Figure 5-3 shows a suggested layout. This chain grate has the advantage of being a partial package-type unit with simple ash-removal arrangement. Ash is removed as it forms with a screw conveyor shown installed in a pit under the conveyor's end.

The amount of ash to be removed from a 90-hp plant, for example, may be about 15 tons/year in a lightly loaded heating plant. But it may be high as 65 tons/year in high-pressure year-round plants. Thus, with today's higher labor costs, mechanical devices for saving labor must be considered.

Q Describe a sprinkler stoker.

A Figure 5-4 shows a forced-draft sprinkler stoker with power-operated shaking grates. Firing doors are fitted to the boiler front beneath the stoker so that the boiler may be hand-fired if the stoker is stopped for any reason. When the stoker operates, a reciprocating feed plate feeds coal from the hopper into a compartment containing a revolving shaft with metal fingers that sprinkle coal over the fire with a spreading action. Rate of feed and sprinkler shaft speed can be adjusted to suit the boiler load. In some types of sprinkler stoker, parts that are exposed to excessive heat are water-cooled.

Q Describe construction and operation of traveling-grate and chain-grate stokers.

A In the *traveling-grate stoker* the small castings forming the grate surface are fastened to heavy transverse bars that are in turn attached to sprocket-

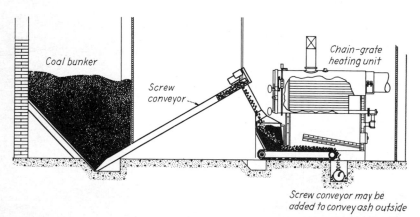

FIG. 5-3.   Chain-grate stoker can be fed by a screw conveyor from fuel bin while ashes are removed by another screw conveyor.

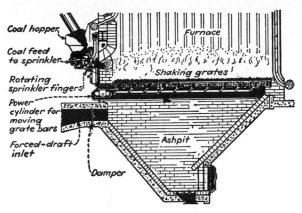

FIG. 5-4.   Sprinkler stoker has shaking grates.

driven chains. In the *chain grate* (Fig. 5-5) the small grate castings are linked together by transverse rods, and the grate is driven by sprocket wheels engaging special sprocket links. Both grates form a continuous chain or band, and the process of combustion is the same in either case. Coal from a hopper is fed onto the front part of the grate through an adjustable door which regulates the thickness of the fuel bed.

The speed at which the grate travels should be regulated so that the coal is entirely burned to ash when it reaches the rear of the grate. The ashes fall over the rear of the grate into an ashpit or ash hopper which is usually emptied at intervals by a conveyor. With induced draft, the ashpit is open and air passes up freely through the grates. With forced draft, the ashpit is closed and the space inside the grate is divided into compartments or zones, thus enabling the air supply to be regulated to different parts of the

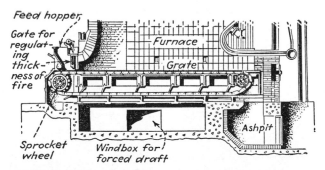

FIG. 5-5.   Chain-grate stoker over forced-draft wind box.

grate to meet combustion needs. A side view of a forced-draft chain-grate stoker is shown in Fig. 5-5.

Q Sketch and describe a front-feed overfeed mechanical stoker.

A Figure 5-6 shows a side view. In this stoker, coal flows by gravity from a hopper onto a dead plate and is fed from there to the grate by a reciprocating pusher. The grate is steeply inclined, and each separate bar has a rocking motion that gradually works the coal down to the bottom of the grate. Ashes pass onto a dumping grate and are dumped at intervals into an ash hopper. The grate bars and pusher-feed plate are operated by an eccentric, and the eccentric shaft is driven by an electric motor or small steam engine or steam turbine.

Q Describe the operation of an underfeed stoker.

A There are a great many different types of underfeed stokers, but practically all operate on the principle illustrated in Fig. 5-7, which shows a sectional view of a forced-draft single-retort underfeed stoker. Coal from a feed hopper is pushed forward into a deep central retort by a power-driven ram and distributed the full length of the retort by auxiliary pushers. The coal spills over the sides of the retort onto inclined side grates, some of which have a reciprocating motion that feeds the coal slowly downward onto dump plates. These are tilted at intervals to deposit ashes in ashpits or ash hoppers. Ashes are removed through doors in the stoker front or, if hoppers are used, discharged into ash cars or conveyors.

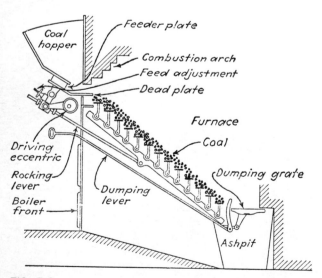

FIG. 5-6. Overfeed inclined-grate stoker is slanted steeply toward the ashpit at the rear end of furnace.

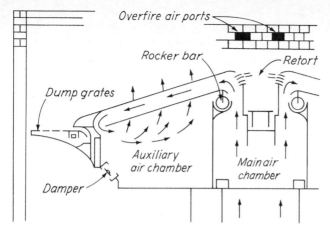

FIG. 5-7. Single-retort stoker is one of many different under-feed types in use today.

The grate bars are hollow, and some of the air for combustion is admitted to the fire through openings in the grate bars at the point of distillation of the gases, that is, just above the fresh green fuel being fed in at the bottom of the retort. The rest of the air passes through the hollow grate bars to auxiliary wind boxes at the lower ends of the grates and is distributed from the boxes up through the fire bars to the fuel on the grates. Air may also be admitted over the fire to assist combustion and prevent smoke.

Figure 5-8 is a side view of a multiple-retort underfeed stoker. It has a number of narrow retorts with narrow grates between. The main grates slope quite steeply from front to rear, and the coal feeds downward to a short grate with reciprocating bars that carry the ashes onto a dump plate. It is tilted periodically by hand or power and the ashes are deposited in an

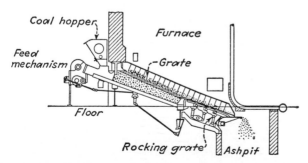

FIG. 5-8. Multiple-retort underfeed stoker has sloping main grate with narrow retorts and grates between.

ashpit or hopper. As in the center-retort stoker, power-operated rams feed the coal into retorts from beneath the grate, and forced draft distributes the air for combustion through the fuel bed.

BURNING PULVERIZED FUEL

Q How is pulverized fuel burned in boiler furnaces?
A The coal is ground to fine dust in a pulverizing mill. An air current blows this dust into the furnace where it burns almost like a gas. Figure 5-9 shows one type of pulverized fuel burner. The fuel-laden air current enters an outer chamber which is shaped to give the current a circular motion. This tends to throw the coal outward; so vanes are placed around the inner circumference to divert the stream toward the center and keep the coal-and-air mixture evenly distributed across the entire area of the pipe. The coal spreader at the burner tip breaks up the mixture into alternate lean and rich layers and causes the stream to assume a conical form.

Secondary air, supplied by a separate fan, is admitted around the tip of the fuel burner. The volume of secondary air can be regulated by dampers,

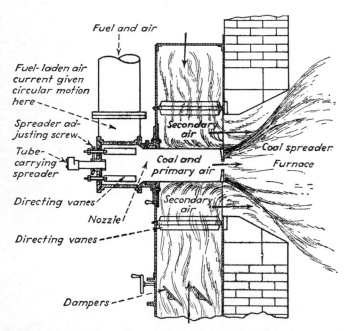

FIG. 5-9. Pulverized-coal burner spreads fuel into furnace.

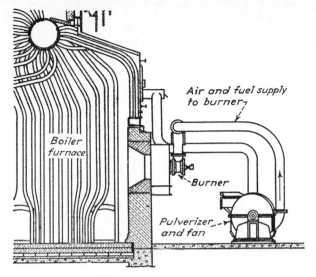

FIG. 5-10. Complete pulverized-coal-burning layout.

and adjustable vanes give it a whirling motion before it enters the furnace. Mingling of the rotating fuel-laden airstream from the burner with the rotating current of secondary air tends to ensure thorough mixing of the air and fuel and thus complete combustion.

The fire is started by inserting an oil or gas torch through the tube in the front of the burner and then turning on the primary air and coal supply Or a torch made from waste dipped in oil may be inserted through an observation opening. The torch should be in a position to ignite the fuel at the instant it is admitted to the furnace. If for any reason the fuel does not ignite, the torch should be withdrawn immediately and the setting allowed to clear before trying again to light up.

The secondary air supply is adjusted after the fire is lit. The shape or angularity of the burning flare can be changed by advancing or withdrawing the coal spreader by means of the tube in the center of the burner. Firing can be switched over to oil by inserting an oil burner in the center pipe.

Q Describe an arrangement for burning pulverized fuel in a boiler furnace.

A In Fig. 5-10, coal from a storage bunker feeds by gravity to a pulverizing mill. A fan draws the fine coal dust from the pulverizer and delivers it to the burners. Another fan supplies the secondary air needed for complete combustion.

Q What are the characteristic effects attending the burning of anthracite, bituminous, and lignite coal?

A *Anthracite,* being mainly composed of fixed carbon and ash, is low in gaseous constituents. Hence it ignites slowly, burns with a bluish flame, and gives off little smoke. For best results with hand firing, don't slice or disturb the fuel bed more than necessary.

Bituminous is more widely used than anthracite for firing power boilers. It varies greatly in composition but always has a higher volatile content than anthracite. In general it burns with a yellow flame. It may be coking or non-coking. Ash content and moisture may vary from very low to very high.

Lignite is usually lower in carbon content than bituminous and fairly high in volatile matter. Its moisture content is notably high. It burns with a smoky yellow flame. Though much lower in heating value than anthracite or bituminous coal, it is a satisfactory fuel when burned on grates and in furnaces of suitable design. Its high moisture content makes lignite hard to store for any great length of time without considerable slacking.

Q For coal firing, describe the various steps in the combustion process from the time fresh coal is thrown on the fire.

A 1. If the coal contains moisture, it is driven off.

2. The dry coal absorbs heat, and the volatile constituents begin to distill off.

3. The volatiles mix with air and burn, their products of combustion passing up the chimney.

4. The fixed carbon left on the grate is now burned until only the ash remains.

Q What furnace conditions are essential to burn coal efficiently and economically?

A 1. Enough air must be admitted to supply the oxygen necessary for the complete combustion of the combustible elements of the fuel.

2. Admission of air must be regulated in such a way that it mixes thoroughly with the combustible gases distilled from the coal and comes in contact with all the carbon left on the grate after the gases are driven off and burned.

3. Temperature of the furnace must be kept at not less than 1800°F.

Q What are the advantages and disadvantages of pulverized fuel compared with stoker-fired solid fuel?

A In general, with pulverized coal it is easier to get proper mixing of fuel and air and to reduce excess air percentages and standby losses. Also, with suitable furnaces, there may be greater flexibility in coal choice.

Disadvantages with certain coals may include slagging of tubes and furnace walls. Pulverized firing adds the cost of pulverizing, and usually discharges more ash from the stack.

Q What are the advantages of oil and gas fuel compared with coal?
A Apart from cost, which varies with the locality, oil and gas possess considerable advantage over coal as boiler fuels. Oil requires less storage space, leaves no ash, and burns better because of more intimate mixing with the air. Furnace and boiler room can be kept cleaner, labor costs lower. Very little excess air is needed to ensure complete combustion. Gas possesses all the advantages of oil, plus elimination of storage space and of steam or mechanical atomizing requirements.

Both oil and gas are smokeless fuels if furnaces and burners are properly designed. Standby losses are small because the fuel supply can be instantly shut off or reduced to a minimum when the load is off.

BURNING FUEL OIL

Q How was liquid fuel originally formed?
A Petroleum as we know it today came from tiny marine plants and animals that lived in the lakes and seas of many million years ago. These lakes have since disappeared, leaving behind decayed matter which the ages have molded into oil. Lying beneath the earth's surface, this oil has been gradually forced by water into large pockets. People have tapped these pockets, drained their contents, and done some molding of their own. We burn the result in our cars and furnaces.

Q What is petroleum?
A Crude petroleum is usually classified in three general bases: paraffin, intermediate (mixed), and asphalt. To make crude petroleum usable, it must be refined. Destructive distillation is a process carried on at relatively high pressures and temperatures. Here are the products produced in order of their boiling point: (1) crude benzine, (2) kerosene distillate, (3) gas oil (diesel oil), (4) lubricating distillates, and (5) residue.

These fractions are further distilled to produce the following marketable products: (1) gasoline and naphtha, (2) kerosene, (3) synthetic gasoline, (4) fuel oil (which remains after the first three fractions are removed), (5) lubricating oils, and (6) waxes, petroleum, asphalt, and coke (obtained from the residue).

Q Name the properties of fuel oil.
A Commercial grades of fuel oil, labeled no. 1 to 6, have various characteristics or physical properties. They should be understood by operating engineers who select and burn oil.

Here are the characteristics, followed by an explanation: (1) viscosity, Saybolt Seconds Universal (SSU) or Saybolt Seconds Furol (SSF), (2) specific gravity, degrees API and Baumé, (3) flash point, (4) specific heat, (5) coefficient of expansion, (6) heating value, (7) sulfur content, (8) moisture and sediment, percent by volume (keep below 1 percent), and (9) pour point.

Q Explain viscosity.

A Viscosity is the internal resistance to flow (opposite of pumpability) of an oil. You should know it because one grade of oil can be atomized thoroughly at only one viscosity.

Viscosity is measured with a viscosimeter (Fig. 5-11) which uses a 60-cc sample of oil heated to a standard temperature of 100, 122, or 210°F. It flows through a tube of given length and diameter. The number of seconds taken for the sample to flow through this tube is the oil's viscosity. This is expressed in SSU or SSF.

Q What is the difference between SSU and SSF?

A The main difference in the two viscosimeters is the tube length that the oil sample flows through. The SSU reading is 10 times the SSF reading. You can find the temperature to heat oil for best combustion by using Chart 5-1 and knowing the SSU or SSF reading and its standard temperature.

For best results pump oil to the burner between 100 and 200 SSU. Here's how to find the exact oil temperature for most efficient combustion. Let's say your oil is no. 6, or bunker C, with a viscosity of 65 SSF at 122°F standard temperature. Follow the heavy dashed line on the chart.

Find point 65 on the vertical scale above 122°F. Then follow the diagonal line downward until you cross the horizontal 155-SSU line. Read the corresponding temperature directly under this point, which is 180°F—the temperature to heat your oil.

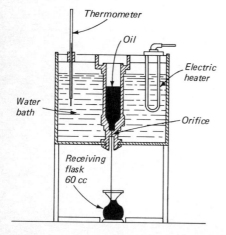

FIG. 5-11. Saybolt Universal viscosimeter tells the oil viscosity by flow through orifice.

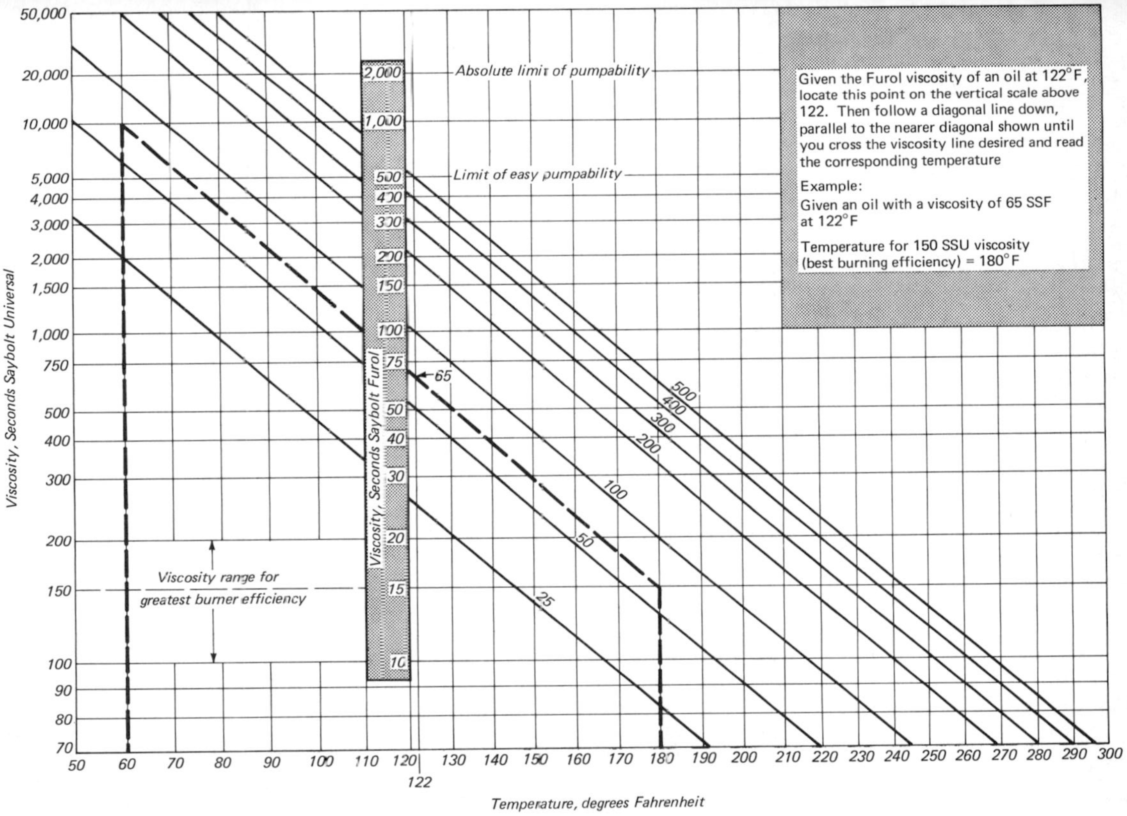
Given the Furol viscosity of an oil at 122°F, locate this point on the vertical scale above 122. Then follow a diagonal line down, parallel to the nearer diagonal shown until you cross the viscosity line desired and read the corresponding temperature
Example:
Given an oil with a viscosity of 65 SSF at 122°F
Temperature for 150 SSU viscosity (best burning efficiency) = 180°F
Absolute limit of pumpability
Limit of easy pumpability
Viscosity, Seconds Saybolt Furol
Viscosity, Seconds Saybolt Universal
Viscosity range for greatest burner efficiency
65
500
400
300
200
100
50
25
50,000
20,000
10,000
5,000
4,000
3,000
2,000
1,500
1,000
750
500
400
300
200
150
100
90
80
70
2,000
1,000
500
400
300
200
150
100
75
50
40
30
20
15
10
50
60
70
80
90
100
110
120
130
140
150
160
170
180
190
200
210
220
230
240
250
260
270
280
290
300
122
Temperature, degrees Fahrenheit

CHART 5-1. Save this chart and use it every time you get a new batch of fuel oil. It's a quick and easy means for finding the right temperature to burn fuel oil efficiently—doing away with all guesswork. It also tells how much to heat oil before it can be pumped from storage tanks to the fuel-oil service pumps.

Here's how to find the minimum temperature for pumping this same 65-SSF oil, at 122°F standard. First find point 65 on the vertical line above 122; then follow this dashed diagonal line upward until you cross the 10,000-SSU line. The corresponding temperature beneath the point is 60°F.

Q Explain specific gravity of a fluid.
A The weight of a unit volume of oil compared with the weight of an equal volume of water, with both the oil and water at 60°F, is called specific gravity.

The specific gravity of standard 60°F oil is found by one of two types of hydrometers. The Bé (Baumé) hydrometer is used mostly in foreign countries, the API (American Petroleum Institute) scales in the United States.

Light oils have a low specific gravity but are high on the Bé and API scales—and vice versa for heavy oils. Here is the relationship between specific gravity and degrees API and Bé:

$$\frac{141.5}{°API = \text{specific gravity at } ^{60}/_{60}°F} - 131.5$$

Q Explain calculations for finding specific gravity.
A When the weight of a unit volume of oil is known and its temperature is 60°F, divide the weight of a unit volume of oil by the weight of the same volume of water at the same temperature, or 60°F. That's what $^{60}/_{60}$ (60 divided by 60) means.

For example, if the weight of 1 gal of oil at 60°F equals 7.986 lb and the weight of 1 gal of water at the same temperature is 8.328 lb, then 7.986 divided by 8.328 equals 0.959, which is the specific gravity $^{60}/_{60}$°F.

$$\text{Specific gravity} = \frac{141.5}{131.5 + °API}$$

$$°Bé = \frac{140}{\text{specific gravity } ^{60}/_{60}°F} - 130$$

$$\text{Specific gravity} = \frac{140}{130 + °Bé}$$

NOTE: For ease of calculations decimals need not be used; we substitute round numbers 140 and 130.

Q How is the heating value of oil found?
A For oils ranging from 18,000 to 19,500 Btu/lb, the approximate heating value is found by this equation: Approximate heating value, Btu/lb = 18,250 + 40 (°Bé − 10)

Since the hydrometer reading increases as the density of the oil decreases, the above equation shows that light oils have a higher heating value than heavier ones, when figured on a weight basis.

Q Explain flash point and the grades of oil.
A The lowest temperature to which fuel oil can be heated so vapor given off flashes momentarily when an open flame is passed over it (Fig. 5-12) is called the flash point. The six grades of oil are: (1) light, (2) medium, and (3) heavy domestic fuel oil; (4) light, (5) medium, and (6) heavy industrial fuel oils. The flash points of grades 1, 2, and 3 range from 110 to 200°F. The industrial grades have a minimum of 150°F.

Q Is fire point important?
A Yes, because it tells where oil starts burning with safety and ease. It's the lowest temperature at which heated fuel oil gives off vapor that supports continuous combustion. This point is usually 10 to 15°F higher than the flash point.

Q Exactly what does specific heat indicate?
A The Btu needed to raise the temperature of 1 pound of oil 1°F is called the specific heat. It varies from 0.4 to 0.5 depending on the oil's specific gravity. The lighter the oil, the higher the specific heat. In short, it shows how much steam it takes to heat oil to the desired temperature, which every boiler operator should know.

Q Explain the coefficient of expansion and what it means to the operator.
A This is the increase or decrease in volume of 1 gal by raising or lower-

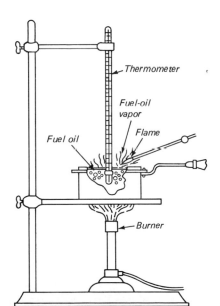

FIG. 5-12. Open-cup test is used to determine flash and fire points of any petroleum.

ing the temperature 1°F. The constant allowed for expansion for each degree increase or decrease above or below 60°F is 0.0004.

The difference in volume of oil at 60 and 120°F is $120 - 60 \times 0.0004 = 60 \times 0.0004 = 0.024$. From this we know that 200 gal of 60°F oil heated to 120°F has a volume of $200 \times 1.024 = 204.8$ gal, an increase of 2.4 percent. So if your 200-gal oil-storage tank is full when that temperature rise takes place, 4.8 gal will spill over. That means not only a nasty cleanup job, but also wasted oil.

Q Why is knowing the sulfur content important?

A The sulfur content of fuel oil is determined in percent by weight. The allowable limit is 4 percent. If you buy oil with excessive amounts of sulfur, sulfurous or sulfuric acids are apt to form in the boiler breaching and uptakes. That corrodes the economizer, air heater, and stack. The EPA is strict about the amount of sulfur you release to atmosphere from your stack.

Q How about moisture and sediment in fuel?

A Moisture or water in fuel oil, if allowed to enter the atomizer (burner), will cause sputtering. This may extinguish the flame, reduce flame temperature, or lengthen the flame. As for sediment, it is troublesome because it clogs strainers and burner sprayer plates. If there's a lot of it, frequent strainer and atomizer clearing will keep you busy or you will lose steam pressure.

If water or sediment or both are present to any great extent, they can be separated from oil by heating. This is done while oil is in settling tanks or by passing it through a centrifuge.

Q Why must plant operators know the pour point of their oil?

A The pour point varies for different grades of oils. It's defined as the lowest temperature for oil flow under set conditions. It indicates the difficulty in handling fuel at its usual minimum temperature. The specific maximum allowable pour point is 15°F.

If you burn oil, you'd better get all this information down pat. It'll help you buy the right oil for your furnace and burn and store it properly and will save you headaches and money in the long run. If you don't burn oil, paste this information in your hat because some day you may work in an oil-fired plant.

FUEL-OIL ATOMIZERS

Q Describe some type of steam-atomizing burner and explain its operation.

A Figure 5-13 shows the tip of a steam-atomizing oil burner with part cut

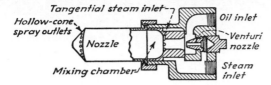

FIG. 5-13.   Steam-atomizing burner-tip details.

away to show the internal construction. Figure 5-14 is a sectional view of
the complete burner installed in a boiler setting.

The fuel oil entering the burner tip meets a high-velocity steam jet
which forces it past a row of smaller steam jets into a mixing chamber.
These small steam jets give the oil a whirling motion that tends to break it
up into very fine particles through the action of centrifugal force. The atomized oil is then blown through small holes in the tip of the burner in the
form of a cone-shaped spray. Air for combustion, controlled by adjustable
louvers, passes through the burner. The burner and oil line can be blown
clean with steam when the burner is not in use.

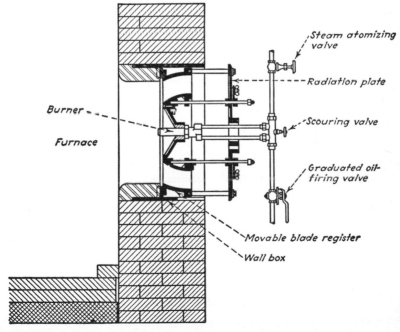

FIG. 5-14.   Steam-atomizing oil burner complete with hand control valves.

Q What is the first step in atomizing oil?

A First step in atomizing oil is heating it enough to get the desired viscosity. This temperature varies slightly for each grade of oil. Lighter oil, numbered 1, 2, and 3, usually needs no preheating. The commercial grades 5 and 6 give best combustion results when heated enough to introduce the oil to the atomizer tip between 150 and 200 SSU (Saybolt Seconds Universal viscosity). Then, after oil is heated to the right temperature, it must be pumped into the burner at the right pressure.

Q Describe a burner in which the oil is atomized by pressure alone.

A This type is usually called a *mechanical* burner to distinguish it from the steam-atomizing burner. The mechanical burner breaks up the oil into very fine particles, by forcing it under high pressure through a number of very small passages in the tip. Oil pressures run from 30 to 300 psi, depending on the burner, and the oil must be heated to temperatures ranging from 100 to 300°F so that it will flow freely. Air for combustion is usually admitted through the burner casing, and the burners are blown

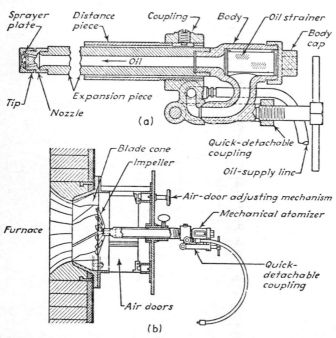

FIG. 5-15. (*a*) Mechanical oil-burner-tip details, and (*b*) the complete mechanical oil-burner installation.

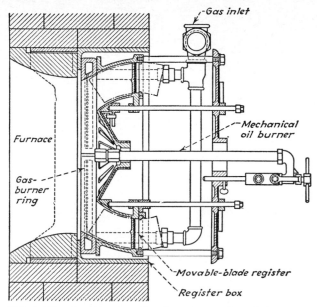

FIG. 5-16.   Combination oil and gas burner installation.

out with compressed air when not in use, to prevent the oil from hardening and clogging the passages.

A mechanical oil-burner tip is shown in section in Fig. 5-15a. In this burner, oil is forced through tangential slots in the nozzle against a sprayer plate; this causes the oil spray to issue from the burner in the form of a hollow cone.

Figure 5-15b shows the complete burner installation. Air for combustion enters through adjustable louvers which give it a rotary motion. This motion is further accentuated by the diffuser, or impeller plate, at the tip of the burner. The diffuser plate also splits up the airstream so that air and oil spray mix thoroughly before burning. Burners of the types in Figs. 5-14 and 5-15b are also made to burn either oil or gas. There are even triple-service burners that burn oil, gas, or pulverized fuel.

Q   Sketch and describe a burner that burns oil or gas as required.
A   A combination oil and gas burner is shown in Fig. 5-16. The oil-burning part is of the mechanical type, somewhat like that of Fig. 5-15. The gas burner is a ring with many small jets on the inside. The air supply is controlled by opening or closing the movable blade registers.

Q   Sketch a method of installing oil or gas burners in a boiler furnace.

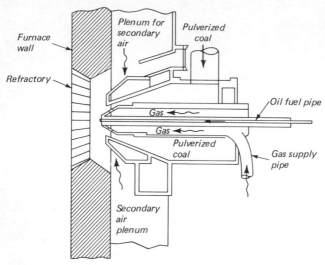

FIG. 5-17.   Coal-oil-gas burner (not to scale; schematic only).

A   Because the burning characteristics of oil and gas are much alike, the same method of installation does for both, and very often combination burners are used that burn either oil or gas.

Loose firebrick checker walls and firebrick baffle walls are sometimes placed in gas- or oil-fired furnaces to assist in the mixing and ignition of the gas-air mixture. When using these, take great care to construct them so that heat is not concentrated on some particular part of the boiler, thus burning or bagging tubes or plates.

Q   Sketch a burner for all three fuels, coal, oil, and gas.
A   Figure 5-17 is a schematic of such a burner, being used more widely today because it enables using the most economical fuel as costs fluctuate.

BURNING GAS

Q   Explain the construction and operation of low- and high-pressure gas burners.
A   The important thing in all gas-burning devices is a correct air-and-gas mixture at the burner tip. Low-pressure burners, using gas at a pressure less than 2 psi, are usually of the multijet type, in which gas from a manifold is supplied to a number of small single jets, or circular rows of small jets, centered in or discharging around the inner circumference of circular air openings in a block of some heat-resisting material. The whole is encased in a rectangular cast-iron box, built into the boiler setting and having

louver doors in front to regulate the air supply. Draft may be natural, induced, or forced.

Figure 5-18a shows a high-pressure gas mixer in which the energy of the gas jet draws air into the mixing chamber and delivers a correctly proportioned mixture to the burner. When the regulating valve is opened, gas flows through a small nozzle into a venturi tube (a tube with a contracted section). Entrainment of air with high-velocity gas in the narrow venturi section draws air in through large openings in the end. The gas-air mixture is piped to a burner. The gas-burner tip may be in a variety of forms. The one in Fig. 5-18b is called a sealed-in tip because the proper gas-air mixture is piped to the burner, and no additional air is drawn in around the burner tip. Size of the air openings in the venturi tube end is increased or decreased by turning a revolving shutter, which can be locked in any desired position.

Q Burning gas as fuel appears simple. Is it?

A *No.* It is really difficult and more dangerous than burning other fuels. Here are the reasons:

1. Low flame luminosity from many types of gas makes it hard to see what's going on in the furnace.

2. Invisibility of unburned gas accumulations presents an explosion hazard.

3. Varying heat content, ranging from 80 Btu/ft^3 for blast-furnace gas to about 1000 Btu/ft^3 for natural gas to 2000 Btu/ft^3 for rich refinery gas.

4. Varying characteristics of flame-propagation speed and ignition temperature.

The ring burner is the most common type. Gas issues from a number of small holes around the inner periphery of the ring, jetting into a stream of

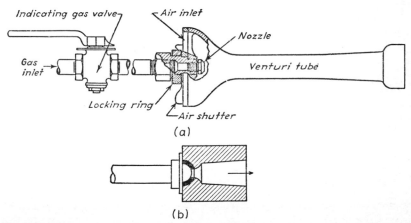

FIG. 5-18.   (*a*) High-pressure gas burner, and (*b*) details of sealed-in burner tip.

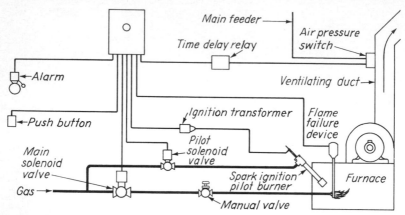

FIG. 5-19.   Purging protection for gas-burning furnace is extremely important.

air passing through the center. Damper vanes regulate the quantity of entering air, usually giving it a spin to promote turbulence.

> CAUTION: Always purge supply headers through a vent line to the roof before attempting to light off the first burner. Leave vent open until first burner is lit because the gas flow out of the vent makes it easier to control pressure in the burner header.

Q   Why should a gas-burning furnace have an air-pressure switch hooked into the system as shown in Fig. 5-19?

A   The air-pressure switch in the ventilating duct in Fig. 15-19 prevents ignition of accumulated gas if the furnace is not properly purged before it can be relighted. The flame failure device shuts off the gas if the flame fails unexpectedly. To prevent furnace explosions, every gas-burning furnace must have a safety hookup for proper purging.

As an added safety precaution, every gas-fired boiler should be checked yearly by the gas supplier or contractor for proper operation and analysis of the flue gas for proper CO_2 content. This is not only a safety precaution, but also is necessary for energy conservation.

TERMINOLOGY USED WITH FUEL

ASH-FREE BASIS: When fuels are delivered on an ash-free basis, it means that the percentage of the ash has been deducted and the other constituents have their percentages recalculated on a 100 percent total without the ash.

BAGASSE: A fuel produced as a by-product of the abstraction of juice from sugarcane. The dried cane is usually fed into a specially designed furnace by means of overfeed stokers.

BLAST-FURNACE GAS: A by-product from a blast furnace which is the result of burning coke in the furnace with limited air. The gas is combustible but of relatively low calorific measurement.

BLOWTORCH EFFECT: In gas- or oil-burning furnaces, when the flame impinges on any surface, such as a tube or refractory wall, that surface is burned as by a blowtorch. This is a combustion condition to be avoided as destructive to the surface.

CAKING COALS: Coals that become soft under the usual furnace temperatures and merge into undesirable masses of coke. These coals can cause considerable trouble in an underfeed or chain-grate stoker by clogging air-supply apertures, or tuyeres.

CALORIFIC VALUE: The heat value of a fuel, expressed in either Btu per pound or calories. NOTE: 1 lb = 453.59 grams approximately.

COAL GAS: A fuel formed by the distillation of coal, usually in a retort or a coke oven. It is the basic gas in some city artificial-gas systems and shows an average calorific value of 500 to 550 Btu/ft^3.

COLLOIDAL FUEL: A mixture of fuel oil and powdered coal.

FLAME DETECTOR: A device that monitors the flame in a furnace that is burning oil, gas, or pulverized-coal fuel. Failure of the flame results in a signal and the actuation of various protective controls on the fuel feed to prevent explosion from unpurged reignition.

FLY ASH: Combustion ash so fine that it is carried up and into the atmosphere by the movement of the flue gases. It can become a neighborhood nuisance by settling on surfaces in the area after it loses its velocity.

FORMULA FOR GRAVITY, API: The American Petroleum Institute (API) has established the formula for calculating the gravity of a fuel oil as

$$°API = \frac{141.5}{\text{specific gravity at } {}^{60}/_{60}°F} - 131.5$$

The symbol ${}^{60}/_{60}°F$ is interpreted as the ratio of the weight of a given volume of oil at 60°F to the weight of the same volume of water at 60°F. For example, assume an oil at 60°F weighing 7 lb/gal and water at 60°F weighing 8.33 lb/gal; then $^7/_{8.33}$ = 0.84. Substituting 0.84 in the equation gives an answer of 37, approximately, putting this oil in the range of Commercial Standard no. 2.

FUEL-OIL BURNERS, TYPES: (1) Pressure-atomizing or gun-type burners are designed to atomize the oil for combustion, under an oil-supply pressure of 100 psig. Equipment typically includes a horizontal air tube with the oil pipe centered in this tube. (2) Rotary burners have a centrifugal thrower ring that mixes the oil and the air. The flame of the rotary burner generally is vertical. (3) Vaporizing burners, also called pot burners, use the heat of combustion to vaporize the oil in the pool or pot beneath the vaporizer ring, and this vapor rising through the ring ignites and maintains combustion in the burner. Fuel enters the pot by gravity through a float valve, and the rate of firing is controlled by the adjustable positions of the float valve.

FUEL OIL, COMMERCIAL GRADES:

Grade	Gravity, API	Btu/gal
1	38–45	137,000–132,900
2	30–40	141,800–135,800
4	12–32	153,300–140,600
5	8–20	155,900–148,100
6	6–18	157,300–149,400

FUME: Any kind of noxious vapor arising from a process of combustion or chemical reactions. Includes smoke, odorous materials, metallic dust.

FUME AFTERBURNERS: Units designed to consume combustible fumes by means of a direct-fired combustion chamber through which the fumes must pass on their way to the stack and the atmosphere. Afterburners usually are fired with gas, although oil may be used in certain instances. Operating temperature is usually from 1200 to 1400°F and is sometimes reached with the help of a platinum catalyst.

GAS, CARBURETED WATER: An artificial gas formed by passing steam through a bed of glowing coke and thereafter enriching the gas so formed with petroleum vapor. This gas is a common one in city gas systems where coal is an economical fuel.

GAS, COKE-OVEN: A coal gas formed as a by-product in the manufacture of coke.

GAS, LIQUEFIED PETROLEUM (LPG): A gas fuel that is stored in liquid form and is converted into gas as it leaves the storage tank by a pressure regulator that steps down the storage pressure on the liquid at the tank outlet and thereby permits the liquid to assume its normal gaseous state at the existing temperature and reduced pressure. Of the two kinds of liquefied petroleum in general use, at atmospheric pressure propane goes from liquid to gas at −51°F, butane at 15°F.

GAS, NATURAL: The constituents of natural gases vary greatly, depending on the gas field from which the gas is obtained. Natural gas is the major fuel in the gas category.

GRINDABILITY: A descriptive term of a characteristic of coal that is important to pulverized-coal systems. The standard from which grindability is measured is the number 100, representing a very soft, easily ground bituminous coal. As the coal samples become harder, the grindability number decreases. It is usually stated on the Hardgrove scale, and the formula is

Grindability $= 6.93\,W + 13$

W in this equation is the weight in grams of material in a 50-gram sample that will pass through a 200-mesh sieve. The other figures are constants. NOTE: Pulverizer mills generally may be expected to show 100 percent capacity with coals having a grindability index of 55. As the index falls to 25 or 30, the mill capacity likewise falls to about 65 percent.

HIGHER HEAT VALUE: A standard recommended by the ASME, the higher heat value of a fuel includes the heat value of the hydrogen in the fuel. A lower heat value, recognized in certain other countries, modifies the total heat in the fuel to reflect losses resulting from the hydrogen burning to form water. This water absorbs sufficient heat to be converted into superheated steam and so is lost.

HOGGED FUEL: Wood that has been chipped and shredded, usually by a machine called a "hog."

KEROSENE: This petroleum product is a liquid fuel having an average latent heat of vaporization of 105 to 110 Btu/lb and a specific heat of 0.50. Sometimes called coal oil.

LIGNITE: A coal of high moisture content and low calorific value, generally less than 8300 Btu/lb. May require predrying before being used as a fuel.

LIQUID-ASH-REMOVAL SYSTEM: An arrangement of piping by which molten ash is removed continuously or intermittently, as desired, from the bottom of a furnace. The operating medium is usually compressed air with pneumatic controls.

LONG-FLAME BURNER: An oil or gas burner in which the mixture of fuel and air is delayed long enough to produce a long flame from the burner nozzle. Can be a source of trouble if the flame impinges on either refractory or tube surfaces.

META-ANTHRACITE: A coal classification of the highest rank with fixed carbon of 98 percent or higher and volatile matter 2 percent or less, when computed on a moisture-free and ash-free basis.

PULSATION: A panting of the flames in a furnace, indicating cyclic and rapid changes in the pressure in the furnace.

PURGE: The evacuation of air or any other designated gas from a duct line, pipeline, container, or furnace. Purging may be done in some instances simply by use of a fan or blower, in others by driving out the air or gas by means of an inert gas, such as nitrogen, under higher pressure.

RETORT: A trough or channel built into an underfeed stoker, through which the stoker ram pushes green coal into the fire. The coal enters the fire from below, hence the name "underfeed."

SECONDARY AIR: Air introduced into a furnace above or around the flames as may be necessary to promote combustion. This air is in addition to the primary air which enters either as a mixture with fuel or as blast underneath a stoker.

SLACK: A coal of fine size, often screenings. Maximum size is not likely to exceed $2^1/_2$ in.

SLAG-TAP FURNACE: A furnace for burning pulverized fuel in which the ash puddles in the bottom of the furnace in a molten state and is removed periodically or continuously, depending on the design of the system, while still in the molten condition.

TORCH: Combustible material on a metal rod, such as oil-soaked rags, used to light off oil and gas burners. The torch is extinguished by being plunged into a prepared receptacle.

TUYERES: Castings appearing as components of underfeed stokers and designed to admit air to the green coal moving through the retorts.

VERTICAL FIRING: Oil, gas, or pulverized-coal burners so arranged that the fuel is discharged from the burner in a vertical direction, either upward from low-set burners or downward from burners in the top of the furnace.

WATER GAS: A basic gas in commercial gas works formed by the reaction of an incandescent bed of carbon when injected with live steam. The gas consists mainly of carbon monoxide and hydrogen, and its calorific value is comparatively low, about 300 Btu/ft^3. General practice is to enrich the gas by mixing with another, frequently petroleum gas if it is available.

WINDBOX: A plenum from which air for combustion, such as primary air, is supplied to a stoker or to gas or oil burners.

SUGGESTED READING

Elonka, Stephen M., and Anthony L. Kohan: *Standard Boiler Operators' Questions and Answers,* McGraw-Hill Book Company, New York, 1969.

Elonka, Stephen M.: *Standard Plant Operators' Manual,* McGraw-Hill Book Company, New York, 3d ed., 1980.

Elonka, Stephen M.: *Standard Basic Math and Applied Plant Calculations,* McGraw-Hill Book Company, New York, 1978.

6

FLUE-GAS ANALYSIS, DRAFT, AND DRAFT CONTROL

Flue-gas analysis tells the operator what goes on inside the boiler's furnace so that any excess or deficiency of the needed combustion products can be corrected. With today's automatic controls, combustion-control equipment is more strictly concerned with two basic functions: adjusting the fuel supply to maintain constant steam flow or pressure under varying boiler loads, and correcting and maintaining the ratio of combustion air to the fuel supply.

Whether the boiler is of the natural-draft type, or has forced draft equipped with power-operated draft controls, or has fully automatic combustion controls, the operating engineer must see that the exact needed draft is supplied at all times.

FLUE GAS

Q What is flue gas?
A It is the name given to the complete and incomplete products of combustion and excess air passing to the chimney. Commonly considered constituents are water vapor (H_2O), carbon dioxide (CO_2), carbon monoxide (CO), and nitrogen (N_2).

Q What constituents are measured in the ordinary commercial analysis of flue gas?
A Ordinary Orsat apparatus cannot measure moisture. It directly measures the percentages by volume of carbon dioxide, oxygen (O_2), and carbon monoxide in dry flue gas. The remainder is assumed to be nitrogen.

Q   What is the flue-gas analysis if carbon is the only combustible element in the fuel and if combustion is complete?

A   When carbon burns completely with one volume of oxygen, it produces one volume of carbon dioxide and no carbon monoxide. Note that air contains 21 percent oxygen by volume and 79 percent nitrogen. The nitrogen goes through the furnace unchanged. Any oxygen that burns completely with carbon produces the same volume of carbon dioxide. The remainder of the oxygen is excess and goes through unchanged. Thus, the flue gas here analyzes 79 percent nitrogen, and the oxygen and the carbon dioxide add up to 21 percent.

Thus, if there is no excess air the analysis is

	Percent
Oxygen	0
Carbon dioxide	21
Nitrogen	79
Total	100

If there is 100 percent excess air, unburned oxygen equals that burned to carbon dioxide, and gas analysis is

	Percent
Oxygen	10.5
Carbon dioxide	10.5
Nitrogen	79.0
Total	100.0

If there is 50 percent excess air, the unburned oxygen is half that burned to carbon dioxide, and the gas analysis is

	Percent
Oxygen	7
Carbon dioxide	14
Nitrogen	79
Total	100

Q   What is the effect on the flue-gas analysis if the fuel contains no combustible element except carbon, and if part of the carbon is incompletely burned to carbon monoxide?

A   The original oxygen now goes in three directions. Part goes through unchanged. Part burns completely to form the same volume of carbon dioxide. The remainder burns incompletely to form twice its volume of carbon monoxide.

Q   What is the effect of hydrogen in the fuel?

A   Hydrogen burns with part of the oxygen to produce water. This condenses and does not show up in the Orsat analysis; so it is equivalent to a shrinkage.

Q   Illustrate the combined effect of all these by an example.

A   Suppose that out of the 21 ft³ of oxygen in 100 ft³ of air, 4 ft³ is excess going through unchanged, 1 ft³ burns to make 2 ft³ of carbon monoxide, 13 ft³ burns to 13 ft³ of carbon dioxide, and 3 ft³ burns with hydrogen to produce water. The 79 ft³ of nitrogen goes through unchanged.

The first column below gives the products of combustion in cubic feet, and the second column in percent:

	Ft³	Percent
Oxygen	4.0	4.1
Carbon monoxide	2.0	2.0
Carbon dioxide	13.0	13.3
Water (disappears)	0.0	
Nitrogen	79.0	80.6
Total	98.0	100.0

In this problem, carbon monoxide is purposely shown higher than would normally occur in boiler operation. Generally the figure is less than 1 percent.

THE ORSAT ANALYZER

Q   Sketch and describe an apparatus for analyzing flue gases.

A   The Orsat apparatus (Fig. 6-1) is a common type of flue-gas analyzer. Its principal parts are a water-leveling bottle, a water-jacketed measuring burette, three pipettes, three containers, various connecting tubes and valves, and a frame to hold all these parts rigidly in position.

The leveling bottle and measuring burette contain pure water. The first pipette (*a*) is packed with steel wool kept wet with caustic-potash solution from the container below. The steel wool in the second pipette (*b*) is wet with pyrogallic solution from the lower container. The third pipette (*c*) contains copper strips wet with cuprous (copper) chloride solution.

Pipettes (*a*), (*b*), and (*c*) are for the absorption of CO_2, O_2, and CO, in the order named. Other chemicals than those indicated are sometimes used for these absorptions.

The Orsat is connected to the sampling tube in the stack or breeching by a length of rubber tubing in series with an aspirator rubber bulb (Fig. 6-2) to pump the gas and (generally) a gas filter (glass wool in a glass tube) to remove dust and tar. The gas line connects to the back side (not shown, Fig. 6-1) of the three-way cock. This cock has three positions: (1) to connect sampling tube to measuring burette; (2) to connect burette to atmosphere; (3) to connect sampling tube to atmosphere. (This last position is *closed* as far as the Orsat is concerned.)

Q Explain how a flue-gas analysis is made, using the indicated type of Orsat.

A The following covers the main steps only. For actual operation always study the detailed instructions supplied with the particular instrument. These show how to avoid contamination of the sample with air or of the chemical with water, or vice versa—also how to ensure easy operation and precise readings.

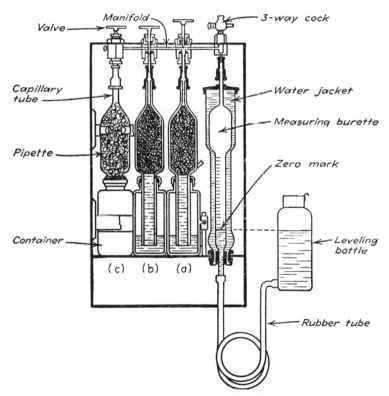

FIG. 6-1. Orsat flue-gas analyzer is an instrument that is a big fuel saver.

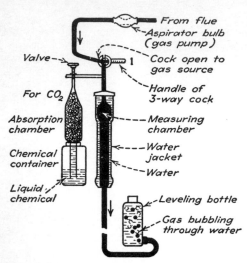

FIG. 6-2. Pumping sample of flue gas into measuring burette is done with aspirator bulb.

Step 1 (Fig. 6-2)

Use a hand aspirator bulb to pump gas through the measuring burette. It bubbles out of the measuring bottle as shown. Continue until you are sure that all air has been displaced.

Step 2 (Fig. 6-3)

With the three-way cock set to connect the burette to atmosphere, raise the leveling bottle slowly until the water is at the zero mark near the bottom of the burette. This is also level with the water in the bottle; so the sample is measured at atmospheric pressure. (All gas measurements must be made at atmospheric pressure, that is, with the water level the same in the burette as in the leveling bottle.)

Step 3 (Fig. 6-4)

To absorb CO_2, close the three-way cock and lift the bottle high. Open the valve to the CO_2 pipette, so the gas displaces the liquid and contacts the caustic solution on steel wool, which absorbs CO_2. Pinch the tube to check the flow when water rises to the mark in the capillary tube above the burette. Allow a few seconds for CO_2 absorption.

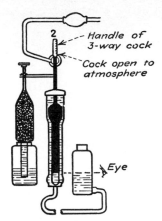

2 -- *Handle of 3-way cock*

-- *Cock open to atmosphere*

Eye

FIG. 6-3.   Bring gas sample to zero at atmospheric pressure.

Step 4 (Fig. 6-5)

Next, keeping the tube pinched, lower the bottle. Then release the rubber tube cautiously until the liquid rises exactly to the mark in the capillary tube on top of the pipette. After closing the pipette valve, raise or lower the bottle until the water level in the bottle is exactly the same as in the burette. Read the graduation at the water level in the burette. This is CO_2 percentage. Repeat steps 3 and 4 to make sure all CO_2 is absorbed.

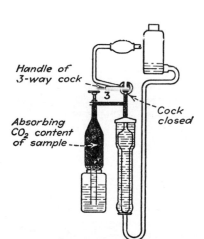

Handle of 3-way cock __

3

Absorbing CO_2 content of sample ---

Cock closed

FIG. 6-4.   Now gas sample is transferred to the first absorption pipette.

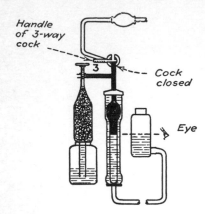

FIG. 6-5.   Returning gas sample to burette to measure percentage of CO_2.

Step 5

Repeat steps 3 and 4 for pipette (b) to absorb the O_2. The *increase* in burette reading is percentage of O_2.

Step 6

Again repeat steps 3 and 4 with pipette (c) to absorb the CO. The *further* increase in burette reading is the percentage of CO.

Q   What are some of the peculiarities of the chemicals used in the Orsat?
A   The caustic to absorb CO_2 acts surely and rapidly. This solution need not be protected from the air and lasts a long time. The CO_2 should always be absorbed first.

The solutions for absorbing O_2 and CO act more slowly, and must be protected from contact with the air. If the O_2 is not completely absorbed in pipette (b), it will be in pipette (c) and will therefore be erroneously reported as part of the CO. Be sure O_2 is *completely* absorbed in (b).

If a second pass through any pipette produces no increase in absorption, it may generally be assumed that the solution is in good shape.

PUTTING READINGS TO WORK

Q   Give an example of readings during an Orsat analysis.
A   Reading after CO_2 absorption, 12.8 percent; after O_2 absorption, 18.7

percent; and after CO absorption, 19.8 percent. Then analysis is

	Percent
CO_2	12.8
$O_2(18.7 - 12.8)$	5.9
$CO(19.8 - 18.7)$	1.17

Q What gas analysis results are reasonable?
A The following comments hold only for solid or liquid fuels (not for gas fuels): The sum of the CO_2, O_2, and half the CO should never be above 21. With coal this sum should never be less than 18.5. With fuel oil, the sum should never be less than 15.5.

Q Assuming there is no CO in the gas, give a table showing the percentage of CO_2 for various percentages of excess air with typical anthracite coal, bituminous coal, fuel oil, and natural gas. Also show for each CO_2 percentage about what the O_2 should be, as a check on the analysis.
A See Table 6-1.

Q Which is the more reliable indicator of excess air, CO_2 or oxygen?
A Table 6-1 shows that oxygen percentage is the more reliable indicator, because it reads about the same for any coal, and almost the same for oil as for coal. For any coal or fuel oil the excess air can be closely estimated from the O_2 by the following rule.

Subtract the O_2 percentage from 21. Divide the O_2 percentage by this difference and multiply by 100 to get the percentage of excess air.

TABLE 6-1 Corresponding Percentages of Excess Air, Carbon Dioxide, and Oxygen

Kind of fuel	Ingredients	Percent excess air					
		0	20	40	60	80	100
Anthracite	CO_2	19.5	16.0	13.8	12.0	10.7	9.6
	O_2	0.0	3.5	6.1	8.0	9.4	10.6
Bituminous	CO_2	18.6	15.5	13.2	11.5	10.1	9.2
	O_2	0.0	3.5	6.0	8.0	9.5	10.6
Fuel oil	CO_2	15.5	12.6	10.6	9.3	8.2	7.4
	O_2	0.0	3.7	6.4	8.1	9.6	10.8
Natural gas	CO_2	12.2	10.0	8.5	7.5	6.5	5.7
	O_2	0.0	4.0	6.5	8.5	9.9	10.9

NOTE: Where gas contains CO, substitute for CO_2 the sum of the CO_2 and half the CO. Thus, if CO_2 is 12% and CO is 2%, equivalent CO_2 is 13%.

EXAMPLE: What is the percentage of excess air for coal or oil if the O_2 in the flue gas is 8.3 percent? Solution: $21 - 8.3 = 12.7$. Then $8.3 \div 12.7 = 0.65$ and $0.65 \times 100 = 65$ percent excess air.   *Ans.*

Q Give some typical flue-gas analyses for bituminous-coal firing under various conditions.
A Analyses are as follows:

	% CO_2	% O_2	% CO	% N_2 (by difference)
Poor	9	10.5	0.5	80.0
Average	12	7.5	0.1	80.4
Good	15	4.0	0.2	80.8

Note (from Table 6-1) that the last condition above corresponds to less than 20 percent excess air. Any further reduction in excess air would undoubtedly raise the CO excessively.

Q What percentage of CO_2 do you expect when burning natural gas under good conditions?
A Because of the high H_2 content of natural gas, 20 percent excess air, which is about the practical minimum, corresponds to about 10 percent CO_2.

Q What percentage CO_2 would you expect when burning fuel oil under good conditions?
A The C content is relatively higher in fuel oil than in natural gas, and so the attainable percentage of CO_2 will be higher (but less than with coal). With 20 percent excess air, CO_2 with oil fuel should be about 12.5 percent.

Q Does a high percentage of CO_2 always indicate good combustion?
A High percentage usually indicates that the excess air is being kept to the minimum, but it does not necessarily indicate the best attainable combustion, as there may also be a considerable percentage of CO. A mere 1 percent of CO may mean a loss of nearly 5 percent of the total heat value of the fuel burned. CO_2 should be high, with little or no CO.

Even when a CO_2 recorder is used, gas should be checked periodically with an Orsat to see whether CO is present.

COMBUSTION CALCULATIONS

Q Give a formula for estimating the approximate heating value of a fuel from the ultimate analysis.

A Dulong's formula, shown below, is probably the most widely used for this purpose. It is based on the known heat values of 1 lb each of pure carbon, pure hydrogen, and pure sulfur. Each of these values is multiplied in turn by the percentage of that particular element in the fuel, and the results are added together.

In the formula, the heat value of 1 lb of hydrogen is taken as 62,000 Btu, the heat value of 1 lb of carbon as 14,600 Btu, and that of 1 lb of sulfur as 4000 Btu. It is assumed that all the oxygen shown in the analysis is already combined with one-eighth its weight of hydrogen in the form of moisture, and the hydrogen so combined is therefore incombustible. The symbols C, H, O, and S represent the percentages of carbon, hydrogen, oxygen, and sulfur. The formula is as follows:

$$\text{Heat value per pound} = \frac{14{,}600C + 62{,}000\left(H - \dfrac{O}{8}\right) + 4000S}{100} \text{ Btu}$$

Q Give a formula for estimating the theoretical amount of air required for the complete combustion of 1 lb of fuel.
A The following formula is based on the known amount of air required for the complete combustion of 1 lb each of pure carbon, pure hydrogen, and pure sulfur. In the formula the figures represent this air in pounds, and the symbols are the percentages of carbon, hydrogen, oxygen, and sulfur.

$$\text{Air required, pounds} = \frac{11.61C + 34.8\left(H - \dfrac{O}{8}\right) + 4.35S}{100}$$

Q Calculate the heat value of a coal having the following ultimate analysis: C, 82.4 percent; H, 4.1 percent; O, 2.3 percent; S, 0.5 percent; ash, 10.7 percent.
A From Dulong's formula:

$$\text{Heat value} = \frac{14{,}600 \times 82.4 + 62{,}000\left(4.1 - \dfrac{2.3}{8}\right) + 4000 \times 0.5}{100}$$

$$= 146 \times 82.4 + 620 \times 3.81 + 40 \times 0.5$$
$$= 12{,}030 + 2362 + 20$$
$$= 14{,}412 \text{ Btu} \qquad \textit{Ans.}$$

Note that the heating value of the sulfur is negligible.

Q Calculate the theoretical amount of air required for the complete combustion of the fuel in the previous question.

$$\textbf{A} \quad \text{Air required} = \frac{11.61 \times 82.4 + 34.8 \left(4.1 - \dfrac{2.3}{8}\right) + 4.35 \times 0.5}{100}$$

$$= \frac{11.61 \times 82.4 + 34.8 \times 3.81 + 4.35 \times 0.5}{100}$$

$$= \frac{957 + 133 + 2}{100}$$

$$= 10.9 \text{ lb of air per lb coal} \qquad \textit{Ans.}$$

NATURAL AND MECHANICAL DRAFT

Q What is *draft?*
A *Draft* is the difference of pressure producing air flow through a boiler furnace, flue, and chimney.

Q Why is draft necessary?
A It supplies air in sufficient *quantity* to ensure complete combustion and under sufficient *pressure* to overcome the resistance offered by the boiler shell, tubes, furnace walls, baffles, dampers, breeching, and chimney lining.

Q In what units is draft measured?
A It is measured in *inches of water column* as indicated by (1) difference between the water-column heights in the two legs of a glass U tube; (2) height of the column in an inclined-glass draft gage; (3) pointer reading in an indicating or recording draft gage.

Q How is draft measurement in inches of water converted into pressure in pounds or ounces per square inch?
A Weight of a cubic foot of water varies with temperature, being 62.43 lb at the point of maximum density, 39.2°F. The drop in weight is very slight over the ordinary range of boiler-room temperatures; so for all practical purposes take it as 62.4 lb/ft³. Since 1 ft³ = 1728 in³., dividing 62.4 by 1728 gives 0.036 psi, the pressure in pounds per square inch indicated by a column of water 1 in. high. This number multiplied by 16 gives 0.576, the pressure in ounces per square inch.

Q Sketch a U-tube draft gage, and explain its operation.
A Figure 6-6 illustrates the principle of the U-tube draft gage (also called a manometer). One end of the glass U tube is open to the atmosphere, the other connected to the boiler furnace, flue, chimney, or other enclosed space where it is desired to measure draft pressure. The U tube is partly filled with water, and when there is no difference in the pressure acting

upon the water surface in the two legs, water level is the same in each leg and the reading on the scale zero.

If the pressure within the boiler setting or chimney is *greater* than the atmospheric pressure, it forces the water down in the leg connected to the setting, and up a corresponding amount in the leg open to the atmosphere. If the pressure within the boiler setting is *less* than the atmospheric pressure, the latter forces the water down in the leg open to the atmosphere and up in the leg connected to the setting. The difference in inches between the levels of the two water columns is the draft measurement. In Fig. 6-6 the scale is divided into inches and tenths of an inch, and the water levels show a reading of 1 in. of draft.

Figure 6-7 shows a more elaborate U-tube gage. Oil is used instead of water, and the scale is calibrated to read directly in hundredths of inches of water, with the aid of a vernier. Connected as shown, the scale reads pressure above atmospheric. Suction can be read by changing the top connections.

When taking a reading on this, or any other instrument where the water level in a tube must be read, note that the water surface is not flat, but concave. The curvature is caused by surface tension, and the shape assumed by the water is called a *meniscus*. The same phenomenon is observed with lighter liquids, such as oils, but when mercury is used the meniscus is reversed, or convex. The bottom of the meniscus is usually read on the scale when water or oil is used and the top when mercury is used.

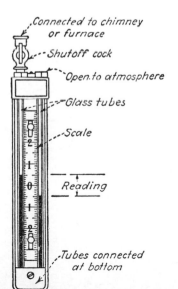

FIG. 6-6. U-tube draft gage has a pen, is open to atmosphere.

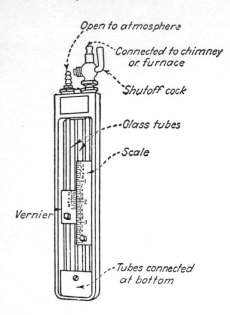

FIG. 6-7. Vernier on this U-tube draft gage gives far more accuracy.

Q Sketch an inclined-tube draft gage and explain its operation.

A This gage is the same in principle as the U tube, but one leg is inclined (Fig. 6-8), and a liquid is used that will not evaporate so readily as water. As the liquid moves through a considerable distance in the inclined tube for a very small change in vertical height, a much larger scale can be used than that in the vertical-tube gage, so that readings are much finer. The gage scale in Fig. 6-8 is registering pressure below atmospheric pressure, the reading being 0.25 in. of draft. Here, the furnace pressure is below atmospheric, and the atmospheric pressure is forcing the liquid down in the inclined tube.

With chimney draft alone, or with chimney plus induced-draft fan alone, the furnace pressure is always less than atmospheric. If a forced-draft fan is added, pressure in the furnace may be greater or less than atmospheric, or equal to it—that is, balanced. If furnace pressure is higher than atmospheric, the connection shown causes the liquid to be forced upward in the inclined tube. Therefore, either the scale or the connection must be reversed.

Q Describe a diaphragm-operated draft gage.

A Figure 6-9 shows the operating principle of this gage. A thin diaphragm of very flexible material is enclosed in a metal case and connected through suitable links or gearing to a pointer that moves over a scale graduated to read in inches of draft. Referring to the sketch, the compartment

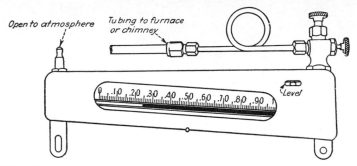

FIG. 6-8.   Inclined-tube draft gage must be absolutely level.

to the right of the diaphragm is airtight and that to the left is open to the atmosphere. The airtight compartment is connected by tubing to the point where it is desired to measure draft pressure. The movement of the diaphragm under this pressure is transmitted to the pointer by the connecting links so that it moves over the scale and indicates the pressure.

When the pressure to be measured is below atmospheric, the same instrument can be used, but the tubing is connected to the left-hand compartment, and the right-hand compartment is left open to the atmosphere.

Recording draft gages act on the same principle as indicating gages, but the end of the pointer carries a pen which is in contact with a moving chart.

Metal expanding bellows, bourdon tubes made in single-tube and spiral- and helical-tube forms, sealed oil columns, and mercury columns operating floats, are also used as pressure elements in indicating and recording draft gages.

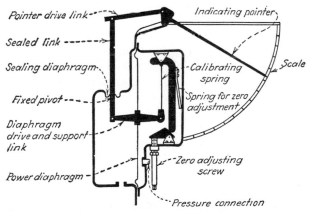

FIG. 6-9.   Diaphragm-operated draft gage is widely used.

Q What is *natural* draft and how is it controlled?

A It is draft produced by a chimney alone. It is caused by the difference in weight between the column of hot gas inside the chimney and a column of cool outside air of the same height and cross section. Being much lighter than outside air, chimney gas tends to rise, and the heavier outside air flows in through the ashpit to take its place. It is usually controlled by hand-operated dampers in the chimney and breeching connecting the boiler to the chimney.

Q What is *mechanical* draft?

A It is draft artificially produced by mechanical devices, such as fans and, in some units, steam jets.

Three basic methods of applying fans to boilers are:

1. Balanced draft where a forced-draft (F-D) fan (blower) (Fig. 6-10) pushes air into the furnace and an induced-draft (I-D) fan (draws) or a high stack (chimney) provides draft to remove the gases from the boiler. Here the furnace is maintained at from 0.05 to 0.10 in. of water gage below atmospheric pressure.

2. An induced-draft fan or the chimney provides enough draft for flow into the furnace, causing the products of combustion to discharge to atmosphere. Here the furnace is kept at a slight pressure below the atmosphere so that combustion air flows through the unit.

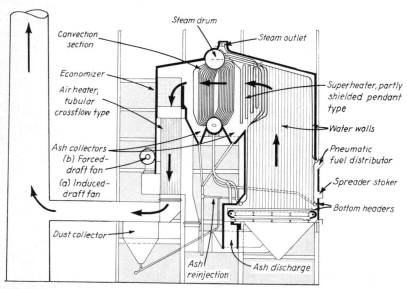

FIG. 6-10.   Induced and forced draft by fans moving air through furnace.

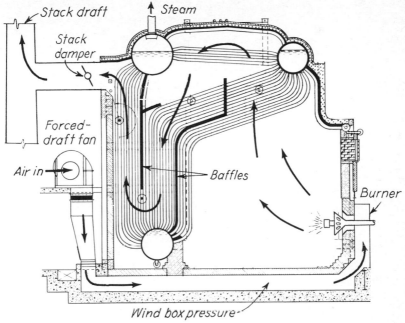

FIG. 6-11.   Forced draft pushes air into wind box for preheating by furnace.

3. The pressurized furnace (Fig. 6-11) uses a blower to deliver the air to the furnace, causing combustion products to flow through the unit and up the stack.

Fans are also used to supply over-fire air (Fig. 6-12) to the furnace in some designs for distributing the flame more evenly throughout the fur-

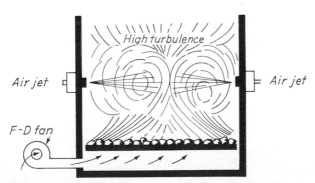

FIG. 6-12.   Air jets provide turbulence over the fuel bed.

nace and thus come in contact with more tube area. Pulverized-fuel-burning boilers have blowers which deliver the coal dust to the furnace.

Q What are the main characteristics of *forced* draft?
A *Forced* draft often (but not necessarily) maintains a pressure above atmospheric in the boiler setting; therefore, flue gas may be forced out into the boiler room through cracks or leaks in the setting. Such leakage should not be important with a boiler setting in good condition. In a few rare cases where boilers are hand-fired, flame may shoot out of firing doors if the draft is not cut out before the doors are opened.

Q Sketch an F-D fan driven by a turboblower.
A Figure 6-13 is a forced-draft propeller fan built into the side wall of a boiler setting. A small direct-connected steam turbine fed by steam from the boiler assures draft in case of electric power failure.

Forced draft is commonly used with underfeed stokers, since considerable pressure is needed to force the air required for combustion up through the deep fuel bed. In general, all types of large boilers operated at high ratings are normally equipped with forced draft.

Forced-draft fans may draw their air supply from the boiler room or from air preheaters, the latter being widely used with large boiler units.

Q What are the main characteristics of *induced* draft?
A *Induced* draft creates a partial vacuum in the boiler furnace so there is no likelihood of furnace gas leaking out through cracks in the setting. If any cracks are present, air is drawn in and it may reduce furnace efficiency considerably. Because of this, induced draft may not be as satisfactory as

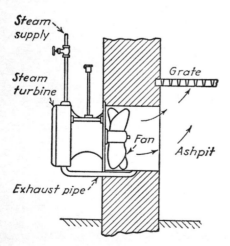

FIG. 6-13. Forced draft using turbine blower beneath fuel bed.

forced draft where solid fuel is fired and the fuel bed is thick, but it may fit the requirements of small or medium-sized plants where furnaces are hand-fired, equipped with overfeed stokers of certain types, or where boiler fuel is oil or gas.

Induced-draft fans are of similar construction to forced-draft fans, but are usually of the single inlet type, and larger than forced-draft fans for the same boiler output, because the hot flue gas is much greater in volume than the air originally fed to the furnace. Induced-draft fans are also likely to deteriorate more rapidly than forced-draft fans as they are exposed to higher temperatures, cinders, and possible corrosive action.

Q   What is *balanced* draft?
A   This term is usually applied to a combination of forced and induced draft, automatically regulated to keep the boiler furnace at approximately atmospheric pressure.

AUTOMATIC DRAFT CONTROL

Q   Discuss *automatic draft control*.
A   It is the automatic regulation of fans and dampers for increasing or decreasing airflow to maintain constant steam pressure as the load changes, and also to maintain good combustion conditions. Devices to control draft-fan speed and damper position frequently employ diaphragms. Changes of air or steam pressure act upon the sensitive diaphragms to open or close the electrical switches, fluid valves, or steam valves that control fan speed or damper position.

Q   What is balanced draft?
A   Balanced draft is any system where the draft is zero (static pressure = atmospheric pressure). At times, it also refers to a unit in which the top of the furnace operates at slightly less than atmospheric pressure.

The balanced-draft regulator is a large diaphragm exposed on one side to furnace pressure conveyed through a draft tube built into the wall of the setting. Movement of this diaphragm operates a pilot valve that admits water or air under pressure to the damper-control cylinder, causing the piston to adjust the damper. When the damper takes up its new position, the furnace pressure comes back to normal, and the diaphragm returns to its original position, no further movement taking place until another change occurs in the furnace pressure. The action of the diaphragm under furnace pressure is opposed by springs that can be adjusted to secure any desired pressure within the range of the regulator.

Fan speed is controlled by a steam-pressure regulator operated by rise or fall of steam pressure in the steam main. If steam pressure falls a little, the regulator opens the fan engine's throttle valve to speed up the fan sufficiently to carry the higher load.

Q Explain the use of both forced and induced draft for one boiler.
A In Fig. 6-14 air is forced into the ashpit by a forced-draft fan, and the waste furnace gases are drawn through an economizer and discharged to the chimney by an induced-draft fan. The boiler feedwater passes through rows of tubes in the economizer and absorbs some of the heat from the waste flue gases before they go to the chimney.

Q What advantages has mechanical draft over natural draft?
A Mechanical draft is independent of wind or temperature changes. After the limit of combustion is reached with natural draft, more fuel may be burned and the boiler capacity increased by installing mechanical draft; also, poorer grades of fuel may be burned than with natural draft. Where a very thick fire is carried, as for underfeed stokers, mechanical draft must be used to force air through the thick fuel bed. Also, when economizers are

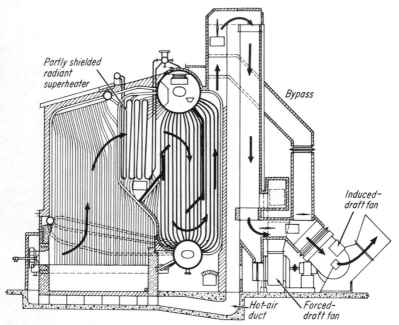

FIG. 6-14. Modern steam generator using both forced and induced draft. (*Courtesy of Power.*)

placed between the boiler and chimney to utilize heat from the waste flue gas, mechanical draft is essential to overcome the resistance offered by the banks of tubes in the economizer.

CHIMNEYS

Q Why is a chimney necessary?

A It creates natural draft where no draft fans or blowers are used and discharges the products of combustion at such a height that they will not be a nuisance to the surrounding community. Where mechanical draft is used, the second reason is the main one.

Q What are the common types of chimneys, and the advantages and disadvantages of each?

A Chimneys are built of steel, brick, or concrete. The steel chimney is usually the cheapest and most easily erected, but it requires more care and attention if it must last a comparable time. Unless the outside is kept well painted, a steel stack deteriorates rapidly from the action of the weather, and the flue gas corrodes the inner surface if it has no protective lining. Also, the excessive heat loss through an unlined steel stack reduces the draft produced; hence practically all permanent steel stacks are lined with firebrick or some other fire-resisting material. A common practice in building this lining is to use firebrick for the base section, where the heat is most intense, and common brick for the rest of the distance to the top.

Steel chimneys may be self-supporting, in which case the base flares out to about twice the upper chimney diameter and is bolted to a substantial concrete foundation. If the chimney is not self-supporting it is kept upright, and braced against wind pressure, by steel guy wires or cables with one end anchored in the ground and the other fastened to a ring on the chimney, about two-thirds of the height from the bottom. Small steel stacks are usually of uniform diameter throughout, but large stacks, especially the self-supporting type, may be tapered gradually from bottom to top.

The brick chimney is more expensive to build than the steel, but it lasts longer and stands weathering much better. It is usually built with a uniform inside diameter from bottom to top but a decreasing outside diameter. In the best construction, an inner lining is used with an annular space between the lining and the outer wall, so that the lining expands or contracts without affecting the outer wall in any way. A hard close-grained brick is used for the outer-wall construction. The inner lining may be all firebrick, or firebrick part way up from the base and common brick the rest of the way. Rectangular bricks can be used for chimney building, but specially shaped radial bricks make a stronger and neater job.

Chimneys are also built of reinforced concrete. This type can be built rapidly, yet it is very strong because of the steel reinforcing. The chimney is practically airtight and less likely to have air leakage than a brick chimney with its multitude of joints. Walls may be thinner than brick, thus giving a lighter construction without sacrificing strength. A concrete chimney should also have an inner refractory lining at least part of the way up from the base.

Q Numerate the principal points to observe when designing a chimney.
A They are:
  1. Height must give the desired draft.
  2. Cross-sectional area must be sufficient for boiler load served.
  3. Foundation must be solid and substantial.
  4. Base must support the weight.
  5. Chimney must resist maximum wind pressure to which it is likely to be exposed.
  6. It must have high resistance to the weathering action of wind, rain, heat, and cold.
  7. A good inner lining should protect the inside of the chimney and prevent excessive heat loss through radiation.

Q Why are chimneys lined, and what materials are used?
A All types should be lined to prevent damage from the effects of unequal expansion due to the difference between flue-gas temperature inside the chimney and that of the outside air. The lining also protects against the corrosive action of the products of combustion passing up the chimney and can be renewed without rebuilding the entire chimney. A space should be left between the lining and the main wall, and the lining should not be bonded to the outer wall. The chimney may be lined completely or only part way up. Concrete, common brick, firebrick, and other refractory materials are used for chimney lining.

DRAFT CALCULATIONS

Q Give a formula for calculating the draft that can be produced by a chimney of a given height.
A The following formula is commonly used. It is based on the fact that air and flue gas expand in volume with increase of temperature, so that the higher the temperature the less they weigh per cubic foot. The formula for the resulting natural draft is

$$D = 0.52 \times H \times P \left(\frac{1}{T_o} - \frac{1}{T_c} \right)$$

where D = draft pressure, in. of water
   H = height of chimney, ft
   P = atmospheric pressure, psi absolute
   T_o = absolute temperature of outside air, °F
   T_c = absolute temperature of chimney gas, °F

NOTE: Absolute temperature is Fahrenheit temperature $+460°$.

Q What draft is produced by a chimney 100 ft high, temperature of chimney gas 500°F, outside temperature 60°F, atmospheric pressure 14.7 psi?
A T_o = 60 + 460 = 520
   T_c = 500 + 460 = 960

$$D = 0.52 \times H \times P \left(\frac{1}{T_o} - \frac{1}{T_c}\right)$$

$$= 0.52 \times 100 \times 14.7 \left(\frac{1}{520} - \frac{1}{960}\right)$$

$$= 764(0.00192 - 0.00104) = 764 \times 0.00088$$
$$= 0.67 \text{ in.} \qquad Ans.$$

Note that this is not the *net* draft the chimney would produce with a boiler carrying a load. Instead it is the *static* draft; that is, the draft the hot gas in the stack would produce if there were no flow of gas up the stack. In actual operation the gas encounters resistance flowing up the stack, and this uses up part of the total draft as computed by the formula just given. This draft loss in the stack increases with the rate of fuel burning and is much greater for a small-diameter than for a large-diameter stack. To keep it within reasonable limits the cross-sectional area of the chimney opening must be increased almost in proportion to the amount of coal burned by the boilers it serves.

Q Is a higher or a lower chimney needed when burning oil or gas fuel than when burning coal?
A Use of oil or gas fuel eliminates the resistance of grates and fuel bed, and hence permits a lower chimney unless (and this is important) the stack must allow for changing over to coal. Excessive draft from a stack higher than necessary can be reduced by dampers.

Q What is meant by *total* draft pressure and *available* draft pressure?
A The draft formula in two previous questions calculates the *total* static-draft pressure. The *available* draft is the draft at any point from the ashpit to the chimney top that is available after the *pressure drop* to that point has been deducted from the total static-draft pressure.
   The total available draft is equal to the total theoretical static draft, less

the pressure used in producing velocity and overcoming friction within the chimney and connecting passages.

Q Give a formula for finding the height of a chimney to produce a given total or static draft.

A With the previous formula transposed,

$$H = \frac{D}{0.52P \left(\dfrac{1}{T_o} - \dfrac{1}{T_c} \right)}$$

Q What chimney height is required to give a total static draft of 0.8 in. of water column with a boiler-room temperature of 90°F, average temperature of chimney gas 500°F, and atmospheric pressure 13 psi?

A $T_o = 90 + 460 = 550$

$T_c = 500 + 460 = 960$

$$H = \frac{0.8}{0.52 \times 13 \left(\dfrac{1}{550} - \dfrac{1}{960} \right)}$$

$$= \frac{0.8}{0.52 \times 13(0.00182 - 0.00104)}$$

$$= \frac{0.8}{0.52 \times 13 \times 0.00078} = \frac{0.8}{0.00527} = \frac{800}{5.27}$$

$$= 152 \text{ ft} \qquad Ans.$$

Q How do you find the chimney area for a given height and coal consumption?

A There are many different formulas for finding chimney area, but all are of a rule-of-thumb nature and give undependable results. The type of chimney construction, method of firing, boiler and furnace design, etc., are among the variable factors that make it impossible to evolve a single formula to suit all conditions.

The only safe procedure is to turn over the actual determination of chimney dimensions, particularly the internal diameter, to an engineer who is able to calculate them directly from a knowledge of the engineering principles involved, and who has full knowledge of the boiler and furnace design, baffling, firing method, load conditions, etc.

MAINTAINING A CLEAN STACK

With strict air pollution regulations in force today, the fireman's job is to maintain a "clean stack." A "dirty stack" not only pollutes the air but wastes fuel by sending unburned carbon and other particulates up into the atmosphere and may result in a heavy fine to the plant owner.

Fly ash and unburned carbon particles are the principal emissions from coal-fired boilers. Some of the more objectionable particulates are nitrogen oxide (NO_x) and sulfur dioxide (SO_2). For removing some of these, cyclone collectors, filters, electrostatic precipitators, and venturi scrubbers are used. We cannot give details here, but the boiler-room operator should know how they work and also how to keep them in repair. To burn fuel efficiently, the person in charge of a boiler plant must first of all know what causes a dirty stack.

Q   What causes heavy black smoke when fuel oil is burned?
A   Insufficient air or excess fuel causes smoke. As air needed for combustion is increased, the smoke lightens until it turns into a light-brown haze; this corresponds to the best operating conditions. A further increase in air will cause a clear stack; dense, white smoke indicates far too much air. Thus air must be regulated until a faint, light-brown haze issues from the stack.

Smoke can also be caused from:

1. Sprayer plates of unequal size in the same furnace.

2. Oil spray on the diffuser, carbon formation on the throat tile, a dirty atomizer, improper atomizer position, a damaged diffuser, worn plates or nozzles, no sprayer plate in the atomizer (firemen often forget to put them back when cleaning burners), a loose nut holding the plate in position, too much throttling of the air-register doors. If carbon forms on the throat tile and cannot be punched off without damaging the throat, remove the atomizer, crack the air-register doors, and let it burn off.

3. The atomizer may be in too far, causing incomplete mixing of fuel and air, and flame fluttering. If the atomizer is out too far, carbon forms on the burner throat. Both conditions cause smoke.

4. Fuel viscosity may be too high (oil not hot enough).

5. Air register and burner throat may not be centered, or the double front may be too deep, resulting in air leaking into the throat without passing through the register doors.

6. Air distribution in the double front may be faulty; check air pressure at various points in this area.

Q   How is smoke density measured?
A   A microRingelmann chart (Fig. 6-15) is used for comparing the smoke's density when no instruments are available. A more accurate method is measuring the stack gas with an in-stack gas analyzer (Fig. 6-16) which continuously measures emissions.

Q   Explain the microRingelmann method.
A   Ringelmann charts were originally used. These have five grids, no. 5

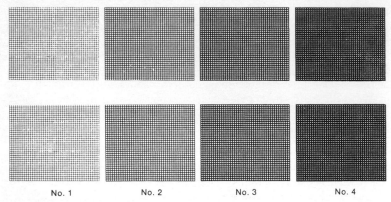

| No. 1 | No. 2 | No. 3 | No. 4 |

FIG. 6-15. MicroRingelmann chart for measuring smoke density. These grids are a direct facsimile reduction of the standard Ringelmann chart as issued by the U.S. Bureau of Mines. *(Copyright 1954 by McGraw-Hill Publishing Company, publisher of* Power, *1221 Avenue of the Americas, New York, N.Y. 10020.)*

being solid black. But they have to be positioned 50 ft from the viewer, making them awkward to use. *Power* magazine omitted the black grid and developed the smaller four-grid microRingelmann chart, which can be read at arm's length. To use:

1. Hold the chart at arm's length and view the smoke through the slot in the chart.

2. Make sure the light shining on the chart is the same as that shining on the smoke being examined. For best results, the sun should be behind the observer.

3. Match the shade of the smoke as closely as possible with the corresponding grid on the chart.

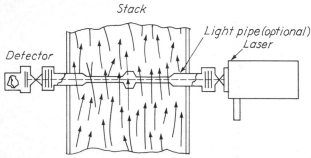

FIG. 6-16. In-stack continuous particulate-emission detector.

4. Enter the density of the smoke and the time of each observation on a record sheet.

5. Make repeated observations at regular intervals of $\frac{1}{4}$ or $\frac{1}{2}$ min.

6. To compute the smoke's density, use the formula: equivalent units of no. 1 smoke $\times$ 0.20 divided by number of observations = percentage of smoke density.

7. Note and record the distance to the stack, the direction of the stack (from where you are standing), the shape and diameter of the stack, and the speed and direction of the wind.

Most major cities have codes regulating the permissible smoke density; so be sure to get this information before fines are imposed on offenders.

Q Describe the in-stack continuous gas analyzer.

A There are various designs, some combining continuous monitoring with instant indication to allow fast response to excessive emissions by alerting the boiler-plant operator to air-pollution violations. Photoelectric detection is the principal measuring technique, with laser beams used in designs (Fig. 6-16) to minimize the effects of light scattering.

The basic monitoring system consists of a lamp and a receiving unit, mounted on opposite sides of the exhaust stack, as the figure shows. The sealed light source projects a beam of energy through particles of fly ash and unburned carbon traveling upward in the stack. The photocell receiver detects energy radiated from the light source and produces a minute electrical signal proportional to the energy detected. Since part of the radiant light energy is deflected or absorbed by suspended particulates, radiant energy detected by the photocell varies with the amount of suspended particulate matter.

The electrical signal is built up through high-gain amplifiers and is transmitted to a continuous-reading recorder, usually located at a central control panel. In most systems, this recorder contains an alarm contact that operates when density exceeds a preset value. This contact can be wired to an external annunciator, signal light, or other type of alarm system.

TERMINOLOGY USED WITH DRAFT

AEROSOL: The term for heterogeneous assemblages of minute particles suspended in the air. These particles may be solid or liquid and as small as 0.01 micrometer or less. Fog, dust, and smoke may be termed aerosols.

BREECHING: The metal duct that carries the smoke and gases of combustion from a furnace to the stack or chimney for ultimate discharge to the atmosphere.

CHIMNEY EFFECT: The upward movement of warm air or gas, compared with the ambient air or gas, due to the lesser density of the warmed air or gas. Chimney effect may be a cause of uneven heating in buildings two or more stories high.

FOULING: A condition of the flue-gas passages in a boiler or furnace that adversely affects the transfer of heat. Usually in the form of soot or scale.

PARTICLE SIZE: In evaluating the efficiency of a filter, electrostatic or others, it is customary to indicate the percentage of particles of a specified micrometer diameter and larger that are caught by the filter and removed from the airstream.

RETARDERS: Also called "spinners," these are helical strips or ribbons of metal centered in horizontal or vertical fire tubes of a fire-tube boiler for increasing the wiping effect on the inner surfaces of the tubes, by the flue gases on their way to the stack. The tubes cannot be cleaned by scraping or brushing until these strips are removed. Retarders also increase the boiler frictional resistance to the flue gases.

SUGGESTED READING

Elonka, Stephen M., and Anthony L. Kohan: *Standard Boiler Operators' Questions and Answers,* McGraw-Hill Book Company, New York, 1969.

Elonka, Stephen M: *Standard Basic Math and Applied Plant Calculations,* McGraw-Hill-Book Company, New York, 1978.

Spring, Harry M., and Anthony L. Kohan: *Boiler Operator's Guide,* McGraw-Hill Book Company, New York, 1981.

7

HEAT, HOW TO USE
STEAM TABLES

Today, more than ever before, heat (energy) is the key to modern civilization.
And our known sources of heat (fuel) are diminishing daily. Since it is the stationary engineer who is charged with operating equipment which releases this
heat and turns it into energy, isn't it logical that the engineer should *know* as
much about heat as possible to do the job efficiently?

Here we explain not only heat but how any boiler operator or operating engineer can use the steam tables to figure out problems involving heating water,
making steam, and superheating for using fuel as economically as possible.

HEAT

Q What is heat?
A It is a form of energy due to molecular motion. The molecules of any
substance containing heat are assumed to be constantly in motion, and the
intensity (or temperature) of heat depends upon the rapidity of this molecular vibration. The temperature of a body will rise as the rate of vibration
increases, and fall as it decreases. To understand this fully, just consider
that at absolute zero ($-459.8°F$), there is *no* molecular action, therefore *no*
heat. That means anything *above* absolute zero has its molecules in constant
motion.

Q What are the effects of adding heat to a body?

A Addition of heat to a body may cause: (1) rise in temperature; (2) change of state—for example, from solid to liquid (ice to water) or liquid to gas (water to steam); (3) performance of external work by expansion of the solid, liquid, or gaseous body to which the heat is added.

All the foregoing effects are seen if heat is applied to ice to melt it into water, thus changing its state without raising its temperature; if addition of heat is continued until the water reaches the boiling point, thus raising its temperature without changing its state; finally, if heat is added until the water turns into steam, which is another change of state without a rise in temperature.

As steam is generated, it exerts pressure on the walls of the vessel in which it is confined. If this vessel is a cylinder containing a movable piston, the steam can do external work by moving the piston.

Q How is heat transferred from one body to another?
A It is convenient to think of heat as flowing like a fluid from one body to another, but strictly speaking there is no transfer of any physical substance. Molecules in the hotter substance are vibrating at a higher rate than those in the colder substance; therefore, when the bodies are brought into contact, the effect is to increase molecular vibration in the colder body and decrease vibration in the hotter body until equilibrium is established. Unless artificially reversed by outside power (as in a refrigerating machine), heat transfer is always from the hotter to the colder body.

Q What is temperature?
A It is a measure of heat *intensity,* or *degree* of hotness or coldness as distinct from *quantity.* A very small body and a very large one may be at exactly the same temperature, but it is obvious that the large body contains a much greater *quantity* of heat than the small body.

MEASURING TEMPERATURE

Q Why is the measurement of temperature important today?
A Quality control is the answer, whether temperature of steam (degrees superheat), heat-treating of metals, sterilization, pasteurizing milk, petroleum refining, safe operation of machinery in every industry, etc., including heating and cooling processes over a broad spectrum.

Q Define the units of temperature.
A See Fig. 7-1. The Fahrenheit temperature scale (°F) divides the temperature interval between the ice point and the steam point of water into 180 parts, or degrees. The ice point is given the value 32°F, and the steam point 212°F.

The Celsius (formerly centigrade) temperature scale (°C) subdivides the temperature interval between ice and steam into 100 parts, or degrees. The ice point has the value of 0°C and the steam point 100°C.

The Kelvin temperature scale (not illustrated) defines the absolute temperature scale. Absolute zero, or the lowest theoretical temperature, 0 K, is that condition where molecular motion ceases and there is therefore no heat. On the Fahrenheit scale this point has the value 459.6°F below zero (−460°F). On the Celsius scale it is −273°C. The ice point is +273 K, or 0°C. The boiling point is +373 K, or 100°C.

Q What is meant by absolute temperature?
A The volume of a perfect gas, under constant pressure, is known to decrease $1/273$ of its volume at 0°C for every degree Celsius fall in temperature. From this it would appear that at 273° below zero on the Celsius scale, volume of the gas would be reduced to zero, and the molecular motion that produces heat would have entirely ceased. This extreme low point is called *absolute zero,* meaning the lowest possible temperature that could be attained. In the same way, 460° below zero on the Fahrenheit scale is the absolute zero for that scale.

Absolute temperatures are reckoned from absolute zero. To reduce any Fahrenheit reading to absolute temperature, add 460°, and to reduce any Celsius reading to absolute temperature, add 273°. Thus 26°F would be 26 + 460 = 486°F abs, and 26°C would be 26 + 273 = 299°C abs.

Q In what units is temperature measured, and what instruments are used to measure it?
A Temperature is measured in *degrees.* Thermometers measure all ordinary ranges of temperatures up to around 1000°F; pyrometers measure very high temperatures beyond the range of thermometers.

Q Describe the construction of a thermometer.
A A thermometer is a glass tube with a very small central bore, having one end blown into bulb form and the other end closed. Bulb and tube are partly filled with a liquid. (Mercury and alcohol are commonly used for this purpose.) The air is exhausted from the remaining portion of the tube, except in very high-temperature thermometers where this space is filled with a gas. Approximate ranges of the common types of glass thermometers are: mercury-filled, from −38 to 750°F; mercury- and nitrogen-filled, from −38 to 1000°F; alcohol-filled, from −95 to 150°F.

When placed in a heated atmosphere or liquid, mercury or alcohol expands and travels upward in the tube, a very small expansion causing quite a noticeable upward movement. A scale of degrees is etched on the glass of a mercury thermometer, marked as follows:

   1. The thermometer is immersed in melting ice at a pressure of 14.7

psia and a mark is made at the top of the mercury column. This is the *freezing point,* called 0° on the Celsius scale and 32° on the Fahrenheit. The two thermometer scales are compared in Fig. 7-1.

2. The thermometer is immersed in boiling water at a pressure of 14.7 psi abs and a mark is made at the top of the mercury column. This is the *boiling point,* called 100° on the Celsius scale and 212° on the Fahrenheit.

3. Distance between the freezing and boiling points is divided into 100 equal parts, or 100°, on the Celsius thermometer, and into 180 equal parts, or 180°, on the Fahrenheit.

The Celsius thermometer has a more logical scale than the Fahrenheit, and is the one commonly used in scientific calculations, but the Fahrenheit is widely used by engineers and others for many everyday purposes.

Q How can thermometer readings in Celsius degrees be converted into readings in Fahrenheit degrees?

A Between the freezing and the boiling points there are 180° on the Fahrenheit scale and 100° on the Celsius scale; hence 180 Fahrenheit degrees equals 100 Celsius degrees. Dividing each by 10, we get a simpler ratio of 18 Fahrenheit degrees equals 10 Celsius degrees. So if we multiply Celsius degrees by 1.8, we get the corresponding number of Fahrenheit degrees. In order, however, that the readings may exactly correspond in position on both scales, we must add 32 because 32 on the Fahrenheit scale is equivalent to 0 on the Celsius scale. The rule is: to convert Celsius temperature to Fahrenheit, multiply by 1.8 and add 32. Thus:

$$22°C = 22 \times 1.8 + 32 = 39.6 + 32 = 71.6°F$$

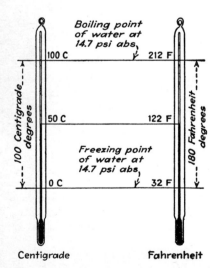

FIG. 7-1. Comparison of Celsius and Fahrenheit scales.

Q   How can you convert readings in Fahrenheit into Celsius degrees?

A   To convert Fahrenheit into Celsius degrees, subtract 32, then divide by 1.8. Thus:

$$105°F = (105 - 32) \div 1.8 = 73 \div 1.8 = 40.5°C$$

Q   What instruments are used for measuring very high temperatures?

A   *Pyrometers* measure temperatures above the range of thermometers. There are a number of types, but those operated electrically are probably most common. Of these the *thermocouple* and *optical* pyrometers are typical.

In the thermocouple pyrometer, two rods or wires of dissimilar metals are joined and sealed in a porcelain tube (Fig. 7-2 illustrates the principle of the thermocouple.). Wires are connected to these rods and to a galvanometer. The tube containing the rods is exposed to the heat at the point where it is desired to measure the temperature. As the rods heat up, an electric voltage is induced at their junction, proportional to the difference in temperature between the hot junction and the so-called cold junction (where the rods connect to the leads). The resulting current flows through the circuit and deflects the galvanometer needle. The dial of the galvanometer is graduated to read in degrees of temperature.

As for the optical pyrometer, it has a telescope containing a tiny filament that glows when an electric current passes through it. In the circuit with this filament is a small battery and a galvanometer. By means of a variable resistance built into the telescope, current flow through the filament is varied until the filament seems to disappear entirely when the telescope is focused on the furnace flame or wall. At this point the temperature is read on the galvanometer dial.

Unlike the thermocouple, no part of the optical pyrometer is exposed directly to furnace heat, and one can be quite a distance from the flame being observed. Operation of this instrument depends upon the fact that there is a definite relationship between color and temperature.

The electrical optical pyrometer has an electric battery, but none is needed with the thermocouple pyrometer.

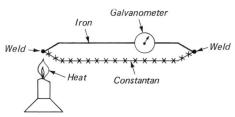

FIG. 7-2.   Thermocouple principle is based on expansion of metals.

Q How is quantity of heat measured?
A Quantity of heat is measured in British thermal units (usually abbreviated Btu). One Btu is the $1/_{180}$ part of the heat required to raise the temperature of one pound of water from 32°F to 212°F, or about the amount of heat needed to raise the temperature of one pound of water through one degree Fahrenheit. The latter definition is close enough for most practical purposes.

SPECIFIC HEAT

Q What is specific heat?
A The amount of heat, expressed in Btu, required to elevate one pound of a material one degree Fahrenheit. The specific heat of water is 1 for all practical purposes, although the exact description is $1/_{180}$ of the heat needed to raise the temperature of 1 lb of water from 32 to 212°F. The specific heat of air is 0.24; steam varies but may be assumed at 0.49 at lower temperatures, 0.50 at 300°F. Refer to the steam tables for exact values.

Q Give the specific heats of some common materials.
A The following short table gives the specific heats of several very common substances:

Material	Specific heat Btu/lb/°F
Water	1.000
Ice	0.49
Iron (cast)	0.13
Copper	0.093
Aluminum	0.218
Brass	0.088
Coal	0.318
Concrete	0.270
Air (at constant pressure)	0.24
Air (at constant volume)	0.17
Flue gas (at constant pressure)	0.240

Figures are approximate, as some materials vary greatly in composition. Also, specific heat changes with temperature.

Q How is heat transmitted from one substance to another?
A By radiation, conduction, and convection.

Q What is radiation?
A It is the giving off of heat from a hot body by ether waves of the same nature as light waves. Radiant heat does not warm the air to any great ex-

tent as it passes through it but is absorbed or reflected by any solid obstruction. In a boiler furnace we have direct heat radiation from the boiler fire to all parts of the boiler that can "see" the fire.

Q   What is conduction?
A   It is the passage of heat through a body by the contact of one molecule with another. For example, if one end of an iron bar is placed in a fire while the other end is held in the hand, in a short time the end in the hand will become unbearably hot because of the conduction of heat through the bar from the red-hot end. Here heat is passed along by a series of collisions; fast-moving hot molecules bump into and speed up the cooler, slower molecules. Heat passes in this way through the tube walls and plates of a boiler to the water on the other side.

Q   What is convection?
A   Convection is the transfer of heat by current flow. As gases or liquids are heated by conduction through the walls of a containing vessel, they tend to expand and rise, and their place is taken by the upper colder layers of liquid or gas which, being heavier than the heated liquid or gas, tend to flow downward. In this way convection currents are set up, and the whole body of gas or liquid is gradually heated to a uniform temperature.

It is by means of convection currents that the air of a room is heated to uniform temperature by a steam radiator. Water in a steam boiler is also heated uniformly throughout by convection currents set up by upward flow of the lighter heated water in contact with the heating surface, and by the downward flow of the heavier colder water above.

Q   What is thermal conductivity?
A   Thermal conductivity refers to the rate at which heat passes through a body. The rate varies widely for different substances, and may (but not necessarily) be stated as the number of Btu that can flow in one hour through a block of the material, one square foot in area and one inch thick, with 1°F difference in temperature between the opposite surfaces. Since thermal conductivity varies with temperature, density, and moisture content, tables of thermal conductivity give only very approximate values. The rate of conductivity of metals usually decreases as temperature rises, but for most other substances the rate increases as temperature rises.

Q   Give the thermal conductivity of some common substances.
A   Table 7-1 gives average values of thermal conductivity of some common materials at the temperatures noted.

Q   What is the coefficient of linear expansion of a solid body?
A   It is the fraction of the body's length that it expands when heated 1°F. Stated another way, it is the amount of expansion per unit length, per degree rise in temperature. For example, if the coefficient of steel is

TABLE 7-1   Thermal Conductivity or Heat Transfer
(Btu/ft^2, 1 in. thick, per °F temperature
difference, per hr)

Substance	Temperature °F	Conductivity
Air	50	1.58
Water	140	4.62
Common brick	70	4.56
Firebrick	2000	12.00
Lead	64	241.00
Cast iron	216	320.00
Steel	212	310.00
Yellow brass	212	738.00
Copper	64	2668.00

0.0000063, a piece of steel expands that fraction of its length for each degree rise in temperature. If it is 1 in. long, it expands 0.0000063 in.; and if it is 1 ft long, it expands 0.0000063 ft. To put it another way, steel expands 6.3 parts in a million for every degree rise.

Q Give the coefficients of linear expansion of some common materials.
A Coefficients of expansion of metals and alloys vary with composition and degree of purity. In some cases the coefficient increases at temperatures above 212°F, but the increase is small, and the values in Table 7-2 give results close enough for all practical purposes.

Q What do we know about the expansion and contraction of liquids?

TABLE 7-2   Coefficients of Linear Expansion
(Average values per °F between 32 and 212°F)

Aluminum	0.000,0128
Brass (cast)	0.000,0104
Bronze	0.000,0104
Copper	0.000,0091
Cast iron	0.000,0059
Wrought iron	0.000,0063
Lead	0.000,0164
Nickel	0.000,0072
Steel	0.000,0063
Tin	0.000,0119
Zinc	0.000,0219
Glass	0.000,0050

Factors of evaporation for various feedwater and steam-pressure conditions are given in Table 7-5. Other factors can readily be calculated from Tables 7-3 and 7-4.

Q What is the developed horsepower output of a steam boiler working at 160 psia steam pressure and evaporating 3400 lb of water per hr from feedwater at 140°F?
A Heat put into 1 lb of steam at 160 psia, with feedwater at 140°F, is 1195.1 − (140 − 32) = 1087.1 Btu. Then

$$\text{Equivalent evaporation} = \frac{3400 \times 1087.1}{970.3}$$

$$= 3809 \text{ lb of water per hr}$$
Boiler hp = 3809 ÷ 34.5 = 110 hp   *Ans.*

You shorten the work by using a table of factors of evaporation; thus the factor of evaporation here, from Table 7-5, is 1.12. Then

Equivalent evaporation = 3400 × 1.12 = 3808 lb of water per hr
Boiler hp = 3808 ÷ 34.5 = 110 hp   *Ans.*

Q What information is required for a simple boiler efficiency test?
A The following data must be secured: (1) temperature of feedwater, °F; (2) steam pressure, psi; (3) heat value of fuel used, Btu/lb; (4) water evaporated during test, lb; (5) fuel burned during test, lb; (6) duration of test, hr.

Q Find the efficiency of a steam boiler, given the following data:
Average feedwater temperature, 160°F
Steam pressure, 125 psig (approx. 140 psia)
Heat value of fuel, 10,200 Btu/lb
Water evaporated in test, 36,000 lb
Coal fired, 6000 lb.
Duration of test, 8 hr
A Water evaporated per lb coal = 36,000 ÷ 6000 = 6 lb.
Total heat put into 1 lb of steam = $h - (t - 32)$
$$= 1193 - (160 - 32) = 1065 \text{ Btu}$$
Total heat put into steam produced per lb of coal fired = 6 × 1065
$$= 6390 \text{ Btu}$$

$$\text{Efficiency of boiler} = \frac{\text{total heat in steam in Btu}}{\text{heat value of 1 lb fuel in Btu}}$$

$$= 6390 \div 10,200 = 62.6 \text{ percent} \qquad \textit{Ans.}$$

Q If for licensing or other purposes, 10 ft² of heating surface is taken as equivalent to 1 boiler hp, what would be the rating on this basis of an hrt boiler having the following dimensions: diameter 60 in., length 16 ft, outside tube diameter 3½ in., thickness of tube wall 0.12 in., number of tubes 60? Take lower half of shell, inner surface (gas-contact surface) of tubes, and two-thirds of area of the tube sheets, less area of tube holes, as heating surface.

A   Heating surface of shell $= \dfrac{5 \times 3.142 \times 16}{2} = 125.7$ ft²

Inside diameter of tubes $= 3.5 - (2 \times 0.12) = 3.26$ in.

Area of tubes $= \dfrac{3.26 \times 3.142 \times 16 \times 60}{12} = 819.4$ ft²

Area of tube sheets

$$= \frac{2}{3} \left(5 \times 5 \times 0.785 - \frac{60 \times 3.26 \times 3.26 \times 0.785}{144} \right) \times 2$$

$$= \frac{2}{3}(19.6 - 3.5) \times 2 = 21.5 \text{ ft}^2$$

Total heating surface $= 125.7 + 819.4 + 21.5 = 966.6$ ft²
Nominal boiler hp $= 966.6 \div 10 = 97$ hp   *Ans.*

Q   What is the specific heat of water?
A   It is the heat in Btu required to raise the temperature of the water 1°F. The mean specific heat at atmospheric pressure is $^1/_{180}$ of the heat required to raise the temperature of 1 lb of water through the 180° from 32 to 212°F. It is sufficiently accurate to consider this as 1. Even at higher temperatures, this is usually close enough for practical purposes.

Q   What is the specific heat of superheated steam?
A   The specific heat of superheated steam is the amount of heat required to raise the temperature of 1 lb of superheated steam, at constant pressure, 1°F. This amount varies with pressure and temperature, and the *mean specific heat* is found for any given set of conditions by dividing the increase in heat in Btu, as found from the superheated steam tables, by the increase in temperature in degrees Fahrenheit. For example, assume we wish to find the mean specific heat of superheat for steam at 200 psia at a temperature of 500°F. From the superheated steam tables, the temperature of saturated steam at 200 psia is 381.79°F, and the total heat is 1198.4 Btu. The total heat of the superheated steam is 1268.9 Btu. Then

Rise in temperature $= 500 - 381.79 = 118.21°F$
Increase in total heat $= 1268.9 - 1198.4 = 70.5$ Btu

Mean specific heat $= \dfrac{70.5}{118.21} = 0.597$ *Ans.*

Q   How much heat is required to produce 1 lb of superheated steam at 300 psia and 600°F from feedwater at 180°F?

A From the superheat table (Table 7-4) final heat content (enthalpy) = 1314.7 Btu. Then

Heat added = 1314.7 − (180 − 32)
= 1314.7 − 148
= 1166.7 Btu/lb   *Ans.*

THE CALORIMETER

Q Explain the operation of the throttling calorimeter in Fig. 7-3.

A Steam from the main steam pipe passes through a nozzle having an opening about $^3/_{100}$ in. in diameter, which reduces it to approximately atmospheric pressure. Since moderately high-pressure saturated steam contains a greater number of heat units per pound (higher enthalpy) than lower-pressure saturated steam, as can be seen from Table 7-3, this excess heat superheats the steam in the calorimeter, thereby raising its temperature above that of saturated steam at the lower pressure. The thermometer in the central well shows the temperature of the superheated steam. The manometer shows the pressure within the calorimeter.

From the observed pressure and temperature, quality of the steam can be calculated from the steam tables by the following formula:

$$q = \frac{H + 0.47(t_s - t) - h}{L}$$

where q = quality of steam

H = total heat in 1 lb steam (enthalpy of saturated vapor) at calorimeter pressure. (To get calorimeter pressure first add the

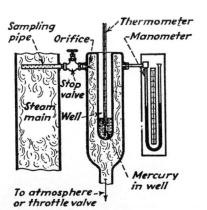

FIG. 7-3.   Throttling calorimeter with sampling tube in steam main.

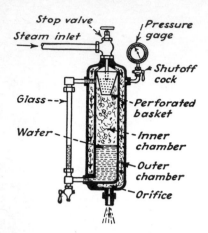

FIG. 7-4. Separating calorimeter takes moisture from steam.

barometer reading to the manometer reading. Multiply this by 0.49 to get calorimeter pressure in psia.)

0.47 = specific heat of superheated steam, atmospheric pressure

t_s = temperature of superheated steam in the calorimeter

t = temperature of saturated steam at calorimeter pressure (from steam tables)

L = latent heat (enthalpy of vaporization) of high-pressure steam in main

h = sensible heat (enthalpy of saturated liquid) of high-pressure steam in steam main

Q Describe the operation of the separating calorimeter in Fig. 7-4.

A When the steam is admitted through the valve at the top of the calorimeter, it flows downward into the perforated cup where its direction of flow is reversed. Moisture in the steam, being heavier than the steam itself, is left in the cup and drains down into the bottom of the inner chamber, where the amount can be measured on the scale. Dry steam passes upward to top of cup, enters the outer cylinder, and finally passes out through the orifice in the bottom. Weight of dry steam flowing can be read on the gage dial, or it can be found by attaching a hose to the outlet at the bottom and leading this hose into a tank of water resting on a scale. The tank's weight and contents are checked before and after the calorimeter is drained into it. The difference in weight gives the weight of the condensed steam. When weights of the moisture and dry steam are found, the percentage of moisture and quality of steam can be calculated:

Percent moisture in steam

$$= \frac{\text{weight of moisture in calorimeter}}{\text{dry-steam weight} + \text{moisture weight}} \times 100\%$$

Quality of steam $= 100 -$ percent moisture

Q Many process industries use great quantities of hot water. It is the operating engineer who must supply the hot water, and the steam for keeping it at the correct temperature. Explain the water-heating system shown in Fig. 7-5.

A This unit keeps the entire storage volume filled with hot water except when stored water is drawn off in excess of recovery capacity. There is no delay in the heating process while strata of cold water rise to the level of the thermostatic bulb. Forced circulation assures that the temperature of the stored water is sensed continually. Whenever necessary, steam is admitted to the heat exchanger to bring stored water up to temperature.

The anticipator system responds immediately to a combination of hot-water delivery rate and incoming-cold-water temperature. It begins to admit steam before any temperature change can be sensed in the storage tank. The steam control valve is a highly sensitive pilot regulator, but overall response is always delayed by the time it takes for the valve to open, for steam to enter the heat exchanger, for steam to condense, and for its heat to be transferred to the water. The fast action of the anticipator control minimizes harmful effects of these built-in response lags.

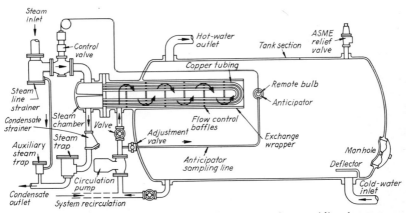

FIG. 7-5. Hot-water heater with control-flow system for providing hot process water automatically. (*Courtesy of Patterson-Kelley Co., Inc.*)

TERMINOLOGY USED WITH STEAM

DEGREES OF SUPERHEAT: Often applied to steam, the term may apply to any gas. Each of these fluids normally has a pressure that is characteristic of its temperature, and when the temperature is higher than is normal for the existing pressure, the increase of temperature is expressed as degrees of superheat.

DRY STEAM: Steam that has a temperature constant with its pressure and contains no more than one-half of 1 percent moisture.

HEAT, LATENT: The heat, in Btu, required for a material to change its state, as ice to water and water to water vapor or steam.

HEAT, SENSIBLE: Heat that changes the temperature of a material but does not change its form.

HEAT-TRANSMISSION COEFFICIENT: A figure that may be used to calculate the total heat transfer through a material and that represents all phases affecting the transfer, such as conduction, radiation, convection, and the surface values. It is often expressed as the number of Btu (or the percentage of 1 Btu) transmitted per degree Fahrenheit difference in temperature per square foot of area, for the given material and its thickness, over a time of one hour. The symbol for this coefficient is U.

ISOTHERMIC: This word describes a condition of change accomplished at constant temperature.

MECHANICAL EQUIVALENT OF HEAT: One Btu = 778.2 ft · lb of mechanical energy.

MEDIUM, HEATING: A substance used to convey heat from the heat source to the heat application. A heating medium may be water, steam, air, gas, or a proprietary liquid such as Dowtherm or Humble-Therm 500.

VAPOR: The gaseous form of a substance that, under other conditions of pressure, temperature, or both, is a solid or a liquid.

SUGGESTED READING

Elonka, Stephen M., and Anthony L. Kohan: *Standard Boiler Operators' Questions and Answers,* McGraw-Hill Book Company, New York, 1969.

Elonka, Stephen M.: *Standard Plant Operators' Manual,* McGraw-Hill Book Company, New York, 1980 (has over 2000 illustrations).

8

BOILERS (STEAM GENERATORS)

Now that we have covered basics (in previous chapters) of heat and how to burn fuels, we are prepared to learn about the equipment used for generating steam and heating water. Today, the steam generated in industrial plants, commercial buildings, service establishments, and central power generating stations has skyrocketed in cost.

Because of mushrooming fuel prices, the operator of boilers is suddenly shoved onto center stage, under spotlights, his every move anxiously observed by management. The information in this chapter will aid not only in passing license examinations and getting through interviews for better-paying jobs but also in operating the many modern steam generators safely and efficiently.

DESIGN DETAILS

Q What is a boiler?
A A boiler (steam generator) is a container into which water can be fed and, by applying heat, evaporated continuously into steam. In early designs the boiler was a simple shell with a feed pipe and steam outlet, mounted on a brick setting. Fuel was burned on a grate within the setting, and the heat so released was directed over the lower shell surface (similar to a kettle) before most of the heat was wasted out through the flue.

Designers soon learned that heating a single large vessel of water was very inefficient. So they then directed the hot combustion products through tubes, surrounded by water, within the boiler shell. Such a fire-tube design not only increased the heat surface exposed to the water, it also helped distribute steam formation more uniformly through the body of water. Then they designed boilers with water inside tubes and fire outside. And the more and smaller the tubes, the more efficient the unit.

Q   What is meant by steam space in connection with steam boilers?
A   A steam boiler is only partly filled with water when in operation. Remaining space is called steam space because it is needed for the disengagement of steam from the water and for storage of this steam until it is drawn off through the steam main.

Q   What constitutes the heating surface of a steam boiler?
A   The heating surface is that area of tubes, furnace, tube sheets, and headers that is *exposed* to the *products* of *combustion*.

Q   What is grate surface?
A   This is the area of the grate upon which fire rests, in a coal- or wood-fired boiler. It is usually measured in square feet. Thus the area of a fire grate 6 ft long by 5 ft wide would be $6 \times 5 = 30$ ft^2.

Q   In connection with steam boilers, what is meant by (1) *water line,* (2) *fire line?*
A   1. The *water line* is the level at which the water stands in the boiler. This level should always be higher than any part of the boiler that is exposed to excessive heat and likely to be damaged if not covered by water.
 2. The *fire line* is the highest point of the heating surface in most common types of boilers, but this definition cannot be applied to all boilers. The fire line is level with the top row of tubes in the horizontal-return-tubular and dry-back marine boilers, at the highest point of the crown sheet in the locomotive boiler, and at the upper tube sheet in the wet-top vertical fire-tube boiler. The fusible plug is so located in each of these boilers that it will give warning of low water before the water level falls below the fire line.
 In the dry-top vertical fire-tube boiler and many types of water-tube boiler, there is no definite line at which water-heating surface ends, and the bottom of the gage glass in these boilers simply indicates the point below which it is deemed unsafe to carry the water level.

Q   What are some of the requirements of a good type of steam boiler?
A   1. Strong and simple construction.
 2. Materials and workmanship of highest standard.
 3. Design that ensures constant circulation of water in boiler, thus distributing heat evenly through the entire body of water and keeping the various parts of the heating surface as nearly as possible at the same temperature.

4. Large area of heating surface to ensure the utmost possible transfer of heat from the hot gases in the furnace to the water in the boiler.

5. All parts of boiler readily accessible for repair, inspection, and cleaning.

6. Ample combustion space so that gases will be completely burned before passing to the chimney.

7. Large steam space so that steam is able to rise freely from surface of water.

Q What is the difference between a tube and a flue?

A Both are cylindrical tubes, but the term *tube* is usually applied to those of small diameter up to about 6 in. Over this diameter, they are called flues. Note that tube sizes refer to *outside* diameters.

FIRE-TUBE TYPES

Q How are boilers classified?

A Boilers may be classified in the following ways: (1) according to direction of axis of shell, *vertical* or *horizontal;* (2) according to use to which they are put, *stationary, portable, tractor,* or *marine;* (3) according to location of furnace, *internally fired* or *externally fired;* (4) according to relative positions of water and hot gases, *water-tube* or *fire-tube;* (5) as special boilers—*electric, once-through, cast-iron* etc.

Q Are fire-tube types classified in other ways?

A Yes, there are various classifications. For example, such boilers may also be classified as shell boilers —that is, water and steam are contained within a single shell housing the steam-producing elements. Noncylindrical sections and flat surfaces are given added resistance to internal pressure by various means: diagonal stays, through bolts, or tubes which are flared at the tube sheets to act as stays.

In such a shell, the force tending to burst it along the length is twice that tending to burst it around the girth. Thus high pressures and large diameters would lead to extremely thick shell plates. Hence there is a definite economic limit on the pressure and capacity that can be reached with shell-type boilers.

Q What is the pressure limitation of fire-tube boilers?

A An operating pressure of 250 psi (pounds per square inch) is considered the practical ceiling, and in the United States, capacity rarely exceeds 25,000 lb of steam per hr—roughly 750 boiler horsepower. In Europe,

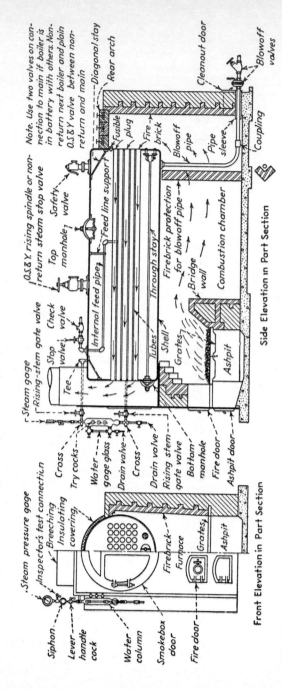

Note. Use two valves on connection to main if boiler is in battery with others. Nonreturn next boiler and plain O.S.& Y. valve between nonreturn and main

Side Elevation in Part Section

Front Elevation in Part Section

FIG. 8-1.   Horizontal-return tubular boiler (hrt) is an older design but still found in many smaller industrial plants.

Q How do you test the water column and water-gage glass to prove that all passages are clear, while the boiler is in operation?

A 1. Close the top valve on the column and the top valve on the glass. Open the drain valve on the glass. If water blows freely from the drain, the water passages from boiler to column and from column to glass are clear.

2. Close the bottom valves on the column and glass, and open the top valves. If steam blows freely from the drain valve at the bottom of the glass, the steam passages from boiler to column and from column to glass are clear.

3. Close the drain valve on the glass and open the drain valve on the column. If the steam blows freely from the column drain, the column itself is clear.

4. Close the column drain valve and open the bottom valves to the column and glass. Note whether water rises quickly to the correct level. If action is sluggish, some obstruction may still be in pipes or valves. Make sure all drain valves are tightly closed and all other valves wide open. Seal the valves on the water-column pipe connections in open position.

Q What points must be observed in fitting up a water gage?

A The gage must be well lighted and so placed that water level can be easily seen at all times. Do not use water glasses or water-glass guards that obscure water level. Automatic ball-check shutoff valves may be used if they comply with the ASME code. Quick-closing valves with lever handles and hanging chains shut off steam and water connections without danger of the operator's being scalded if a glass breaks. It is important to set the visible bottom of the gage glass at the correct distance above the highest point of the boiler heating surface that might be damaged by low water.

On very high boilers the water gage glass is sometimes set with its top tilted outward so that it may be more easily seen, and special forms of gages are sometimes used with lamps and mirrors to project gage readings to floor level. Flat glasses are used in gages for very high steam pressures and may be constructed to make water appear black and steam white.

Q What are gage cocks and where are they used?

A Gage cocks are small globe valves with side outlets and wheel or lever handles. They are a check on the water gage or a temporary means of finding water level when a gage glass breaks by showing whether water or steam blows out when a cock is opened. When the gage glass is placed directly on the boiler head or shell, gage cocks are also directly attached. When a water column is used, gage cocks are placed on the side of the column, as in Fig. 10-6.

According to the ASME code, each boiler must have three or more gage cocks located within the visible length of the water glass, except when a boiler has two water glasses independently connected to the boiler at least

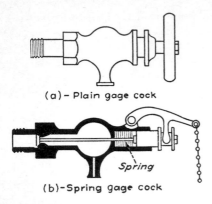

(a) - Plain gage cock

(b) - Spring gage cock

Spring

FIG. 10-7. Two types of gage cocks.

2 ft apart. Firebox or water-leg boilers with not more than 50 ft² heating surface need have only two gage cocks. The bottom gage cock is placed level with the visible bottom of the water glass; others are spaced vertically at suitable distances.

Q Sketch two types of gage cock.
A Figure 10-7*a* is a simple gage cock with a wheel handle for opening and closing, while Fig. 10-7*b* is a spring-lever-handle type in which the valve is kept closed by a strong spring and opened by a chain-operated lever.

WATER ALARMS

Q Describe the high- and low-water alarm shown in Fig. 10-8.
A This form of high- and low-water alarm has a whistle placed on the water column and operated by rods and levers attached to floats or weights within the column. Movement of these weights or floats as the water level falls below or rises above the danger points causes the warning whistle to blow.

Figure 10-8 shows a high- and low-water alarm of the solid-weight type. It depends for its action upon the well-known fact that a body weighs less in water than in air. The upper weight is in the steam space, and the lower weight is wholly submerged in water. Water in this design is about at normal operating level. In this position the weights are balanced and the whistle valve is closed.

If the level falls to a point where the lower weight is no longer submerged, the balance is disturbed. The lower weight, now unsupported, moves downward. This causes a corresponding upward movement of the upper weight, which blows the warning whistle.

If the water rises so high that the upper weight is submerged, its downward pull is lessened and the balance disturbed, so that the lower weight again moves downward and the upper weight rises, blowing the warning whistle as before.

The rods by which the weights are suspended are so attached to the short cross levers at the top that upper and lower weights must always move in opposite directions. Either low or high water moves the weights farther apart, and the resulting turn of the levers opens the steam valve to the whistle.

Q   Explain why in the high-pressure water gage (Fig. 10-9) the water space appears green in color and the steam space red, color changes which make it possible to read the gage more easily.

A   The bicolor water gage works on the principle that a refracted (bending) ray of light differs as it passes indirectly through different materials.

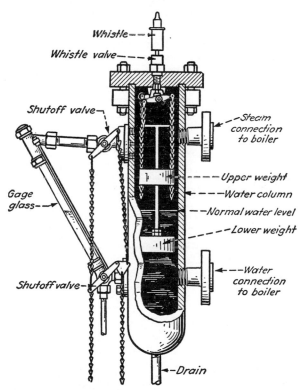

FIG. 10-8.   High- and low-water alarm with gage glass at angle.

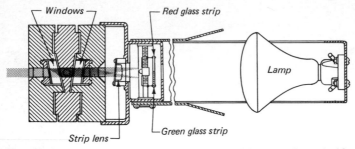

FIG. 10-9.   Cross section of bicolor water gage with steam shown inside the gage.   (*Courtesy of Diamond Power Specialty Corp.*)

Example: When the beam passes indirectly through a column of steam, the amount of bending to which it is subjected is not the same as when it passes through a similar column of water.

The bicolor water gage in Fig. 10-9 has parallel strips of red and green glass in which the colors are permanently fused. When steam occupies the space between the windows, the index of refraction is such that the green light beam is bent out of the field of vision. The red beam is bent so that it emerges from the gage into the line of vision for the observer, who then sees a red glass. Today bicolor gages are used with industrial television cameras for remote reading of water level.

Q   Explain the remote water-level indicator in Fig. 10-10.

A   This indicator operates on the differential-pressure principle; it has connecting tubes leading to the spaces in the boiler drum above and below the water level. One tube has a fixed static head, the other a head which corresponds to changing water level. As fluctuations of water level in the boiler drum overhead vary in pressure, the indicating element registers the corresponding level.

LOW-WATER CUTOFFS

Q   Explain the low-water cutoff.

A   This device is separate from the programming sequence control, and is extremely important to general safety. Its job is to shut down the boiler immediately if water drops to a dangerously low level. There are three basic types (Fig. 10-11). In (*a*), the *float-magnet* design, a nonferrous sleeve encloses a ferrous plunger on a float rod which stays out of the permanent magnet's field when the water level is normal. A switch on the magnet holds the burner circuit closed. When lowered water drops the float, the magnet swings in to the plunger, which tilts the mercury switch and opens the burner circuit.

In the *float-linkage* type (*b*), the float is connected through a linkage to a plate supporting a mercury switch. On a horizontal plane when the water level is normal, the switch holds the burner circuit closed. When lowered water drops the float, the linkage tilts the plate and attached mercury switch, opening the burner circuit.

In the *electrode-probe* type (*c*), an electrode, insulated from its grounded

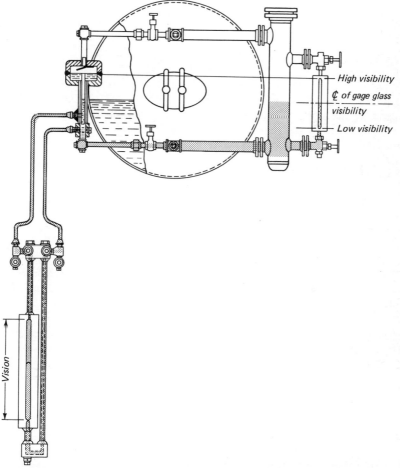

FIG. 10-10.   Remote boiler water-level indicator details.   (*Courtesy of Reliance Gage Column Co.*)

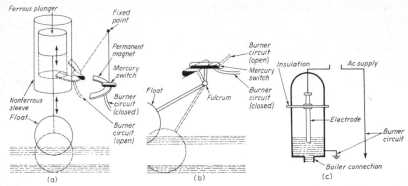

FIG. 10-11.   Basic low-water cutoff types. (*a*) Float magnet; (*b*) float linkage; (*c*) electrode probe.

housing, is partially submerged when the water level is normal. With the burner circuit in series, current flows from the electrode through the water to the housing to hold the burner circuit closed. Lowered water bares the tip, breaking the circuit to the burner.

Q   Explain the location of the low-water cutoff (Fig. 10-12) for heating boilers.

A   Whether the unit is a steel boiler with independent water columns or a cast-iron boiler with a water glass in the first section, it's important to connect the water space-equalizing line at a point about 6 in. below the water glass (or above the boiler's firebox) where there is a large volume of water and slow circulation.

Q   Water-column gage glasses are hard to read from a distance, especially if the glass tube is not clean and the water not clear. Explain one way to solve the problem.

A   Figure 10-13 shows a gage glass with a strip in back, painted with black and white angles. The water level stands out because the strips run in opposite directions starting at water level. The amount of strip distortion through the water-filled tube section depends on tube size and distance from tube to painted card. A little experimenting soon shows the best strip width and distance from the glass.

SAFETY VALVES

Q   What is the function of the boiler safety valve?

A   It prevents boiler pressure from rising above a certain predetermined pressure by opening to allow excess steam to escape into the atmosphere when that point is reached, thus guarding against a possible explosion through excessive pressure.

If the boiler has more than 500 ft² of heating surface, it should have two or more safety valves. In any case, safety-valve capacity must be such as to discharge all the steam the boiler can generate without allowing pressure to rise more than 6 percent above highest pressure at which any valve is set, and in no case more than 6 percent above the maximum allowable working pressure. All safety valves must be the direct spring-loaded pop type.

Q Explain the operation of the spring-loaded pop safety valve.

A It opens suddenly with a popping sound when the steam pressure reaches the point at which the valve is set to blow, hence the name "pop." This quick action is obtained in all pop valves by some method of suddenly increasing the upward force on the spring that holds the valve down on its seat. For example, on some types of safety valves, an extension or lip on the

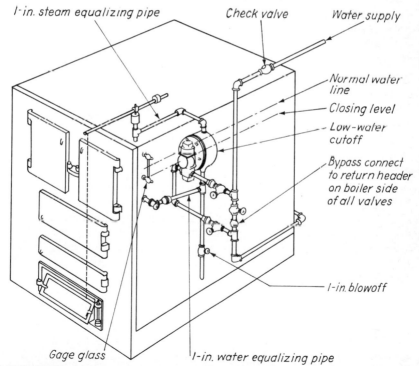

FIG. 10-12. Low-water cutoff correctly installed has water and steam space-equalizing lines connected properly to get true boiler water level in float chamber.

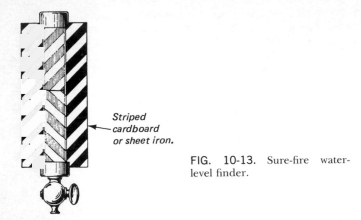

Striped
cardboard
or sheet iron.

FIG. 10-13. Sure-fire water-
level finder.

valve presents a greater area to the upward pressure of escaping steam as
soon as the valve opens.

It is held open until the pressure drops a few pounds. Then it closes just
as suddenly as it opened. This quick opening-and-closing action prevents
"wiredrawing" of the steam and probable grooving of valve and seat.

Q What would cause a safety valve to stick to its seat?

A Most likely corrosion of valve and valve seat as a result of the valve's not
being opened for a long period. To avoid this most dangerous condition,
allow the valve to blow off periodically by using the hand lever to raise it
from its seat or, preferably, by raising the steam pressure to the popping
point.

Q Describe and sketch a spring-loaded pop safety valve.

A A common cast-iron-body type is shown in Fig. 10-14. The valve disk is
held firmly on its seat by a heavy coil spring. The point at which the valve
lifts and relieves the pressure is set by screwing the adjusting nut up or
down and so decreasing or increasing compression of the spring. A lock nut
keeps the adjusting nut from moving once it is set. The cap on top of the
valve may be sealed, if desired, to prevent access to the adjusting nuts. A
hand lever lifts the valve from its seat if it is thought to be sticking, and a
blowdown adjusting arrangement regulates the number of pounds that the
valve blows down before it closes. Figure 10-14 shows an enclosed spring,
which is used when high steam temperatures might affect proper operation of the spring.

Q What is meant by *blowdown* of a safety valve, and how is it regulated?

A Difference between the pressure at which the valve pops and that at
which it closes is the *blowdown*. The valve remains open until the pressure

falls a few pounds below the original popping point because when it does open, a greater area of disk surface is presented to the upward rush of escaping steam and further upward pressure is exerted on the valve lip. In Fig. 10-14 the amount of blowdown may be varied by raising or lowering the adjusting ring on the outside of the valve seat, thus decreasing or increasing the area of the small ports through which the steam escapes from the "huddling" chamber.

The ASME code requires blowdown adjustment to be made in the factory and sealed. Changes which are required in the pressure at which the valve pops or in the amount of blowdown should be made by the boiler inspector or another competent person.

Q What is the reason for allowing blowdown on a safety valve?

A It prevents "chattering," that is, the rapid opening and closing of the

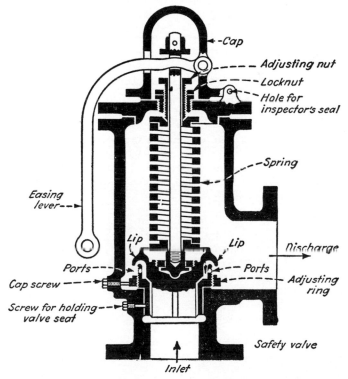

FIG. 10-14. Pop type safety valve for steam boiler must be thoroughly familiar to all operators in order to keep in operating shape.

valve that would take place if the steam were allowed to escape freely and there were no upward pressure on the spring other than that exerted on the bottom of the valve disk.

Q What are some points to consider in attaching a safety valve to a steam boiler?

A Connect the valve as close to the boiler as possible with no unnecessary pipe or fitting between. No other valve of any description may be placed between boiler and safety valve, nor on any escape pipe from it to the atmosphere. If a discharge pipe is used, it must have an open-ended drain to prevent water from lodging above the valve. Locate all safety valves so that the discharge is clear of running boards, working platforms, or any other place where escaping steam is likely to scald anyone. Connect every safety valve in an upright position with the spindle vertical. The valve must not be connected to an internal pipe and the hole in the boiler shell must be at least equal in area to the combined areas of all safety valves connected to the boiler. A muffler's construction must be such that it puts no back pressure on the valve.

Q How is the capacity of a safety valve tested?

A Tests are made under steam in the factory to determine the capacity, and the procedure to be followed in these tests is laid down in the ASME Code for Power Boilers.

If a test is required to be made on a boiler in operation, there are several ways in which this may be carried out. Probably the best check on safety-valve capacity is by an accumulation test. This is made by shutting off all steam outlets from the boiler and forcing the fires to the maximum. Under these conditions, the safety valve or valves should prevent the pressure from rising more than 6 percent above the highest pressure at which the valve or valves are set to pop and in no case more than 6 percent above the maximum allowable working pressure.

Q What is the total upward force on a valve disk 3 in. in diameter when the boiler steam pressure is 150 psi?

A Area of valve = diameter squared × 0.785
     = 3 × 3 × 0.785 = 7.07 in.²
   Total force = area of valve × pressure, psi
     = 7.07 × 150 = 1060 lb   *Ans.*

Q If the downward force exerted by the spring of a safety valve is 850 lb and the diameter of the valve disk is $2^1/_2$ in., at what pressure will the valve open?

A Area of valve disk = 2.5 × 2.5 × 0.785 = 4.91 in.².
   Opening steam pressure = spring pressure ÷ valve area
     = 850 ÷ 4.91 = 173 psi   *Ans.*

FUSIBLE PLUGS

Q What is a fusible plug?

A This device protects the boiler from damage through low water. It is a bronze plug having a fine thread on the outside and a tapered hole in the center filled with a soft metal alloy, which is almost pure tin with a melting point of around 450°F.

Plug is screwed into the boiler plate at some specified point above the highest point of the boiler heating surface that might be damaged by low water. In normal operation, one end of the plug is always covered by water while the other end is exposed to fire or hot gas. The tin filling does not melt as long as the plug is submerged, but if falling water level uncovers the plug, the tin melts and steam blows out of the central hole, thus warning of low water.

A fusible plug is not practicable in many modern boilers where pressures and temperatures are high and shells of drums are often thick, but it is still a valuable accessory in low- and medium-pressure boilers, especially in fire-tube ones. The ASME code gives the correct location of the fusible plug in most well-known types of boilers.

Q Sketch and describe some approved forms of fusible plugs.

A Those recommended in the ASME code are shown in Fig. 10-15. The water-side plug (*a*) is screwed into the plate from the water side of the boiler; (*b*) shows a fire-side plug, so called because it screws into the plate from the fire side; (*c*) is a fire-side plug with the end recessed for a plug wrench to reduce the amount projecting beyond the plate surface. Plugs (*a*) and (*b*) have hexagonal ends and are screwed in with an ordinary wrench. The plug must project at least $^3/_4$ in. beyond the plate on the water side and should not project more than 1 in. on the fire side. The boiler pressure is always on the large end of the tapered soft-metal core.

Q What care must a fusible plug receive?

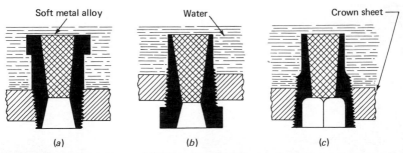

FIG. 10-15. Three types of fusible plugs used for steam generators today. (*a*) Water-side plug; (*b*) fire-side plug; (*c*) fire-side plug.

A　Examine it every time the boiler is shut down for washing or inspection and clean all soot and scale off the ends. It must be renewed at least once a year.

Q　How would you refill a fusible plug?
A　The ASME code recommends that used casings not be refilled. But if it is necessary to refill a plug, heat it to melt out the old filling, clean the hole well, and tin it. Next, place a small plug of putty in the small end of the hole, set the plug on a smooth, level surface with the small end of the tapered hole down, and pour in some molten tin. When the plug cools, smooth off the end with a file, and it is again ready for use.

When melting the tin, take care not to overheat it.

Q　Where should the fusible plug be placed in the following boilers: (1) hrt, (2) water-leg, (3) horizontal-drum water-tube?
A　(1) Hrt boiler: in the rear head not less than 1 in. above the upper row of tubes. Measurement is taken from the upper surface of the tubes to the center of the plug; (2) water-leg boiler: in the highest part of the crown sheet; (3) horizontal-drum water-tube boiler: in the upper drum not less than 6 in. above the bottom of the drum over the first pass of the products of combustion.

SUGGESTED READING

Elonka, Stephen M.: *Standard Plant Operators' Manual,* McGraw-Hill Book Company, New York, 1980 (has over 2000 illustrations).

11

FEED SYSTEMS AND BLOWOFF ACCESSORIES

Automatic feedwater regulators, blowoff valves, and related fittings do their job unfailingly *only* if the operator is familiar with them, uses them as intended, and keeps them in perfect working condition. Even simple valves used in boiler-room piping systems cannot be used for just any service, as their marking indicates. Here we explain the proper uses of these fittings.

It is important to remember, incidentally, that boilers may not always be blown down and discharged into the sewer, since some local ordinances forbid the practice. That is why blowoff tanks are used.

Operators who are proud of their licenses and boiler rooms usually run a "tight ship" and keep their equipment "shipshape"—provided, of course, that management cooperates.

FEED LINE

Q Where should feedwater be introduced into the boiler?

A It should be introduced at a point where it will not discharge directly against any riveted joint or surface exposed to high-temperature gas or direct radiation from the fire. When feedwater must be introduced close to riveted joints or heating surface, use an internal baffle plate to deflect the entering water from the joint or heating surface. In some boilers internal pipes lead the water to the most suitable point of discharge inside the boiler. Some fire-tube and water-tube boilers have internal feed pipes for this purpose. Actual discharge of feedwater should always be below the minimum water level.

Q What does the ASME Code for Power Boilers say regarding means of supplying feedwater?

A Two means of supplying feedwater to the boiler should be provided for all boilers having more than 500 ft² of water-heating surface. These may be two pumps, two inspirators, two injectors, or any suitable combination. If another source of supply is available at a pressure at least 6 percent higher than that at which the safety valve is set, it may be considered as one means of supply.

Q How should the feed pipe be attached to the boiler shell?

A Figure 11-1 shows an approved form of threaded flange connection. This provides for both internal and external feed pipes. Both pipes should be so threaded that they will make tight joints without actually butting the ends together.

Q What valves are necessary on the boiler feedwater line at the boiler shell, and what are their functions?

A A stop valve must be placed next to the boiler shell so that the entire line can be shut off from the boiler during any emergency work on the check valve without shutting the boiler down. At all other times the stop valve is left wide open. The check valve is placed next to the stop valve. It is so constructed that the feedwater passes through it freely into the boiler, but, if feedwater pressure falls below boiler pressure, the check valve closes and prevents steam or water from the boiler from backing up into the feed line.

Q Describe an arrangement of feedwater connections to a fire-tube boiler.

A The connection in Fig. 11-2 consists of a swing-check valve, gate-stop valve, cross with brass plugs, flanged and threaded pad, and internal feed pipe. A *gate* valve is superior to a *globe* valve in that it provides a *straightway* opening, but the seats and disks are more easily renewed in the globe valve. If a globe valve is used here, connect it so that the inlet is *under* the valve disk. Then, if the valve stem breaks, the feedwater pressure will hold the valve open and allow water to enter the boiler.

The swing-check valve is simply a hinged disk that swings open as the feedwater forces its way through and drops back on its seat when feedwater

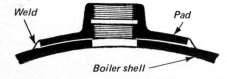

Weld *Pad*

Boiler shell

FIG. 11-1. Pad needed for feed-pipe connection to boiler shell.

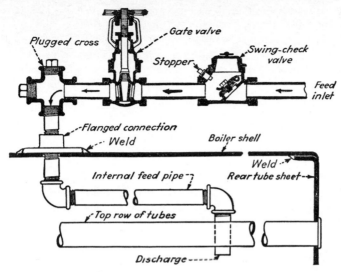

FIG. 11-2. Types of feedwater connections needed for fire-tube boiler.

pressure falls below boiler pressure. The swing-check valve also gives a straightway passage.

BOILER BLOWDOWN

Q What is the purpose of the blowoff valve?

A It has several functions. When necessary it empties the boiler for cleaning, inspection, or repair. It blows out mud, scale, or sediment when the boiler is in operation and prevents excessive concentration of soluble impurities in the boiler.

Q What grade of pipe and types of valves should be used on boiler blowoff lines?

A Use extra-heavy pipe and only valves that offer an unrestricted passage. Plug cocks, especially those that can be lubricated, and quick-opening valves are suitable, as well as certain other Y and angle slow-opening valves that are made for this purpose. Do not use ordinary globe valves on blowoff lines as they tend to prevent passage of mud or scale and cannot be cleared, when plugged, by passing a rod or heavy wire through the blowoff pipe and valve. Gate valves, although they offer a straightway passage, are unsuitable for blowoff because scale can lodge beneath the gate and prevent its closing.

The ASME code specifies that all boilers carrying over 100 psi working pressure, except those used for traction or portable purposes, shall have two blowoff valves on each blowoff pipe. These may be two slow-opening valves, or one slow-opening valve and one quick-opening valve or plug cock. Traction and portable boilers must have one slow- or quick-opening blowoff valve.

Q   What is the proper sequence in opening and closing blowoff valves when blowing down a boiler?

A   When a boiler is equipped with both a blowoff valve and cock or quick-opening valve (Fig. 11-3) in the same blowoff connection, always open the cock or quick-opening valve first and the blowoff valve second. To close, always close the blowoff valve first and the cock or quick-opening valve second. Caution the boiler operator to open and close blowoff valves and cocks slowly, to reduce shock as much as possible. And *never* take your hands off the blowoff valve while it is open, nor your eyes off the gage glass.

Q   When and how often should boilers be blown down?

A   Boilers should be blown down only during periods when steam production is at a minimum. The reason is that circulation in some boilers is very sensitive. Blowing down when steam production is at a maximum could upset the circulation so badly that serious damage could be done to some parts of the boiler, especially the tubes.

Blowdown valves on water walls serve primarily as drain valves. Never blow down water walls when the boiler is in operation. If difficulties arise that require blowing down water walls, do this only under banked conditions or only in accordance with the boiler manufacturer's instructions.

Q   What are some typical blowoff valves?

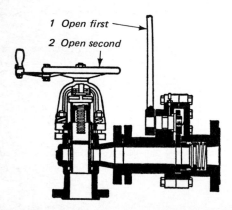

1 Open first

2 Open second

FIG. 11-3. Angle and quick-opening straightway blowoff valves and sequence for opening.

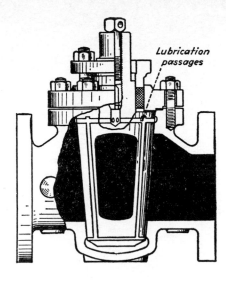

FIG. 11-4. Plug cock installed for boiler blowdown.

A Figure 11-3 shows a typical blowoff arrangement: an angle valve in series with a quick-opening valve.

Figure 11-4 shows details of a plug cock. Any plug cock used for blowoff must have a guard or gland to hold the plug in place, and the end of the plug must be marked in line with the passage through the plug so that anyone can see whether the valve is open or closed.

Q When is a blowoff tank necessary, and how is it constructed?

A Necessary when there is no open space available into which blowoff from the boilers can discharge without danger of accident or damage to property. For example, where the blowoff must be discharged to a sewer, the sewer would probably be damaged by blowing hot water under high pressure directly into it, and water and steam might possibly back up into other sewer connections.

The blowoff tank always stands nearly full of water, as in Fig. 11-5. When the boilers are blown down, the cooler water in the bottom of the tank overflows to the sewer as it is displaced by the entering hot water. The large open vent prevents pressure from building up in the tank, and a small pipe runs from the top of the overflow into the vent pipe to prevent any siphoning action that might empty the tank.

Q List the precautions to take with blowoff valves.

A 1. Avoid accidental opening of a blowoff valve, especially of the quick-opening type. Guard against this by removing the handle or locking it in a closed position when the valve is not in use.

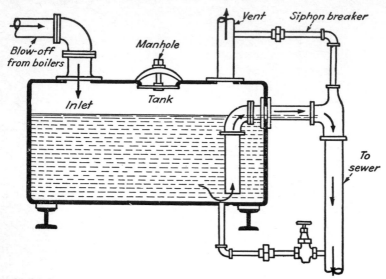

FIG. 11-5.   Blowdown tank must be piped properly for safe operation.

2. Open and close the valve slowly to prevent water hammer and possible rupture of pipes, valves, or fittings. Double blowoff valves protect against this trouble and avoid the need to shut down immediately if one of the valves fails to close properly.

3. If a blowoff valve appears to leak when closed, don't try to force it onto its seat, but open it again so that boiler pressure will wash the valve and remove whatever may be lodging on the seat. Forcing only damages the valve, whereas the extra blow-through will probably remove the leakage cause.

VALVES FOR BOILERS

Q What types of stop valves should be used on steam boilers and steam mains?

A Fit each main or auxiliary discharge steam outlet, except the safety valve and superheater connections, with a stop valve placed as close to the boiler as possible. When outlet size is over 2 in. (pipe size), valves must be the outside-screw-and-yoke type to indicate by the position of the spindle whether the valve is open or closed.

When two or more boilers are connected to a common steam main, the steam connection from each boiler having a manhole must have two stop valves in series, with an ample free-blowing drain between them, the discharge of the drain to be in full view of the operator when opening or closing the valves. Both valves may be of the outside-screw-and-yoke type, but it is preferable that one be an automatic nonreturn valve. This should be placed next to the boiler so that it can be examined and adjusted or repaired when the boiler is off the line.

Q   Describe an *outside-screw-and-yoke* valve.

A   Commonly called an OS&Y valve (Fig. 11-6), its screwed part of the valve stem works in a bushing *outside* the main valve body. This bushing is fastened solidly to the handwheel and turns in a socket at top of the outside yoke. The threaded valve stem can rise and fall but cannot turn. When the handwheel is turned, the threaded bushing also turns, thus raising or lowering the valve stem and opening or closing the valve. Very little of the threaded stem can be seen above the handwheel when the valve is closed, but a considerable length shows when the valve is open. Thus the amount of valve stem in sight above the handwheel is a positive indication of whether the valve is open or closed.

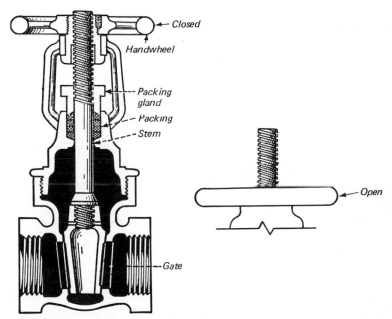

FIG. 11-6.   OS&Y gate valve tells at a glance if open or shut.

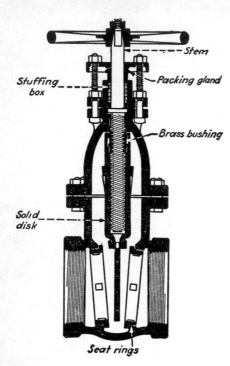

FIG. 11-7. Inside-screw gate valve.

Q Describe an *inside-screw gate* valve.
A The stem of the valve shown in Fig. 11-7 rotates with the handwheel but does not rise or fall. Thus rotation screws the wedge up or down without any outside indication of the valve being open or closed.

Q Explain the construction and operation of an *automatic nonreturn* valve.
A This valve is usually placed on the main steam outlets of boilers in battery with others, in addition to the OS&Y stop valve. The automatic nonreturn valve shown in Fig. 11-8 can be closed by screwing down the outside stem but can be opened only by boiler-steam pressure, as the outside stem is not attached to the valve. The dashpot on top of the valve spindle cushions the valve movement and prevents chattering.

When a boiler is about ready to cut in, the OS&Y stop valve is opened. As soon as boiler pressure rises a little above the pressure in the steam main, it raises the nonreturn valve and automatically puts the boiler on the line. During operation, if for any reason the boiler pressure falls below the main header pressure, the nonreturn valve closes and cuts the boiler out. It can also be used in this way to cut out the boiler when it is being taken off the line for cleaning or repair. It really acts as a check valve, allowing steam flow from the boiler to the main but preventing steam flow from the main to the boiler.

Q What is the function of a *double-acting automatic nonreturn valve?*

A Like the single-acting nonreturn valve shown in Fig. 11-8, it allows steam to flow from the boiler to the main as long as boiler pressure is higher than pressure in the main. It closes when the boiler pressure falls below the steam main pressure, but in addition, it closes against the boiler pressure if the steam main pressure falls below a certain point. It thus protects against heavy steam flow from the main into the boiler in the event of some accident, such as the bursting of a tube. It also prevents a heavy flow of steam from the boiler into the main if some accident to the main or its connections suddenly lowers the steam pressure.

The double-acting automatic nonreturn valve can also cut the boiler on or off the line in the same way as the single-acting nonreturn valve.

Q What are the commonest types of valves for steam, water, air, or gas service?

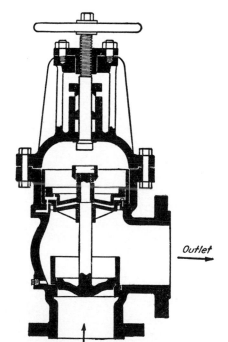

Outlet

From boiler

FIG. 11-8. Automatic nonreturn valve can be opened only by steam inside the boiler.

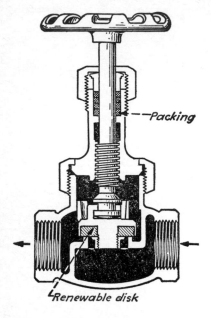

FIG. 11-9. Globe valve with disk and renewable seat used for cool liquids.

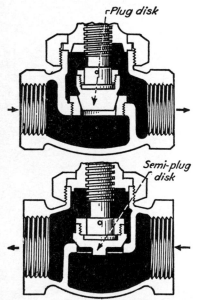

FIG. 11-10. Globe valves may have tapered plug or semiplug disk.

A The two commonest are *gate* and *globe* valves. Gate valves give a straight way opening and are therefore suitable for pipelines carrying heavy fluids, as they offer little resistance to the flow. Globe valves, so called because of their globular shape, do not give a straightway passage and are therefore not so suitable as gate valves where free and unrestricted passage is particularly desirable. However, globe valves are preferred to gate valves on most small pipelines because disks and seats can easily be renewed. Sectional views of small globe valves are shown in Figs. 11-9 and 11-10. Two types of gate valve are shown in Fig. 11-11.

Q Sketch some form of *reducing* valve and describe its construction and operation.

A A *reducing* valve is simply an automatic throttle valve for use where low-pressure steam for heating or process is taken from high-pressure mains. Figure 11-12 shows a weighted-lever reducing valve for medium pressures. Referring to the figure, when there is no pressure on the low-pressure side,

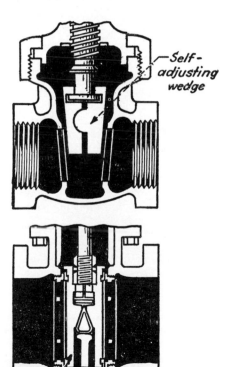

FIG. 11-11. Gate valve seats come in various designs depending upon service.

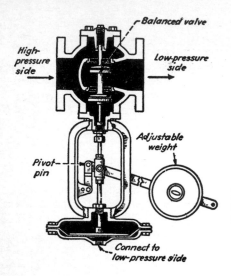

FIG. 11-12.   Weighted-lever reducing valve is set by moving weight.

the weights are so adjusted as to keep the balanced valve open. As pressure builds up on the low-pressure side, it acts upward on the thin metal diaphragm attached to the valve stem, to lift the weight and close the valve. Any desired reduction in pressure within the limits of the valve is accomplished by adjusting the position of the weight on the lever.

Stop valves should be placed on both sides of the reducing valve to cut it out if it should get out of order. Run a bypass around the valve to provide a means of continuing the supply of steam temporarily. A pressure gage and a low-pressure safety valve should be placed on the low-pressure side.

Springs are used instead of levers on some makes of reducing valves. Plungers, operated by steam, air, oil, or water pressure, are also used to control the amount of valve opening in various forms of reducing and regulating valves.

Q   What are automatically controlled valves?
A   They are valves fitted with some means of opening and closing other than, or in addition to, the handwheel or lever. These may be a solenoid or electric motor, or compressed-air or hydraulically operated motors, or plungers. Very large valves that cannot be opened or closed easily by hand are generally equipped with power-operated mechanisms. Sometimes small valves also are fitted with an automatic device for opening and closing by remote control. See Figs. 11-13 to 11-17.

Q   What does the marking *200 WOG* on a valve body mean?
A   A body marked *200 WOG* (water, oil, gas) means the valve is safe for no more than 200 psi, whether for *cold* water, oil, or gas. A valve marked

125 S, for example, means that the valve is for 125 psi saturated *steam,* at its equivalent temperature of 344°F (steam tables give the temperature). For cold service, this 125 S valve may be used up to 300 psi. This shows how dangerous it is to use a *cold* valve for *hot* service. Always check the valve body before installing valves in a system.

Q Explain metals used for boiler-room valves.
A Metals for valves installed in boiler piping systems are usually brass, bronze, iron, or steel. Almost any of these metals is good for steam, cold water, oil, or gas. But each has certain limits. For example, never use brass above 500°F. Bronze can stand 350 psi pressure and temperature up to 700°F. Bronze and brass valves usually come in smaller sizes, up to and including 2 in. Above this size, use iron up to 250 psi and 500°F. Steel is for temperatures above 550°F, and is also used when the valve must withstand high internal pressure. Of course, metals for valves and piping in supercritical pressure systems (3206.2 psi, 705.4°F) and for nuclear power plants are more specialized, as covered in the boiler code.

FEEDWATER-CONTROL SYSTEMS

Q How many basic feedwater-control systems are there?
A Feedwater-control systems are divided into four major classifications, each defined by the number of measurements utilized in the overall control action.

Q What is an automatic feedwater regulator?
A This device automatically regulates feedwater supply according to load, and so does away with hand operation of valves on feed lines. Most feedwater regulators are controlled by temperature, their action depending upon expansion and contraction of some metal part of variation in vapor pressure

Q Describe some type of automatic feedwater regulator operated by expansion and contraction of metal.
A Figure 11-13 diagrams a feedwater regulator operated by expansion and contraction of an inclined metal tube, or thermostat in line with the water gage, and connected to the steam and water spaces, as shown. One end of tube is fixed; the other end is free to move. The free end operates the control valve in the feedwater line through a system of levers. The water level rises and falls in the expansion tube just as it does in the boiler.

The upper end of the tube, being filled with steam, is at approximately steam temperature, but the lower end, which is full of water, is but little warmer than the boiler room. With boiler water at half glass, the tube will

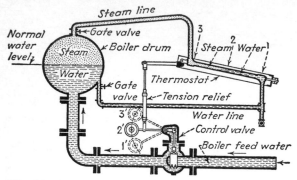

FIG. 11-13. Thermostatic automatic feedwater regulator. (*Courtesy of Copes-Vulcan Div., Blaw Knox.*)

also be half full of water, as shown at *2*, the control valve in the feed line being half open, as at *2'*.

When the load on the boiler increases, the water level falls, exposing more of the tube to steam. This expands the tube, opening the control valve to position *1'*. When the load decreases and the water level rises, the tube contracts, moving the control lever to *3'*, thus closing the control valve, which is balanced to operate with minimum effort.

Q Describe the single-element, self-operated feedwater regulator in Fig. 11-14.

A In this unit, two connections are made in the drum to allow installation of a *generator*. This generator adjusts the vapor pressure within the closed hydraulic system, which in turn furnishes operating power for the feedwater valve. The lower the level of water in the drum, the higher the resulting vapor pressure and the farther the feedwater valve is opened.

This type of regulator establishes a relation between water level in the drum and valve opening. Thus for each rate of steam flow, a slightly different water level will be maintained. This control is not applied to large high-capacity boilers or to those with severely and frequently changing load demands.

Q Describe the single-element, air or electrically operated system in Fig. 11-15.

A Where compressed air or electronic control is used, great improvement can be made in a single-element system. A level recorder is used, thus adding to it a transmitting device and the components of a control system as shown. The water flow into the boiler is regulated from water-level measurement only. A cam in a positioner mounted on the feedwater valve provides a means for matching the valve-flow characteristic to the system requirement.

This system has several advantages over the self-operated type. It holds a constant level at all loads, provides a means for remote manual control of level, and may be converted into a two-element or three-element system if operating conditions require it.

Q Describe the two-element feedwater-control system in Fig. 11-16.

A This system functions primarily from steam-flow measurements but is readjusted from water-level measurement to assure holding the desired level. The steam flow anticipates changes in level and opens the feedwater

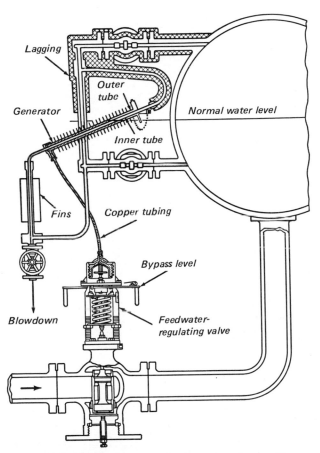

FIG. 11-14.   Thermohydraulic feedwater regulator.   (*Courtesy of Babcock & Wilcox.*)

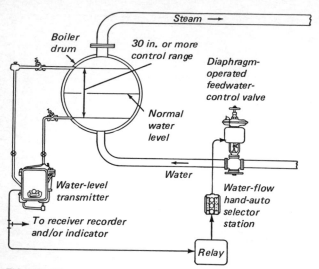

FIG. 11-15. Single-element air-operated feedwater-control system. (*Courtesy of Babcock & Wilcox.*)

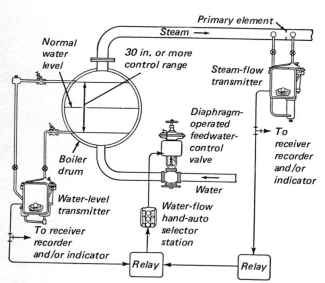

FIG. 11-16. Two-element air-operated feedwater-control system. (*Courtesy of Babcock & Wilcox.*)

valve proportionately. A cam in the positioning relay mounted on the feedwater valve provides a means for matching the valve-flow characteristic to the system requirement.

Such a system minimizes the effect of swell and shrinkage within the drum. It responds quickly, since it does not wait for level to indicate a need for water-flow change. Any boiler installation which has or anticipates frequent changes in load should incorporate the steam-flow element in the feedwater-control system.

Q Explain the three-element feedwater-control system in Fig. 11-17.

A This system functions primarily from steam-flow measurements but is readjusted from the water-level measurement to make certain that the level is held at the desired location. The steam-flow element acts as a continuous load index and positions the control valve immediately on load change. To assure feedwater flow that is always proportional to the steam-flow output, the water-flow measurement is compared with the steam-flow measurement. If pressure variations ahead of the regulating valve cause a change in water flow, the flow ratio repositions the control valve to maintain the proper flow before any effect on level is experienced.

This system may be adjusted to permit the water level in the drum to change during normal operation in accordance with swell, or shrinkage corresponding to the load. This maintains a more nearly constant weight of water in the boiler.

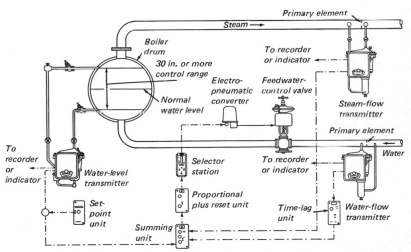

FIG. 11-17. Three-element solid-state electronic feedwater-control system. (*Courtesy of Babcock & Wilcox.*)

Q Figure 11-18 illustrates the control hookup of a vertical hot-water heating boiler common in smaller commercial buildings. Makeup water is by hand control, which requires an attendant. Common practice is to install a high-temperature limit control on the boiler. This shuts off the burner when the water gets too hot. But if the limit control fails, the burner keeps running. If no operator is on watch to open the makeup valve, the water in the heating system evaporates into steam, which escapes through the relief valve, causing the boiler to burn up. Sketch an automatic regulation system of feedwater for this unit.

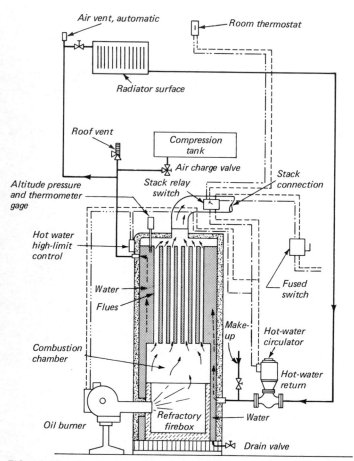

FIG. 11-18.   Circulator-equipped hot-water boiler acts like a forced-circulation boiler, yet makeup isn't automatic.   (*Courtesy of* Power.)

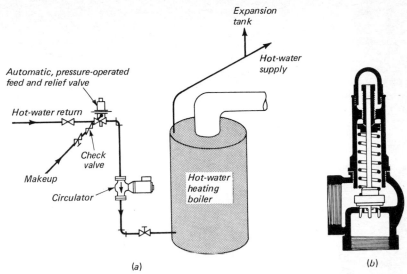

(a)   (b)

FIG. 11-19. (*a*) Automatic regulation of feedwater in boiler return line ensures a steady supply of makeup water as needed. (*b*) Relief spring-loaded valve protects boiler and heating systems against any overpressure dangers. (*Courtesy of* Power.)

A To avoid troubles, install automatic or mechanical regulation (Fig. 11-19*a*). An automatic feed and relief valve in the return line to the boiler has makeup water tied into it. This valve keeps proper pressure on the boiler and heating system, plus adding makeup as needed.

NOTE: Unless the valve can be adjusted to compensate for water head, this system is not adaptable to buildings over three stories in height.

BEWARE: Pressure adjustment on the relief valve (Fig. 12-8*b*), which is designed to meet ASME requirements, should not be made unless the boiler and other equipment have been designed to withstand the pressure increases. Valve location is also important—if possible, it should be level with top of the boiler, because its height affects relieving pressure.

SUGGESTED READING

Elonka, Stephen M., and Alonzo R. Parsons: *Standard Instrumentation Questions and Answers,* McGraw-Hill Book Company, New York, 1962. Reprint 1979, Robert E. Krieger Publishing Company, Inc., Melbourne, Fla. Reprint 1976 (in Portuguese), Editora McGraw-Hill do Brasil, Ltda, São Paulo, Brasil.

12

AUTOMATIC COMBUSTION CONTROL

Steam boilers can be operated with indicators and recorders alone, but few units can be operated efficiently by manual control. Many of the dampers and valves that control flow of air, fuel, and water demand great forces for their rapid manipulation. Thus controls relieve the boiler operators of these tasks and permit them to devote time to other duties, such as burner pattern. Today, manual operation is fading away because of the many packaged flexible control components and systems that do the job more efficiently.

BURNER CONTROLS

Q What basic controls are needed for automation?
A Basically five: (1) fuel flow, (2) combustion-air flow, (3) feedwater flow, (4) furnace draft or pressure, and (5) steam temperature. In addition, regulation is provided in some installation for auxiliary firing, attemperation, gas recirculation, and air-heater protection.

Q Explain the fuel-management system in Fig. 12-1.
A This fuel-system automation permits fuel equipment to be placed in service without supervision by the operator. Such a system will recognize the level of fuel demand to the boiler, will know the operating range of the fuel equipment in service, will reach a decision concerning the need for starting up or shutting down the next increment of fuel equipment, and

will select the next increment based on the firing pattern of burners in service. Such demands for the start-up or shutdown of fuel-preparation and burning equipment can be initiated by the management system shown, without the immediate knowledge of the operators, and their attention may in fact be diverted from the firing and fuel-preparation equipment at the time.

Q Figure 12-2 diagrams a control system for oil- and gas-fired boilers, also for high-temperature water generators. Briefly explain this system.

A Either the boiler steam pressure or the temperature from a water generator actuates the master actuator. The resulting impulse moves a shaft of linkages for controlling (1) the forced-draft-fan damper, through (2) the airflow regulator, and through (3) the power unit and relays, (4) the burner-register louvers, and (5) the oil supply to the burners (by adjusting the orifice-control valve), and (6) the gas supplied to the burners by adjusting the orifice gas valve.

The atomizing steam pressure is controlled from the oil supply applied to the burner, and by the differential steam-control valve. The solenoid valve actuates the gas supply to the pilot. These automatic burners have safety controls which sequence the start-up and shutdown and thus override the operating control to close the fuel valves if any of the components fail. Safety controls shown shut the unit down when there is (1) low water, (2) flame failure, (3) low gas or oil pressure, (4) overpressure, and (5) high water temperature.

Also, an option is provided of recycle or nonrecycle for restarting the unit once it has tripped out because of failure. The recycling control attempts to restart the burner while a nonrecycle control allows the burner to remain out of service until the start switch is closed by hand.

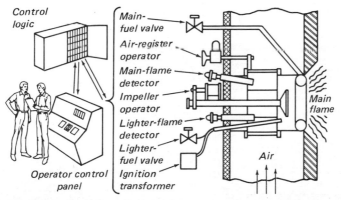

FIG. 12-1. Fuel-management system. (*Courtesy of Babcock & Wilcox.*)

NOTE: Insurance restrictions should be checked when using recycle control. Remember: Automatic start-up sequence and timing depend upon types of burners, specific conditions, and the use of oil, gas, or a combination of the two.

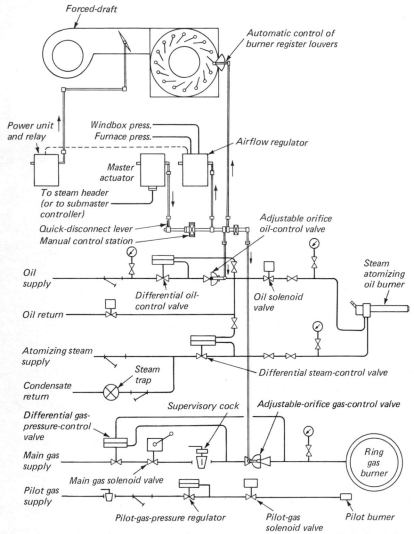

FIG. 12-2. Automatic gas- and oil-fired unit schematic. (*Courtesy of Coen Company, Inc.*)

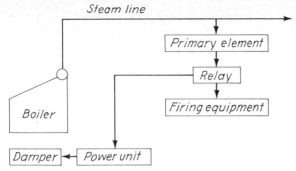

FIG. 12-3. On-off control system has low load range.

Q Explain control systems for modern packaged boilers.
A Such systems generally include safety shutoff valves, limit switches, equipment for ignition proving, combustion control, and flame failure, plus start-up and shutdown programming. The control system senses and makes the boiler respond to changes in steam demand to the point of shutting the unit down when steam needs drop to zero. And if an unsafe condition arises, the boiler locks out.

Automatic regulation of fuel feed, in the proper ratio to the air supply, is the basic job for combustion control equipment. Regulation is initiated in response to a steam-pressure change.

Q What are the three basic combustion control systems?
A (1) On-off, (2) positioning, and (3) metering. An *on-off* system (Fig. 12-3) works between two steam pressure levels to start and stop both air supply and fuel feed. The simplest control, it has a minor drawback in that during the off period, natural draft through the furnace carries away heat, lowering operation efficiency.

In a *positioning* system (Fig. 12-4) response to steam pressure is based on the assumption that, for example, a given damper position will provide sufficient air for a given fuel flow to hold a constant fuel-air ratio throughout the load range. Yet, at any position, airflow may be influenced by voltage variations at the blower motor and by draft loss through the boiler caused by soot or slag and other variables. Thus, although a positioning system may hold steam pressure constant, combustion efficiency may not stay at a high level.

A *metering-type* control (Fig. 12-5) goes a step beyond the positioning setup by actually measuring fuel and air flows and draft loss. These values are balanced against signals for more or less steam pressure. The result is compensation for variables which tend to influence the positioning system's operation.

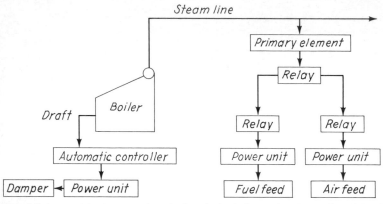

FIG. 12-4. Positioning type has fuel and air tied together in predetermined ratio.

Briefly, metering control yields top precision, and this higher precision is important when a boiler is operated under fluctuating loads for long periods. Generally, both positioning and metering systems include an on-off feature, extending the below-load range beyond that of modulating.

Today's combustion controls are available with hydraulic, electric, or pneumatic relays to transmit impulses to control points.

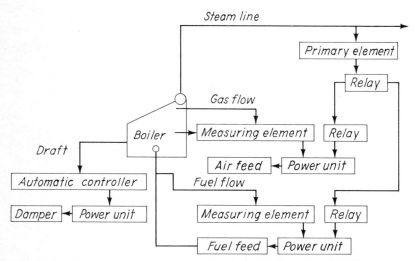

FIG. 12-5. Metering control measures fuel and air flow against signals that indicate steam-pressure fluctuations due to firing and load changes in plant.

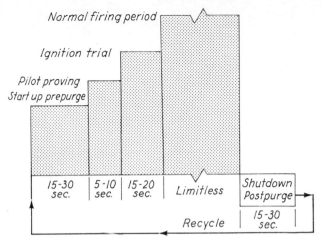

FIG. 12-6. Programming-sequence safeguards boiler through-out its entire operating cycle. (*Courtesy of* Power.)

Q Explain programming sequence.
A Programming sequence (Fig. 12-6) safeguards the boiler throughout the operating cycle. On start-up, purge operation forces air through the furnace and gas passes to clear out any combustible gas pockets. Ignition is next. Depending on the fuel, the ignition system may be electric, spark, light-oil-electric (in which a spark ignites a thin oil which in turn touches off the main fuel), or gas-electric (spark ignites a gas pilot).

A flame scanner is on the job during ignition. The control system won't let the main gas or fuel-oil valve open unless the scanner indicates ignition is normal (ignition proving). Even after lightoff, the scanner continues to supervise the burner. If flame is lost, the scanner shuts the unit down. Following shutdown, the blower again purges the furnace and gas passes before the ignition cycle recurs.

WHY BOILERS FAIL

NOTE: Table 12-1 shows a list of accidents in 1979 to power and heating boilers. All persons working with boilers should study the types and incidence of accidents that occur most often, then study their equipment and set up a system of inspection and operation to prevent such accidents to units under their care.

Q With the many safety devices and automatic controls on present-day boilers, why were there over 3000 boiler accidents in 1979?

TABLE 12-1 Accidents Reported by National Board Members and Other Authorized Inspection Agencies (For the period January 1, 1979–December 31, 1979)

Power boilers:	Accidents	Injuries	Deaths
Tube rupture	281	2	
Shell rupture	33	1	1
Furnace explosions	78	13	1
Flarebacks	6		
Low-water failures	404		
Miscellaneous overheating failures	88		
Piping failures	68	14	3
Poor maintenance of controls	73		
Unsafe practice	23		
Construction-code violation (welds)	1		
Dry fired	102	2	
Tube-sheet crack	61		
Total	1218	32	5
Heating boilers:			
Tube-sheet crack	74		
Dry fired	73		
Gas explosion	8	3	
Shell rupture	107	4	
Furnace explosions	59	2	
Furnace overheated (bagged)	41		
Low-water failures	269		
Runaway burner	7		
Inadequate safety valve	2		
Construction-code violation	1		
Domestic water heater	31		
Total	672	9	0
Cast-iron boilers:			
Gas explosion (furnace)	49	3	
Failure of low-water cutoff	731		
Pilot-light failure	13		
Malfunction of burner control	8		
Operator error	19		
Cracked section	408		
Total	1228	3	0
Total, all sections	3118	44	5

A   The main reason is the lowering or complete absence of operating standards. Too many owners have only one thought in mind when buying boilers provided with automatic controls—they hope to operate without the benefit of qualified operators. Until such time as all interested parties (these include government and insurance company inspectors, boiler manufacturers, code-making organizations, and engineers) can convince boiler owners that they cannot shed all responsibilities toward safe operating practices by substituting automatic devices, boiler accidents will continue at the present alarming rate.

Q   What is the most common type of boiler accident occurring today?
A   The greatest accident producer today is low water, which usually results in overheating and loosening of tubes, collapse of furnaces, and in
some cases, complete destruction of the boiler. In certain classes of boilers a
low-water condition can set the stage for a disastrous explosion that can
cause serious loss.

Q   Why is low water responsible for most boiler accidents in the past
decade?
A   Primarily because of a lack of boiler-operating and maintenance standards. This is true despite the fact that most boilers, especially heating
ones, are provided with automatic feed devices as one of the many automatic controls. But it is these devices that lull the owners into a false sense
of security. Many owners feel that the boiler is fully automatic and completely protected from accidents. Not understanding fully how potentially
dangerous a fired pressure vessel can be, the owner (or anyone else) does
not seem to take the slightest interest after installing one of these so-called
automatic boilers.

The fact is that automatic feed devices, like all other automatic devices,
will work perhaps a thousand times, perhaps many thousand times. But at
some time they will probably fail, usually with disastrous results. That is
why it's the duty of everyone concerned with boilers to realize that unless
proper operating and maintenance standards are instituted, accidents are
certain to occur.

Q   Isn't it true that, in practically all cases, low-water fuel cutouts are installed on all automatically fired boilers to protect the boiler from overheating, in the event of failure of the feeding device? If such is the case, what
causes the failure?
A   Practically all boilers today are provided with low-water fuel cutouts.
What most people don't realize is that in most accidents, regardless of the
types of cutouts, a *series* of failures occurs. Thus, the basic fault may be failure of the automatic feed device to operate. Then we experience failure of
the low-water fuel cutout. The net result is overheating and burning of the
boiler metal. The failure of the automatic feed device and the subsequent
failure of the low-water fuel cutout to operate stem from the same basic
cause—lack of operating and maintenance standards on the part of the
owner.

Q   How should a low-water cutout be tested?
A   The only positive method of testing a LWCO is by duplicating an actual
low-water condition. This is done by draining the boiler slowly while under
pressure. Draining is done through the boiler blowdown line. We find that
many heating boilers are not provided with facilities for proper draining
—an important consideration.

Many operators mistakenly feel that draining the float chamber of the cutout is the proper test. But this particular drain line is provided only for blowing out sediment that may collect in the float chamber. In most cases the float will drop when this drain is opened because of the sudden rush of water from the float chamber. Every boiler inspector can tell you of numerous experiences of draining the float chamber and having the cutout perform satisfactorily. But when proper testing was done by draining the boiler, the cutout failed to function.

Q What percentage of boiler losses are caused by low water?
A Approximately 75 percent.

Q Various articles have been published on the subject of safety valves, particularly the low-pressure type with a setting of only 15 psi. If it is true that all boilers furnished today are provided with ASME-approved valves, why should anyone take exception to them and why shouldn't they work?
A In the first place, most people have a misconception of the term *ASME-approved*. To set the record straight, the ASME itself does not approve a type of safety valve. By referring to the ASME code, Section I—Power Boilers and Section IV—Low Pressure Heating Boilers, you will see that both codes contain limited design criteria. The codes also require a manufacturer to submit valves for testing. Such tests are solely for pressure-setting and relieving capacity.

In brief, the ASME symbol on a safety valve attests to the fact that the limited design criteria and the materials outlined in the code have been supplied by the manufacturer and that the relieving capacity and set pressure stamped on the valve have been proved.

Getting back to the first part of the question, in some cases experience indicates that a particular type of safety valve has an inherent design weakness. After a short period of operation, the disk may be subject to sticking due to close clearances. This condition will render the valve useless, and the boiler will be without the benefit of overpressure protection.

In regard to the second part of the question, the failure of safety valves to work is usually caused by buildup of foreign deposits that results in "freezing." This is an indication that the valve has not been regularly tested or examined. One of the greatest causes of foreign deposit buildup is a "weeping" or leaking condition. The only way to be sure that a valve is in proper operating condition is to set up and adhere to a regular program of testing the valves by hand while the boiler is under pressure. Also, any weeping or leaking valve should be immediately replaced or repaired. This is important.

Q How often and in what manner should boiler safety valves be tested to make certain they are in proper working order?
A Low-pressure (15 psi) safety valves should be lifted at least once a

month while the boiler is under steam pressure. The valve should be opened fully and the try lever released, so the valve will snap closed. For boilers operating between 16 and 225 psi, the safety valves should be tested weekly by lifting the valves by hand. On these higher-pressure boilers it is good practice to test the safety valves by raising the pressure on the boiler. This can usually be done when the boiler is being taken off the line. Then, if the valve "feathers" from improper seating, it can be corrected when the boiler is cold.

Q What can be done to prevent boiler failures?
A Boiler failures can at least be greatly reduced if boilers are placed under the custody of properly trained operators. This means that boiler owners must use sound judgment when employing boiler operators. Everyone with an interest in boilers must be encouraged in educating boiler owners and operators in proper operating procedures.

A very important step is establishing a regular program for testing of controls and safety devices, then faithfully following through.

Further, a program must be established for periodic maintenance of controls and safety devices. First thing to realize is that providing boilers with the most modern proved controls and safety devices is no guarantee that you will not have boiler failures. Any control or safety device is only as good as the testing and maintenance it receives. You cannot ever relax on these two.

Q What can a boiler owner do to assure that he has taken all possible steps to prevent boiler failure?
A First, purchase the best equipment available for a given service. Second, make certain that the boiler is properly installed and equipped with all the necessary appurtenances and safety devices. Third, before taking final acceptance, specify that the installation be inspected by a commissioned inspector in the employ of an insurance company or the state or municipality. By doing so, the owner will be assured that the equipment and the installation meet the legal requirements of the particular state or municipality. Fourth, the owner should provide the operator with a log book and a set of preventive maintenance and testing procedures. He should insist that such procedures be followed religiously and that the results of the test and maintenance be recorded and be made a permanent part of the boiler-room log.

FIRE SAFETY SWITCH

Q What is the purpose of the ignitable cord on the fire safety switch, found on many horizontal rotary atomizing oil burners?
A These burners have as one of the protection controls a fire safety switch

which shuts down the entire system instantly in the event of a flareback. The switch is held closed by an ignitable cord which will burn quickly, thus releasing the switch, which snaps open and breaks the electric circuit, thereby shutting down the burner.

> CAUTION: Do *not* replace this cord with a wire or any nonburnable cord, as is often done.

SUGGESTED READING

Elonka, Stephen M., and Julian L. Bernstein: *Standard Electronics Questions and Answers,* Vols. I and II, McGraw-Hill Book Company, New York, 1964. Reprinted in Polish by Wydawnictwa Naukowo-Techniczne, Warszwa, Poland, 1968.

13

STAYS, JOINTS, OPENINGS, AND CALCULATIONS

So long as boilers contain flat surfaces, stays to keep them from bulging and bursting will be used. Knowledge of the types of stays and their construction, methods of inspecting the condition of stays when a boiler is opened, and calculations related to stays and stayed areas are all vital to the operating engineer's work.

Whether joints are riveted as in older boilers or welded as in modern units, the forces working to part them must be calculated before repairs are made. Such calculations are required in many license examinations. Completing the calculations in this chapter will give the operating engineer a better understanding of (and respect for) heated surfaces under pressure encountered in daily work—and, we hope, result in safer boiler operation.

STAYS

Q Why are stays necessary in boiler construction?

A Flat surfaces exposed to pressure tend to bulge outward; hence they must be supported by stays. Cylindrical or spherical surfaces do not tend to change their shape under pressure and so do not require staying.

Q Sketch a through stay and explain where it is used.

A Through stays are used to support widely spaced parallel flat surfaces, such as front and rear heads of return-tubular and dry-back boilers. They are long round bars, threaded on the ends with a fine thread, and fitted with nuts and washers to secure the ends of the stay to plates and make a steamtight and watertight joint. When the outer nut would be exposed to excessive heat (as, for example, on the rear head of a return-tubular boiler), an eye is forged on the end of the stay; it is fastened to the rear head by a pin and angle irons. To avoid exposing a double thickness of metal to the furnace heat at this point, the angle iron may be separated a few inches from the head by distance pieces riveted to the angle iron and to the head. Both methods are shown in Fig. 13-1.

Q Sketch and describe a diagonal stay and explain where it is used.
A It is used for the same purpose as the through stay—that is, to stay the flat portions of cylindrical-shell heads that are not supported by the tubes. It is not so direct as the through stay, and it throws stress on the shell plates as well, but it leaves more room above the tubes for inspection, repair, and cleaning. A common form of diagonal stay is shown in Fig. 13-2.

Q Sketch and describe a gusset stay.
A It is a form of diagonal stay in which a plate is used instead of a bar. As illustrated in Fig. 13-3, it consists of a heavy plate fastened by rivets and angle bars to the head and shell. It is more rigid than the diagonal stay, takes up more room, and interferes to a greater extent with water circulation. Gusset stays are used very little in modern boiler construction.

Q Sketch and describe a girder stay.
A The girder stay is still used to support tops of combustion chambers in boilers of the scotch marine type. It consists of a cast steel or built-up girder with its ends resting on the side or end sheets of the firebox or combustion chamber, and supporting the flat crown sheet or top sheet of the combustion chamber by means of bolts, as shown in Fig. 13-4.

Q How are curved crown sheets usually supported in older boilers?
A They are supported by long threaded rods called *radial stays* (Fig. 13-5). These rods are screwed through both crown sheet and wrapper sheet and the ends are riveted over.

Q Sketch and describe some forms of stay bolts.
A They are short stay bars to support flat surfaces that are only a short distance apart, such as inner and outer sheets of the water legs in a locomo-

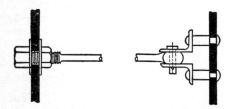

FIG. 13-1. Methods of attaching through stays to boiler plate.

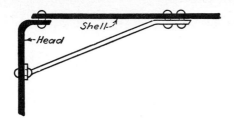

FIG. 13-2.   Diagonal stay retains two flat right-angle surfaces.

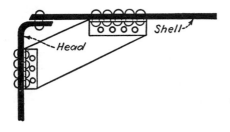

FIG. 13-3.   Gusset stay is much stronger than diagonal stay.

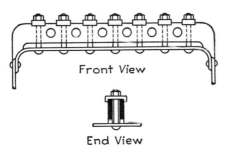

FIG. 13-4.   Girder stay for crown sheet of wet-back scotch marine-type boiler is very rugged.

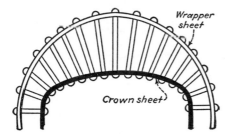

FIG. 13-5.   Radial stays for curved crown sheet of firebox-type boiler.

tive boiler. A plain stay bolt (Fig. 13-6a) is simply a piece of round iron or steel bar, threaded its entire length with a fine thread (12 threads per inch) screwed through both sheets and riveted over on each end. The stay bolt in Fig. 13-6b has the threads turned off on the center part, or sometimes bolt

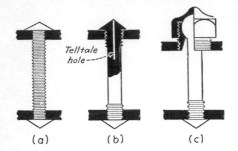

FIG. 13-6. Stay bolts for fire-tube boilers come in various designs.

ends are upset to give enough extra diameter to form a thread and leave the center part smooth. This type is a little more flexible than the type in Fig. 13-6*a* and should be more durable because the bottom of a sharp V thread is a very likely place for a crack to start. It is also claimed to be less liable to damage by corrosion than the all-threaded stay bolt. A still more flexible form, with ball and socket joint on one end, is shown in Fig. 13-6*c*.

Small telltale holes are sometimes drilled in stay-bolt ends to a depth greater than the thickness of head and plate, so that broken stay bolts can be detected by the escape of steam and water. In the case of solid stay bolts, broken bolts can be detected by tapping the heads with a hammer and noticing the sound, but this requires some practice.

Q Sketch some methods of connecting inner and outer plates at the bottom of water legs and openings around fire doors or burner openings in some types of boilers.

A Figure 13-7 shows four methods of connecting plates in such cases: (*a*) an ogee connection; (*b*) a solid ring between the plates with connecting rivets passing through ring and plates; (*c*) plates flanged inward and welded; (*d*) plates flanged inward but with one flange lapped over the other and welded as shown.

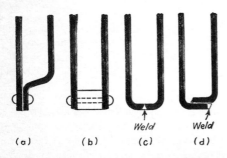

FIG. 13-7. Four methods of connecting plates in the leg-type fire-tube boilers of older designs.

RIVETED JOINTS

Q Boiler plates today are welded, but many boilers in use have riveted joints. What forms of riveted heads are allowed in riveted boiler repairs?
A Various acceptable forms of rivet heads are shown in the ASME Code for Power Boilers, and also in Fig. 13-8.

Q What forms of riveted joints were used in boiler construction?
A Lap joints and butt joints. In the lap joint, edges of plates overlap. In the butt joint, plate edges meet or butt together so that cover straps are used.

Q Describe and sketch a single-riveted lap joint.
A In this joint, plate edges are lapped over and secured by one row of rivets. Figure 13-9a shows a side view and end-sectional view of a single-riveted lap joint.

Q Describe and sketch a double-riveted lap joint.
A In this joint, plate edges are lapped and secured by two rows of rivets, Fig. 13-9b.

Q Describe and sketch a double-riveted butt joint with double straps.
A In this joint, plate edges are butted together and cover straps are placed inside and outside. Figure 13-9c shows a double-riveted double-strap joint with straps of equal width.

Q Describe and sketch a triple-riveted butt joint with double straps of unequal width.

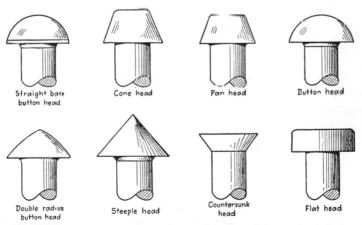

Straight base button head   Cone head   Pan head   Button head

Double radius button head   Steeple head   Countersunk head   Flat head

FIG. 13-8.   Boiler rivet heads are formed in these eight standard shapes.

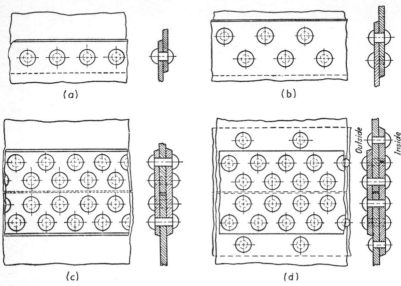

FIG. 13-9. Four basic types of riveted joints used in older steam generators.

A In this joint (Fig. 13-9d) plate edges butt together and cover straps are placed inside and outside, but the outside strap is narrower than the one inside. Outer rows of rivets pass through the inner strap and shell plate but not through the outer strap, and alternate rivets are omitted in these outer rows.

Q Why is a butt joint preferable to a lap joint?
A In the butt joint, shell plates form a true circle, and there is no tendency for the pressure to distort the plates in any way. In the lap joint, plates do not form a true circle at the lap, and internal pressure tends to pull them to the position shown in Fig. 13-10a. In time, this bending action may cause "grooving" or cracking of the plate along the calking edge and thus create a dangerous condition. Properly designed butt joints also have much higher efficiencies than lap joints with the same number of effective rows of rivets, because in the butt joint all or most of the rivets pass through three plate thicknesses and so are in *double shear,* while the rivets in lap joints are all in *single shear.* According to the ASME code, rivets in double shear have twice the strength of rivets in single shear.

Q What preparation of plates and butt straps is necessary to ensure good riveted joints?
A Plates and straps must be rolled or formed to the proper curvature by pressure only, and calking edges planed or milled to an angle not sharper

than 70° with the plane of the plate. Plates and straps should fit closely before riveting so as to avoid undue stress on the rivets or excessive calking to make a joint tight.

Q Are rivet holes punched or drilled?

A A drilled hole is best because it is parallel and exact in size, and the metal around it is not weakened to any great extent by the drilling process. The ASME code permits punching holes at least $1/8$ in. less than full diameter in plates not over $5/16$ in. thick, and at least $1/4$ in. less than full diameter in plates over $5/16$ in. and up to $5/8$ in. Drilling or reaming to full size removes metal around the hole that may have been crushed by the action of the punch.

Final drilling or reaming of rivet holes is done with the joint assembled and held in place by tack bolts. The parts are then separated and all burrs around holes, chips, and cuttings removed before the joint is reassembled and the rivets driven. Tack bolts and barrel pins keep the holes in proper line while the riveting is being done.

Q What is calking and why is it necessary?

A Calking refers to the upsetting or burring up of the edge of the plate or strap after riveting so as to make the edge press down tightly on the plate beneath and thus form a water- and steamtight joint. This is necessary because inequalities in the plate surface make it practically impossible for riveting alone to make a joint absolutely tight. In calking, use a blunt-nosed tool and take great care not to damage or nick the bottom plate. A blunt or round-nosed tool thickens the plate edge, as in Fig. 13-10b, but a sharp tool may actually force the plates apart, as in Fig. 13-10c, and so do more harm than good. After calking and applying the hydrostatic test, certain seams may be seal-welded as a further precaution against leakage.

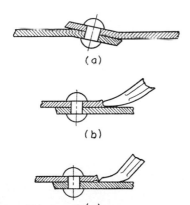

(a)

(b)

(c)

FIG. 13-10. (a) Distortion of single lap joint; (b) correct calking method; (c) incorrect calking opens lap.

Q Which is stronger and safer, riveted or welded seams for pressure vessels?

A Experience and observation of burst tanks indicate that riveted seams in tanks should be avoided. Usually the joint or seam is the weakest part of the tank, whether a longitudinal or girth seam or a riveted head attached to the body. Bursting rivet heads and bodies cause a "shrapnel" effect, which is so forceful that the pieces can easily penetrate concrete walls and even pass through steel plating.

A welded tank conforming to ASME specifications (or to certain specifications of other societies) is generally considered safe. This also applies to boiler drums, although such riveted drums are rarely made today. In view of the damage that can occur just from flying rivet heads, it is best to always use welded drums that have been stress-relieved and built according to the applicable code.

Tanks (or flasks) holding compressed air or gas or liquid air or gas are probably the *most* dangerous should there be a fire or excess heat tending to increase the internal pressure. This applies especially to tanks that are only partially filled with volatile liquids; in such a case, the liquids more readily flash into gas than if the tank were completely full.

EFFICIENCY OF RIVETED JOINTS

Q What is meant by the *efficiency* of a riveted joint?

A This is the ratio of the strength of a unit section of the joint to the same unit length of solid plate. Unit length usually taken is the pitch of the rivets (distance from center to center) in the row having the greatest pitch.

Q What are the various ways in which a riveted joint may fail?

A It may fail by (1) tearing the plate along the centerline of the rivets, (2) shearing the rivets, (3) crushing plates and rivets.

Q What symbols are commonly used in riveted-joint calculations?

A Following are the symbols commonly used in riveted-joint calculations and their meanings: P = pitch of rivets, in., in the row having the greatest pitch; d = diameter of rivet after driving, or diameter of rivet hole, in.; a = cross-sectional area of rivet after driving, in.^{2}; t = thickness of plates, in.; b = thickness of butt strap, in.; s = shearing strength of rivet in *single shear*, usually taken as 44,000 psi; S = shearing strength of rivet in *double shear*, usually taken as 88,000 psi; c = crushing strength of mild steel, usually taken as 95,000 psi; T_s = tensile strength of plate metal, usually taken as 55,000 psi; n = number of rivets in *single shear;* N = number of rivets in *double shear.*

Q What is meant by *single shear* and *double shear*?

A When two lapping plates are riveted, as in a lap joint or a single-strap butt joint, there is only one section of each rivet opposing the tendency to shear the rivets through at right angles to their length. In this case, rivets are said to be in *single shear*. When there are three plate thicknesses in the joint, as in a butt joint with inner and outer straps, two sections of each rivet oppose the shearing stress; so the rivets are said to be in *double shear*.

Q How would you find the efficiency of a single-riveted lap joint?

A Figure 13-11a shows this joint. There are four calculations to be made, using the symbols given previously (the calculations refer to a length of joint equal to P, the rivet pitch):

    1. Strength of solid plate $= P \times t \times T_s$

    2. Strength of plate between rivets $= (P - d) \times t \times T_s$

    3. Shearing strength of *one* rivet in *single shear* (since there is only one rivet in one pitch in this case) $= s \times a$

    4. Crushing strength of plate in front of *one* rivet $= d \times t \times c$

Whichever one of 2, 3, or 4 gives the lowest value is divided by no. 1 to give the joint efficiency.

Q What is the efficiency of a single-riveted lap joint having a rivet pitch of 3 in., a rivet diameter of $1\frac{1}{4}$ in., and a plate thickness of $\frac{1}{2}$ in.?

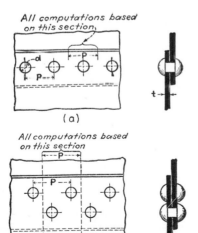

FIG. 13-11. Letter P indicates pitch in single-riveted joint (a) and double-riveted (b) lap joints.

A   Using substitutes in the formulas given in the preceding answer, the following are obtained (carrying each computation to nearest 100 lb):

1. $P \times t \times T_s = 3 \times 0.5 \times 55,000 = 82,500$ lb

2. $(P - d) \times t \times T_s = (3 - 1.25) \times 0.5 \times 55,000 = 48,100$ lb

3. $s \times a = 44,000 \times 1.25^2 \times 0.785 = 54,000$ lb

4. $d \times t \times c = 1.25 \times 0.5 \times 95,000 = 59,400$ lb

Since calculation 2 gives the lowest value, efficiency of joint = 48,100/82,500 = 0.583, or 58.3 percent.   *Ans.*

This means that plate will fail by tearing the "ligament" between rivet holes and that the strength of this ligament is 58.3 percent of the strength of the undrilled plate.

Q   How do you find the efficiency of a double-riveted lap joint?

A   Here (Fig. 13-11*b*), calculations are similar to those for the single-riveted lap joint, except that the value of *n* is 2 instead of 1, because there are *two* rivets in one pitch in a double-riveted lap joint. Thus,

1. Strength of solid plate = $P \times t \times T_s$

2. Strength of plate between rivet holes = $(P - d) \times t \times T_s$

3. Shearing strength of *two* rivets in *single* shear = $2 \times s \times a$

4. Crushing strength of plate in front of *two* rivets = $2 \times d \times t \times c$

Q   What is the efficiency of a double-riveted lap joint having a rivet pitch of 4 in., a rivet diameter of 1 in., and a plate thickness of $3/8$ in.?

A   Using formulas given in preceding answer and evaluating to nearest 100 lb

1. $P \times t \times T_s = 4 \times 0.375 \times 55,000 = 82,500$ lb

2. $(P - d) \times t \times T_s = (4 - 1) \times 0.375 \times 55,000 = 61,900$ lb

3. $2 \times s \times a = 2 \times 44,000 \times 0.785 = 69,100$ lb

4. $2 \times d \times t \times c = 2 \times 1 \times 0.375 \times 95,000 = 71,300$ lb

Since calculation 2 (strength of plate between rivet holes) is lowest, efficiency of joint = \$61,900/82,500 = 0.75, *or* 75 percent.   *Ans.*

Q   How do you find efficiency of a triple-riveted butt joint with two unequal straps and alternate rivets omitted in outer rows?

A   This (Fig. 13-12) is the equivalent of five rivets in one pitch, four being in double shear and one in single shear. Since there are a greater number of rows of rivets than in the previous riveted-joint examples, some extra calculations have to be made, as follows:

1. Strength of solid plate $= P \times t \times T_s$

2. Strength of plate between rivet holes in outer row $= (P - d) \times t \times T_s$

3. Shearing strength of *four* rivets in *double shear* and *one* rivet in *single shear* $= 4 \times S \times a + s \times a$

4. Strength of plate between rivet holes in second row, plus shearing strength of one rivet in single shear in outer row $= (P - 2d) \times t \times T_s + s \times a$

5. Strength of plate between rivet holes in second row, plus crushing strength of butt strap in front of one rivet in outer row $= (P - 2d) \times t \times T_s + d \times b \times c$

6. Crushing strength of plate in front of four rivets, plus crushing strength of butt strap in front of one rivet $= 4 \times d \times t \times c + d \times b \times c$

7. Crushing strength of plate in front of four rivets, plus shearing strength of one rivet in single shear $= 4 \times d \times t \times c + s \times a$

Q What is the efficiency of a triple-riveted butt joint with unequal straps and alternate rivets omitted in outer rows, if the outer pitch is 7 in., rivet diameter 1 in., thickness of plates $^7/_{16}$ in., and thickness of butt strap $^3/_8$ in.?
A 1. $P \times t \times T_s = 7 \times 0.4375 \times 55,000 = 168,400$ lb

2. $(P - d) \times t \times T_s = (7 - 1) \times 0.4375 \times 55,000 = 144,400$ lb

3. $4 \times S \times a + s \times a = 4 \times 88,000 \times 0.785 + 44,000 \times 0.785 = 310,900$ lb

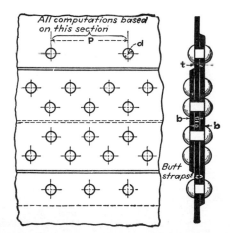

FIG. 13-12.   Letter P indicates pitch in triple-riveted butt joint.

4. $(P - 2d) \times t \times T_s + s \times a = (7 - 2) \times 0.4375 \times 55,000 + 1 \times$ 44,000 $\times$ 0.785 = 154,900 lb

5. $(P - 2d) \times t \times T_s + d \times b \times c = (7 - 2) \times 0.4375 \times 55,000 + 1 \times$ 0.375 $\times$ 95,000 = 155,900 lb

6. $4 \times d \times t \times c + d \times b \times c = 4 \times 1 \times 0.4375 \times 95,000 + 1 \times 0.375$ $\times$ 95,000 = 201,900 lb

7. $4 \times d \times t \times c + s \times a = 4 \times 1 \times 0.4375 \times 95,000 + 1 \times 44,000 \times$ 0.785 = 200,800 lb

Since calculation 2 gives the lowest result, efficiency of joint = 144,400/168,400 = 0.857, or 85.7 percent.   *Ans.*

Q What will be the maximum allowable working pressure in psi on a cylindrical boiler shell if the inside diameter of the weakest course is 6 ft, plate thickness is $^9/_{16}$ in., efficiency of riveted joint is 85 percent, maximum allowable unit stress on steel is 55,000 psi, and maximum temperature to which plate will be subjected is 650°F?

A From the conditions given, the ASME Code for Power Boilers will allow a maximum unit working stress of 13,750 psi, and the maximum allowable working pressure is calculated from the code formula

$$P = \frac{0.8SEt}{R + 0.6t}$$

where P = maximum allowable working pressure, psi

    S = maximum allowable unit working stress, psi

    E = efficiency of longitudinal joint

    t = minimum thickness of shell plates in weakest course, in.

    R = inside radius of weakest course of shell, in.

  Substituting the values given,

$$P = \frac{0.8 \times 13,750 \times 0.85 \times 0.5625}{36 + 0.6 \times 0.5625} = 144.7, \text{ say } 145 \text{ psi}   Ans.$$

Q In a fire-tube boiler 72 in. in diameter and 18 ft long, the distance from the top of the tubes to the shell is 23 in. What is the area of the segment (Fig. 13-13) of the head above the tubes to be supported by staying?

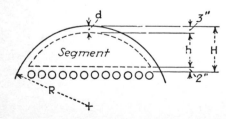

FIG. 13-13. Data needed for calculating the area of boiler segment.

A It is usually assumed that a distance of 2 in. above the top of the tubes is supported by the tubes and that a distance of 3 in. from the shell is supported by the shell. The tables of areas of segments given in the ASME Code for Power Boilers are based on these assumptions, though the code also gives formulas for finding the distance d supported by the shell. These formulas are

(1) d = outer radius of the flange, not exceeding 8 times the thickness of the head

or (2) $d = \dfrac{80t}{\sqrt{P}}$

where t = thickness of the head, in.
d = unstayed distance from shell, in.
P = maximum allowable pressure

Assuming in this case that $t = 0.5625$ and $P = 145$, then $d = 80 \times 0.5625/\sqrt{145} = 3.74$ in., but for all practical purposes, 3 in. may be taken in most cases. The height of the segment to be stayed is $h = H - d - 2$, and the complete formula for finding the area is

$$\text{Area of segment} = \frac{4}{3}(H - d - 2)^2 \times \sqrt{\frac{2(R - d)}{H - d - 2}} - 0.608$$

We can simplify this by substituting h for $H - d - 2$, so $h = 23 - 3.74 - 2 = 17.26$. Then

$$\text{Area} = \frac{4 \times 17.26^2}{3} \times \sqrt{\frac{2(36 - 3.74)}{17.26}} - 0.608$$

$$= \frac{4 \times 298.5}{3} \times \sqrt{\frac{64.52}{17.26}} - 0.608$$

$$= 398 \times \sqrt{3.74 - 0.608} = 398 \times 1.77$$

$$= 704.5 \text{ in.}^2 \qquad Ans.$$

Q Calculate the number and size of through stays required to support the segment in the previous question.
A As we cannot have two unknown quantities in our problem, we must assume either the size of the stays or the number of stays to begin with. Let us assume that $1\frac{1}{2}$-in. stays are to be used, having an allowable unit stress of 8500 psi, and that gage pressure is 145 psi.

Allowable load on one stay = cross-sectional area $\times$ 8500
$$= 1.5^2 \times 0.785 \times 8500 = 15{,}013 \text{ lb}$$
Total load on segment = area in in.^{2} $\times$ gage pressure
$$= 756 \times 145 = 109{,}620 \text{ lb}$$

Number of stays $= \dfrac{109{,}620}{15{,}013} = 7.3$, say 8 stays $\qquad Ans.$

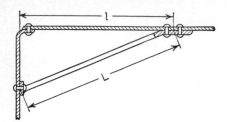

FIG. 13-14. Dimensions needed for finding the size of diagonal stay.

Q Assuming that this segment is to be supported by diagonal stays instead of through stays, what would be their cross-sectional area if the stays are 40 in. long and the distance l in Fig. 13-14 is 35 in.?

A The ASME Code for Power Boilers formula for finding the area of a diagonal stay is

$$A = \frac{aL}{l}$$

where A = cross-sectional area of diagonal stay
  a = cross-sectional area of through stay
  L = length of diagonal stay
  l = length of line drawn at right angles to boiler head, or surface supported to center of palm of diagonal stay

Then, in this case,

$$\text{Area of diagonal stay} = \frac{1.767 \times 40}{35} = 2.02 \text{ in.}^2$$

and, if the stay is a round bar,

$$\text{Diameter} = \sqrt{\frac{2.02}{0.785}}$$

$$= 1.6 \text{ or } 1^5/_8 \text{ in. to nearest eighth of an inch}\qquad Ans.$$

CALCULATING WORKING PRESSURE

Q Figure 13-15 shows a form of cylindrical flue for internally fired boilers known as an Adamson ring. If the diameter of the flue is 36 in. and the length of the section is 48 in. between ring centers with a plate thickness of $^5/_8$ in., what will be the allowable working pressure on this flue in psi?

A The allowable working pressure on an Adamson ring is found from the formula

$$P = \frac{57.6(300t - 1.03L)}{D}$$

where t = thickness of plate, in.
$\quad$ D = outside diameter of the furnace, in.
$\quad$ L = length of furnace section, in.
In this case,

$$P = \frac{57.6[300(0.625) - 1.03(48)]}{36}$$

$\quad$ = 220.9 psi $\qquad$ *Ans.*

Q What is the maximum allowable working pressure on the concave side of an unstayed seamless head, dished to the form of a segment of a sphere of radius 42 in.? The plate is $5/8$ in. thick and the maximum allowable unit working stress on the plate is 12,000 psi.
A The ASME Code for Power Boilers formula for plate thickness is

$$t = \frac{5PL}{4.8SE}$$

where t = thickness of plate, in.
$\quad$ P = maximum allowable working pressure, psi
$\quad$ L = radius to which head is dished, measured on concave side, in.
$\quad$ S = maximum allowable working stress, psi
$\quad$ E = efficiency of weakest joint forming head (efficiency of seamless heads is 100 percent)
Transposing the above formula for P and substituting the values given,

$$P = \frac{4.8SEt}{5L} = \frac{4.8(0.625)(12,000)(1)}{5(42)}$$

$\quad$ = 171.4 psi $\qquad$ *Ans.*

Q What working pressure should be allowed on stayed flat plates, $3/8$ in. thick, with stay bolts pitched 5 in. horizontally and vertically?

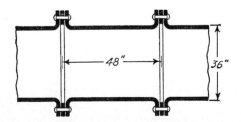

FIG. 13-15. Construction of Adamson ring-type combustion chamber.

A Working pressure is found from the formula in the ASME Code for Power Boilers:

$$Wp = \frac{SCt^2}{p^2}$$

where S = allowable stress, psi, for plate material
$\quad C$ = a factor for staybolts = 2.1 for screwed stays and plates not over $^7/_{16}$ in. thick
$\quad t$ = plate thickness, in.
$\quad p$ = pitch of stays, in.

Allowable stress for plate materials is assumed to be 13,750 psi. Then

$$Wp = \frac{SCt^2}{p^2} = \frac{13,750(2.1)(0.375)^2}{(5^2)} = 162.4$$

Q A flat plate, $^3/_8$ in. thick, is to carry 160 psi working pressure; what should be the pitch of the stay bolts?
A Using the same formula as in the previous question but transposing it for p,

$$p = \sqrt{\frac{SCt^2}{Wp}} = \sqrt{\frac{13,750(2.1)(0.375)^2}{160}} = 5.038 \text{ in.} \qquad Ans.$$

Q A steam drum in a bent-tube boiler is 40 in. in diameter and made from $^5/_8$-in. carbon steel plate. A $^1/_2$-in. reinforcing plate is riveted over the ligament between the tubes, and the ligament has an efficiency of 45 percent. The longitudinal seam of the drum is a double-riveted butt joint with double straps and has an efficiency of 82 percent. The allowable unit working stress on the plate is 9500 psi. What is the maximum allowable working pressure on this drum?
A Using the ASME Code for Power Boilers formula, Wp calculated on efficiency of the joint

$$= \frac{SEt}{R + 0.6t} = \frac{9500 \times 0.82 \times 0.625}{20 + 0.6 \times 0.625}$$

$$= \frac{95 \times 82 \times 0.625}{20,375} = 239 \text{ psi}$$

Wp calculated on efficiency of the ligament

$$= \frac{9500 \times 0.45 \times 1.125}{20 + 0.6 \times 1.125} = \frac{95 \times 45 \times 1.125}{20.675}$$

$$= 233 \text{ psi}$$

Taking the lower value, the allowable working pressure is 233 psi.   *Ans.*

Q Calculate number and pitch of 1-in. stay bolts required to support the flat side sheet of a water leg measuring 30 by 42 in. Boiler pressure is 175 psi, and allowable unit stress on stay bolts is 7500 psi.
A Area of sheet = 30 × 42 = 1260 in.²
Total pressure on sheet = 1260 × 175 = 220,500 lb

Assuming 1-in. stays are used, 12 threads per inch, U.S. standard,

Diameter of stay at root of thread = outside diameter
$$- \text{(pitch} \times 1.732 \times 0.75)$$
$$= 1 - (^1/_{12} \times 1.732 \times 0.75)$$
$$= 1 - 0.108 = 0.892 \text{ in.}$$
Cross-sectional area of one stay = $0.892^2 \times 0.785$
$$= 0.625 \text{ in.}^2$$

Total area of stays = $\dfrac{\text{total load}}{\text{allowable load per in.}^2}$

$$- \frac{220,500}{7500} = 29.4 \text{ in.}^2$$

Number of stays = $\dfrac{29.4}{0.625} = 47,$

say 48 to get even spacing of six rows of eight stays in each row   *Ans.*
Vertical pitch = $^{30}/_6$ = 5 in.
Horizontal pitch = $^{42}/_8$ = 5¹/₄ in.   ⎦   *Ans.*

Q What weight of water will be contained in an hrt boiler, 18 ft long by 72 in. in diameter with seventy-two 4-in. tubes, when it is filled with water for a hydrostatic test?
A Volume of shell = $6^2 \times 0.785 \times 18$ = 508.68 ft³

Volume of tubes = $\dfrac{72 \times 4^2 \times 0.785 \times 18}{144}$ = 133.04 ft³

Volume of water = 508.68 − 113.04 = 395.64 ft³
Weight of water = 395.64 × 62.4 = 24,688 lb   *Ans.*

SHELL OPENINGS

Q What is the common form of manhole opening in flat surfaces?
A It is usually elliptical and flanged *inward* from solid plate (Fig. 13-16). When the plate edge is less than ¹¹/₁₆ in. thick, the gasket-bearing surface is increased by shrinking a narrow ring around the flange and facing off both ring and plate edge to form a smooth joint.

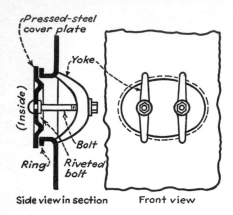

Side view in section   **Front view**

FIG. 13-16.   Design of manhole in flat surface according to code.

Elliptical manholes must not be less than 10 by 16 in. or 11 by 15 in. inside. Circular manholes must not be less than 15 in. inside diameter.

Q   What is the common form of manhole in curved surfaces?
A   The manhole is usually elliptical, with long axis at right angles to the axis of the shell, and is reinforced by a flanged ring welded to the shell (Fig. 13-17). A ring is also shrunk around the flange of the reinforcing ring if more gasket bearing surface is required.

Q   How are manhole cover plates and yokes made, and what materials are used in their construction?
A   Manhole cover plates and yokes are made from rolled, forged, or cast

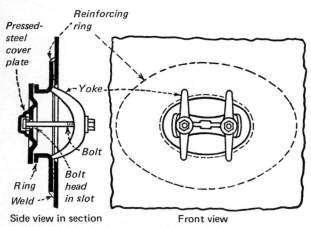

FIG. 13-17.   Manhole and cover details for curved surface.

steel for pressures over 250 psi and temperatures above 450°F. Below this temperature and pressure they may be made from cast iron.

The cover plate in Fig. 13-16 is made from steel plate. The bolts are riveted to the plate, and the yokes or dogs are cast steel or pressed steel.

The cover plate (Fig. 13-17) is also pressed from steel plate, but the bolt heads fit loosely in slots instead of being riveted solidly to the plate. There is no possibility of leakage past the bolt heads with this construction, and probably less liability of bolt breakage, but the riveted-bolt manhole cover is more easily handled and does not often give any trouble.

Q Why are manholes and handholes usually elliptical in shape?
A The elliptical cover plate can easily pass through the manhole when it is necessary to take it out or insert it in place. This shape of manhole also affords easier access to the boiler with the minimum area of opening, and so does not weaken the plate to the same extent as openings of larger area. Figure 13-18 shows a handhole opening and handhole cover plate.

Q What type of gasket is used on manhole and handhole cover plates?
A An asbestos gasket, molded to the correct shape, is commonly used. It should not be over $^{1}/_{4}$ in. thick when compressed.

Q How would you replace a manhole cover plate after washing out a boiler?
A It is not always necessary to replace the gasket, as a good one may be satisfactory after the joint has been broken several times. However, if there is any doubt that the gasket will make a tight joint, it should be discarded for a new gasket.

After cleaning off the joint faces, place the new gasket on the plate and smear the face of the gasket with a paste made from graphite and cylinder oil so that the joint will break readily the next time the boiler is opened.

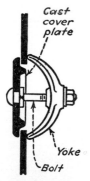

FIG. 13-18.   Handhole with cast cover plate.

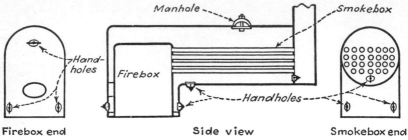

Firebox end   **Side view**   **Smokebox end**

FIG. 13-19.   Handholes and washout plugs needed in firebox water-leg type of

Insert the plate through the manhole opening. Hold it in place with one hand while positioning the yoke, and screw on the nut with the other hand.

Then put the second yoke on, if two are used, and tighten both nuts, making sure that the plate remains central in the manhole opening and that the gasket has not shifted. Large manhole covers in the ends of horizontal drums are sometimes supported on hinges inside the boiler head. These covers are not removed when the boiler is opened up but simply swung back clear of the opening.

Q   Show, by sketches, the number and arrangement of washout plugs or handholes in a firebox water-leg type of boiler.

A   The boiler should have not less than six handholes or washout plugs, located as in Fig. 13-19: one in the rear head below the tubes; one in the front head in line with the crown sheet; four in the lower part of the water leg. When possible, a seventh one should be near the throat sheet.

Handholes should be at least $2^{3}/_{4}$ by $3^{1}/_{2}$ in. and washout plugs not less than 1-in. pipe size. Larger sizes are desirable, since the purpose of these openings is to permit visual inspection of the interior of the boiler and the insertion of washout hose and cleaning tools. Plugs of brass, bronze, or other nonferrous metal may be used where pressures are not over 250 psi. They can be removed more easily than steel plugs and are less likely to corrode or rust in place.

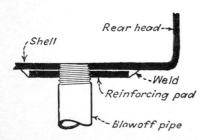

FIG. 13-20.   Reinforcing pad is required for blowoff pipe connection.

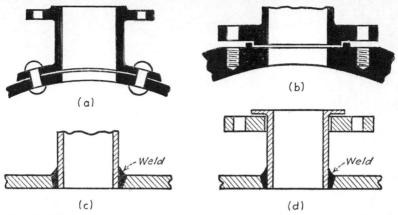

FIG. 13-21.   Four common types of boiler nozzles installed on steam boilers.

Q How are threaded connections attached to boiler shells?
A When the plate is fairly thick and the threaded opening is *very small*, the plate is not greatly weakened by the hole; so the pipe is simply screwed into the plate. If the opening is large and the plate too thin for a strong and leakproof threaded joint, a reinforcing pad may be riveted or welded to the shell and a continuous thread run through both pad and shell plate. Attachment of a blowoff pipe to an hrt boiler by this method is shown in Fig. 13-20. All such connections may be seal-welded either inside or outside the shell, though this is not particularly desirable in the blowoff-pipe connection, which may have to be renewed frequently.

For boiler pressures over 100 psi, the maximum size of threaded pipe connections is 3 in.

Q Sketch several acceptable types of boiler-nozzle construction.
A Figure 13-21*a* shows a nozzle riveted to the shell; Fig. 13-21*b* shows a nozzle attached by studs screwed into the shell plate; Fig. 13-21*c* shows a plain nozzle, welded to the shell plate; and Fig. 13-21*d* shows a welded nozzle with loose flange.

SUGGESTED READING

Elonka, Stephen M.: *Standard Basic Math and Applied Plant Calculations,* McGraw-Hill Book Company, New York, 1978. Reprint, 1979 (in Portuguese), Editora McGraw-Hill do Brasil, Ltda, São Paulo, Brasil.

14

CONSTRUCTION MATERIALS, WELDING, AND CALCULATIONS

In addition to the materials covered in this chapter, metals for higher pressures and temperatures are being developed. Such high-quality alloys must contain expensive alloys for the superheater and reheater sections of large, modern water-tube steam generators.

Today, low-carbon steel is used in most water-tube boilers operating over a wide temperature range. Convection temperatures run between 500 and 700°F. Medium-carbon steel, with 0.35 percent maximum of carbon, permits higher stress levels than low-carbon steel at temperatures up to 950°F, for example. But for superheater tubes, which must resist temperatures above 950°F, alloy steels are required. These may contain chromium, chromium-molybdenum, and chromium-nickel.

Although repairs on the pressure components of boilers must be made only by certified welders, we cover a few questions on welding and calculations that every operator should know.

CONSTRUCTION MATERIALS

Q List the principal materials used in steam-boiler construction and name some of the parts for which each is especially suitable.
A Wrought steel (also called *low-carbon* steel): For boiler plates, bolts, tubes, rivets, stays, pipes, reinforcing rings, manhole door frames, mud drums, nozzles, manhole and handhole covers.

Cast steel: For high-pressure fittings, valves, tube headers, manhole and handhole cover plates, supporting lugs.

Wrought iron: For tubes, pipes, stay bolts, rivets, door-frame rings.

Cast iron: For valves, pipes, fittings, and water columns when pressures do not exceed 250 psi and temperatures are not over 450°F. Cast-iron fittings on blowdown lines are limited to 100 psi.

Malleable cast iron: For pipe fittings, water columns, valve bodies.

Brass: For small valves, gage glass fittings, parts of gages, pipe.

Bronze: For safety-valve seats and same purposes as brass.

Copper: For plates and tubes.

Alloy steels (steels containing small percentages of other metals, such as nickel, chromium, molybdenum): Commonly used when greater strength is required and metal is subjected to high temperatures.

Q Explain briefly how wrought steel, cast steel, wrought iron, cast iron, malleable cast iron, chilled cast iron, brass, bronze, copper, and alloy steels are manufactured.

A Wrought steel: Composed primarily of the element iron, combined with small percentages of other elements, notably carbon (generally less than 0.3 percent). First step in the manufacture of iron and steel is the smelting of iron ore in a blast furnace, with limestone as a flux and coke as fuel. The molten metal is cast in narrow molds as "pigs." The process of manufacturing wrought steel consists of melting this pig iron again in a suitable furnace, usually an "open hearth," burning off the carbon and other impurities, then adding whatever amount of carbon may be required for the desired grade of steel. The finished steel usually contains small percentages of manganese, sulfur, and phosphorus as well as carbon. Molten metal from the steel furnace is run off into ingot molds, and these ingots are rolled, forged, or pressed into plates, bars, angles, and other shapes.

Cast steel: Cast directly from the steel furnace into molds of any desired shape and not formed by rolling or any other process.

Wrought iron: Made by melting pig iron in a "puddling" furnace and working it in such a way as to remove practically all carbon. The pasty mess that is left is almost pure iron.

Cast iron: Made by melting pig and scrap iron in a cupola furnace, then running the molten metal into molds. Iron castings made in this way may contain a large amount of free carbon, that is, carbon that is not combined with the iron but is in the form of graphite. Cast iron high in free carbon is called gray cast iron because of its gray appearance when machined. It machines easily and is rather weak in structure. Cast iron that is low in free carbon is somewhat more like steel. It has a white appearance when machined and is harder and stronger than gray cast iron.

Malleable cast iron: Produced by annealing "white-iron" castings. Packed with oxide or iron in cast-iron boxes, the castings are heated to red heat.

They are kept at this temperature for a considerable time, then allowed to cool slowly. This makes the castings more or less malleable—that is, capable of being bent or worked to some extent.

Chilled castings: Made in special molds that cool or chill the outer surface of the casting rapidly, thus making it very hard.

Copper: Copper ore is first crushed fine, then washed, screened and concentrated, rewashed, and smelted to a *copper matte.* The matte goes through a further refining process to burn out such impurities as sulfur and iron. The resulting *blister copper,* about 99 percent pure, is further refined before being manufactured into wire, rods, and bars.

Brass: Principal ingredients are copper and zinc. Some brasses contain small amounts of tin and lead as well. They can be made soft or hard by varying the proportions and the amount of cold working.

Bronze: Contains copper with either zinc or tin, or both. Many modern bronzes contain such elements as aluminum and nickel.

Alloy steels: Made by the same processes as wrought and cast steel, but containing small percentages of other metals, such as chromium, nickel, and manganese. In boiler work, alloy steels are used for plates and castings for high temperatures (750 to 1100°F).

Q   The following eight terms refer to the physical properties of materials. Explain their meaning briefly: ductility, elasticity, malleability, hardness, toughness, homogeneity, tenacity, resilience.

A   1. Ductility: Ability to withstand drawing out or other deformation without breaking.

2. Elasticity: Ability to return to normal shape after a deforming force has been removed.

3. Malleability: Capable of having the shape changed by rolling, hammering, or bending, without cracking or breaking.

4. Hardness: Ability to withstand surface wear or abrasion.

5. Toughness: Measure of a material's ability to stand up without fracture under repeated twisting or bending.

6. Homogeneity: Refers to the internal structure of a material. When broken, homogeneous material shows uniform grain or fiber.

7. Tenacity (high tensile strength): Ability to stand large pulling force.

8. Resilience: Ability to store up energy under stress and give it back when stress is removed.

PHYSICAL PROPERTIES

Q   What are the physical properties of low-carbon steel, wrought iron, cast iron, cast steel, and brass?

A   Only a very general answer can be given, since all these metals vary widely in quality because of different manufacturing methods, varying proportions of alloying metals, and impurities that cannot be entirely eliminated. Broadly speaking:

Low-carbon steel is ductile, malleable, tenacious, tough, elastic, and fairly homogeneous.

Wrought iron possesses much the same physical properties as low-carbon steel but has lower strength. It is tougher and more ductile and its internal structure is fibrous.

Cast iron is hard, brittle, less uniform in structure than wrought iron and steel, and also much weaker in tension.

Cast steel is far stronger than cast iron, especially in tension and resistance to impact.

Brass varies from a hard and brittle material to a ductile tenacious one, depending on the proportions of the various metals in the alloy.

Q   What is the main difference between iron and steel?
A   Chemically pure iron is an element, or simple substance, that cannot be divided up into anything but particles of the same substance. In practice, it is almost impossible to refine iron so that it is free from all foreign material, but a high-grade charcoal iron really contains few impurities. Steel is an alloy of iron and carbon. The carbon is in solution with the iron, increasing its strength and changing its physical properties to a marked extent.

Steel contains only a very small percentage of carbon, which is combined chemically with the iron. Cast iron, on the other hand, may contain quite a substantial amount of carbon, mostly in a "free" state, that is, simply *mixed* with the iron in the form of graphite.

Q   What is the difference between low-carbon steel and high-carbon steel?
A   Low-carbon steel contains a very small percentage of carbon, from 0 1 to 0.27 percent in soft malleable steel, such as is used for plates, bolts, and rivets, to as much as 0.6 percent in harder steels used in manufacturing rails, hard steel wire, etc.

High-carbon steel contains from 0.6 to about 1.5 percent. A 1 percent steel is suitable for such heavy-duty tools as chisels, punches, and rock drills. Turning tools, drills for metal, reamers, and dies are made from steel containing from 1 to 1.3 percent carbon.

Low-carbon steel is easily welded but cannot be hardened. High-carbon steel can be hardened but may not be so readily welded.

Q   What is an alloy steel?
A   An alloy steel contains some other metal or metals in addition to carbon and iron. Common alloying metals are chromium, nickel, manganese, silicon, vanadium, molybdenum, and tungsten. *Chromium* increases hardness and tenacity without reducing toughness. It is used extensively in high-speed tools and rustless steels; *nickel* greatly increases tensile strength, duc-

tility, elasticity, and corrosion resistance; *manganese* increases hardness and toughness; *silicon* increases hardness and magnetic permeability, making steel particularly suitable for electromagnets; *vanadium* increases resistance to fatigue; *molybdenum* increases hardness and offers greater resistance to heat changes; and *tungsten* hardens and increases heat resistance. It is the main alloying constituent in self-hardening tool steels.

STRENGTH OF MATERIALS

Q What is meant by *load, stress,* and *strain?*
A *Load* is an *external* force acting on a body. *Stress* is the *internal* resistance that the particles composing a body offer to the action of an external load. *Strain* is a change of length or shape caused by an external load. There are three kinds of simple stress and strain: (1) tensile stress and tensile strain; (2) compressive stress and compressive strain; (3) shearing stress and shearing strain. Tensile or compressive strain is measured by the fraction: change of length ÷ original length.

Q If an iron bar 6 ft long and 2 in.² in cross section is subjected to a pull of 20,000 lb and is stretched 0.024 in. in length, what are the load, unit stress, and strain?
A Load = 20,000 lb
  Unit stress = load ÷ area = 20,000 ÷ 2 = 10,000 psi

$$\text{Strain} = \frac{\text{increase in length}}{\text{original length}} = \frac{0.024}{72} = 0.00033 \text{ in./in.}$$

Note that this last answer means there has been an extension in length of 0.00033 in./in. of original length.

Q Explain what is meant by the following terms: *tensile unit stress, compressive unit stress,* and *shear unit stress.*
A *Stress* is the force acting within the material. *Unit stress* is the stress per square inch of cross section. It equals the external load divided by the area that carries the load. Thus, *tensile unit stress* in a rod under tension is equal to the total tension in pounds divided by the square inches of cross section. Likewise, *compressive unit stress* in a short rod under compression is the total compressive force in pounds divided by the cross-sectional area in square inches.

  Shear unit stress is the total shearing force divided by the area resisting that shear. *All unit stresses,* whether tensile, compressive, or shear, are measured in pounds per square inch (psi). In Fig. 14-1a, b, c, and d the unit stress is the total force divided by the area subjected to tension, compression, or shear. Although actual failure is pictured in each case, this merely suggests the type of failure that would occur if the type of stress listed were indefinitely increased. In Fig. 14-1c and d, note the difference between single and double shear. In the latter case, the area carrying the load is double the cross-sectional area of the rivet; therefore, each rivet does double duty.

Q What is meant by the *elastic limit* of a material?
A If a tensile or pulling load is applied to a bar of metal, the bar stretches more and more as the load is increased, and for a while in exact proportion to the load. Finally, a load is reached at which the bar starts to stretch faster for a given increase in load. The unit stress at this point is called the *elastic limit* of the material. If a bar is loaded to less than the elastic limit, it returns to its original length after the load is removed. If it is stretched beyond its elastic limit, it retains a permanent "set" after the load is removed.

Q What is the ultimate strength of a material?
A It is the total load (in tension, compression, or shear, as the case may be) required to pull the piece apart, to crush it, or to shear it, divided by the square inches of area resisting such failure. Since the cross section changes before a piece fails in tension or compression, the material's ultimate strength is customarily figured as the breaking load divided by the *original* cross section.

Q Is there any difference between the strength of a piece and the strength of the material of which it is made?
A If a certain steel has a tensile strength of 60,000 psi, a particular bar of 4 in.² cross section has a strength of 240,000 lb. The same applies to elastic limit and to actual loads. In each case we may consider the *total* loading for

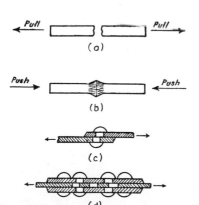

FIG. 14-1. Unit stresses, (a) tensile; (b) compressive; (c) and (d) shear, and effect on riveted joints.

an *object,* such as a bolt or a rivet, but only the unit stress (loading per square inch) in the case of a *material.*

Q What is meant by the *safe working strength* of a material?
A This is the maximum unit stress deemed safe for a material to carry under ordinary working conditions. To provide a sufficient margin of safety, it is always far less than the elastic limit, which in turn is much less than the ultimate strength.

Q What is the *factor of safety?*
A It is the ultimate strength of the *material* divided by the *actual unit stress,* or it is the ultimate strength of the *piece* divided by the actual total load (tension, compression, or shear) on the piece. Thus, it is merely a number showing how strong the material or object is in terms of the imposed load use. If the factor of safety is 5, the load would have to be multiplied 5 times to break the piece.

Factor of safety must allow for uncertainties in the uniformity of the material, for corrosion and wear, for bumps and jerks, for fatigue, for overloading, and for imperfections in the method of figuring.

Thus, in practice, for each kind of work and material, the factors used are those which experience shows to be reasonably safe without undue waste of material from making parts overly strong.

Q What working steam pressure, in pounds per square inch, is allowed on a steam boiler whose bursting pressure is calculated to be 1260 psi if the safety factor is 5.6?
A Note that the stress in the metal is unknown as far as this problem is concerned. Only the bursting steam pressure is given. However, stress in the metal is proportional to the steam pressure; so you merely divide 1260 by 5.6 to get 225 psi, the safe working steam pressure.

Q A steel bolt carries a load that induces a unit stress of 6000 psi. What is the ultimate strength of the bolt material if a safety factor of 8 is used?
A Ultimate strength = working load × factor of safety
             = 6000 × 8 = 48,000 psi

Q If steel bars with an ultimate tensile strength of 56,000 psi are used to make boiler stays and the maximum permissible stress is 8000 psi, what factor of safety is being used?
A Factor of safety
       = ultimate tensile strength ÷ maximum permissible stress
       = 56,000 ÷ 8000 = 7

Q What is the ultimate strength of a round steel bar, 1^1/$_2$ in. in diameter, if the steel's breaking strength is 60,000 psi?
A   Area of bar = 0.785 × 1.5 × 1.5 = 1.77 in.²
        Ultimate strength = 60,000 × 1.77 = 106,200 lb

Q What is the stress in six 1-in. bolts that support a load of 10 tons? Diameter at bottom of threads is 0.84 in.

A Total area of bolts $= 6 \times 0.84 \times 0.84 \times 0.785 = 3.32$ in.^{2}

Stress $=$ load $\div$ area $= (10 \times 2000) \div 3.32 = 6024$ psi

Q What tests are applied to material that is used to manufacture a steam boiler, and what are the purposes of these tests?

A Tension test: To find ultimate tensile strength of a material and amount of elongation when subjected to varying degrees of tensile stress.

Crush test: Applied to boiler tubes to see whether they can stand crushing longitudinally without cracking.

Bend test: To find whether metal stands bending without breaking or cracking.

Flattening test: Ability of part, when it is under test, to stand flattening without cracking.

Transverse test: Resistance of metal to bending.

Homogeneity test: To find whether the internal structure of metal is uniform by examining the broken edges of a fractured specimen.

Fracture test: Same purpose as homogeneity test.

Nick bend test: Same purpose as homogeneity test.

Etch test: Examining structure of material by polishing the surface of the test specimen and etching the polished surface with acid.

Hardness test: For surface hardness.

Hydrostatic test: Water-pressure test applied to pipes, tubes, and castings to detect weak parts in structure; also applied to finished boilers and other pressure vessels to find weaknesses in material or defects in workmanship.

X-ray test: To examine castings and welds for possible flaws.

Q Why are the principal parts of boilers subjected to internal pressure usually cylindrical in form?

A In a cylindrical vessel pressure is exerted equally in every direction on the cylindrical surface, and thus no part tends to become distorted through excessive pressure. See Fig. 14-2a. In Fig. 14-2b, pressure is shown acting on the flat head of a cylindrical shell. Edges of the flat plate are supported by the shell plate, but the center of the plate is not supported in any way and tends to bulge outward in an elliptical form, as the dotted line shows. Elliptical-shaped heads, such as those in many types of water-tube boiler

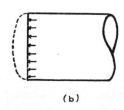

(a)   **(b)**

FIG. 14-2. Effect of pressure on cylindrical surface (*a*), and how it distorts flat surface (*b*).

drums, do not require support because they are already in the shape to which a flat head would be distorted by excessive pressure. On the other hand, flat plates *must* be stayed.

HEAT TREATMENTS AND PROCESSES

Q How is carbon tool steel hardened and tempered?

A Carbon tool steel is hardened by heating it to about 1450°F, then cooling it rapidly by plunging it into water or oil. The steel is not quite so hard and brittle when cooled in oil as when cooled in water, but in either case it is usually too hard and must be *tempered*—that is, some of the hardness drawn out. If the hardened steel's surface is polished and the steel reheated, colors appear, ranging from light straw to deep blue, as temperature increases. The darker the color, the higher the temperature and the softer the temper. If the tool is heated until the blue color appears, practically all hardness has been drawn from the steel. The tempering process can be stopped at any desired point by quenching the steel in oil or water.

Q List some common carbon-steel tools and their tempering colors.

A In the following tools (when made from carbon steel) the temper ranges from hardest to softest in the order given:

Light straw: Scrapers, files, lathe and planer tools for hard materials.
Straw: Milling cutters, lathe tools, taps, dies, reamers, drills.
Brown: Cold chisels for hard materials, woodworking tools.
Brownish purple: Cold chisels for soft materials, woodworking tools.
Dark blue: Springs, wood saws.

Q What is meant by *annealing*?

A *Annealing* is the process of softening metal and relieving internal strains by heating to a temperature between the recrystallization temperature and the lower critical temperature. Rapid heating and short holding time prevent excessive grain growth. It is customary to cool the parts in air. If surface protection is desired, controlled cooling temperatures are used.

Q What is *case hardening*?

A This is a method of hardening the surface of those irons and steels that cannot be hardened by simply heating and then cooling rapidly. One method is to heat the metal to a cherry red, coat the surface with potassium cyanide or potassium ferrocyanide, and then chill suddenly by dipping in water. The cyanide carbonizes the surface of the metal, really converting it into a steel that can be hardened by heating and sudden cooling.

Q What is *soldering*?

A This is a process of joining two pieces of metal by covering their surfaces with a molten alloy of lead and tin, called *solder*. Parts to be joined are first cleaned to the bright metal, then coated with flux and "tinned" by spreading molten solder thinly over the surface with a soldering copper, blow torch, or other source of heat. More flux is applied and surfaces to be joined are clamped together and heated while additional solder is flowed into the joint. Parts are allowed to cool before pressure is released.

The flux helps the solder to flow and to form a firm bond with the metal being soldered. Common fluxes include rosin for tinned steel, rosin and tallow for lead, borax for iron, sal ammoniac for copper and brass. Chloride of zinc is a good general-purpose flux. In addition, several good prepared paste and liquid fluxes are on the market.

Q What is *brazing?*
A *Brazing* is somewhat similar to soldering, except that the strength and melting point of the joining material are much higher. While commonly called *spelter* by practical brazers, the material for high-temperature brazing is actually granulated brass with a melting point ranging from 1450 to 1800°F, according to its composition. So-called "low-temperature" brazing is done with silver alloys ranging in melting point from 1175 to 1600°F. Usual heat source is an oxyacetylene torch. Borax or other flux must be used. A brazed joint is very much stronger than a soldered joint.

Q What is *forge welding?*
A This is the joining of two pieces of metal by heating to their plastic point, then placing in contact, and hammering or pressing them together before they cool. When welding heat is reached, the metal is sprinkled with a flux to remove oxide from the molten surfaces and thus ensure a clean joint. A mixture of sand and borax or sal ammoniac is commonly used for this purpose. Strength of the weld depends on the cleanness of surfaces to be joined and the extent to which they really fuse. If temperature is not up to correct welding heat, metals will not fuse. Heated above welding heat, they will burn. A good weld is practically as strong as the solid metal.

BOILER REPAIRS WELDING

Q Is it permissible for any person to weld pressure parts of a boiler?
A No. All materials and constructions used in repair work must meet ASME code requirements. Repairs to pressure parts are made under the guidance of the National Board Inspection Code, published by the National Board of Boiler and Pressure Vessel Inspectors, Columbus, Ohio. This code covers problems of inspections and repairs to boilers and auxiliary equipment that are not otherwise covered in the ASME code. It suggests laws and regulations for inspection of pressure vessels, and rules for repairs by fusion welding. These rules are acceptable in most states.

No repairs by welding are to be made to a boiler without the approval of an authorized inspector. Where the strength of the structure depends on the strength of the weld, the repair must be made by a qualified operator.

For repair work the welder need not be qualified by the manufacturer. The only requirement is that the inspector be satisfied with the welder's qualifications.

Q What is a "qualified welder"?

A Before a welder is permitted to work on a job covered by a welding code or specification, he must become certified under the code that applies. Because many different codes are in use today, the specific code must be known when taking a qualification test. Qualification of welders is too technical to be fully covered here. See Section IX of the ASME Boiler and Pressure Vessel Code.

For some repairs where the strength of the structure does not depend on the strength of the weld, as in building up corroded surfaces, the National Board rules permit the use of an "approved operator." The requirements for an approved operator are somewhat less rigid than those for a qualified operator. Repairs defined by the National Board Inspection Code include (1) replacement of sections of tubes, provided the remaining part of the tube is not less than 75 percent of its original thickness, (2) seal welding of tubes, (3) building up of certain corroded surfaces, and (4) repairs of cracked ligaments of drums or headers within certain definite limits.

Q Using sketches, explain what is meant by a *lap weld* and a *butt weld* (both forge welds).

A In *lap welding,* ends to be welded are first heated and "scarfed," as in Fig. 14-3a. They are then heated to welding temperature and welded by hammering, with ends overlapped as shown. In *butt welding,* ends are heated to welding temperature, joined by simply butting them together, as in Fig. 14-3b, and finished by hammering.

Q What is *fusion welding?*

A Any process that joins metals while in a molten state without hammer-

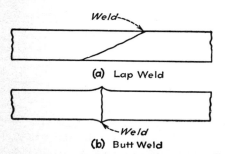

(a) Lap Weld

(b) Butt Weld

FIG. 14-3. Forge welding of (a) lap joint, and (b) butt joints.

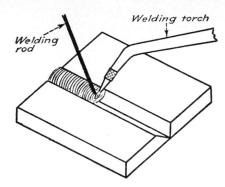

FIG. 14-4. Proper method of oxyacetylene welding.

ing or other pressure is called *fusion welding.* Oxyacetylene and arc welding are examples of fusion welding.

Q   What is *thermit welding?*
A   Chemical action obtained by igniting a mixture of aluminum powder and iron oxide, called *thermit,* produces molten steel at a temperature of around 5000°F. This molten steel is run into a mold surrounding the parts to be welded, melting the parts, and solidifying with them as the steel cools. It is essentially a casting process.

Q   Explain *oxyacetylene welding.*
A   In the *oxyacetylene process* of fusion welding, the joint between parts to be welded is thoroughly cleaned and in many cases cut out in V shape. This vee is filled in by melting a rod of similar metal into the joint and, at the same time, fusing the edges of the metal to be joined with the metal of the filler rod. If fusing is properly done, the joint is practically as strong as the original metal. Heat required to melt the metal is applied by means of a blowpipe burning a mixture of oxygen and acetylene gas that produces an intensely hot flame. The flame must be kept moving with a circular or "weaving" motion to spread the heat and avoid burning the metal. Figure 14-4 shows how an oxyacetylene weld is made.

Q   What are the principal methods of *electric welding?*
A   The three common processes are:
Resistance welding: Parts to be welded are raised to fusing temperature by the passage of a heavy electric current while they are in contact with each other. Spot welding of light metal parts is one form of resistance welding.
Carbon-arc welding: Parts to be welded are connected to one leg of an electric circuit; the other leg is connected to a carbon rod held in a hand receptacle. When the carbon rod touches the metal surfaces to be welded, and is then moved back a fraction of an inch, an arc is formed. Heat of the

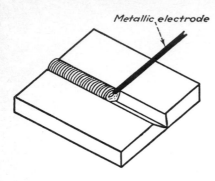

Metallic electrode

FIG. 14-5. Electric welding method.

arc melts the metal, and a filler rod of similar metal held in the arc fills up the joint.

Metallic-arc welding: A wire electrode is used instead of the carbon rod, and the arc is formed between the rod and the metal to be welded. The electrode itself melts and fills the joint. Electrodes are bare wires or, more commonly, metal rods coated with some flux. Figure 14-5 shows this method of electric welding on a prepared V joint. The electrode is constantly moved in a weaving pattern during the welding process. The pattern varies with type of joint.

CODE BOILER REPAIRS

Q What kind of high-pressure boiler repairs may be made by electric or oxyacetylene welding?

A Factory welds can be tested and flaws readily detected by x-ray and other forms of testing apparatus. These testing methods cannot be applied easily to the common run of plant repair jobs; so exercise great care and judgment when carrying out welding repairs on high-pressure boilers. In general, it is inadvisable to weld long cracks in plates that are exposed to intense heat or to high tensile stress, but in many other cases risk of failure is not so great, and welding is more convenient than other forms of repair. Because much more depends upon the human factor in welding than in other classes of mechanical repair work, it should be done only by skilled, experienced welders.

Some common defects that may be welded satisfactorily are: short cracks in internally fired furnaces; cracks in stayed surfaces, such as side sheets of water legs; cracks in tube sheets; cracks in headers.

Always ask the advice of boiler-inspection authorities before attempting any welding repairs, and follow their instructions closely.

Q When a damaged plate is to be repaired by welding in a new piece, how are the plate edges and patch piece prepared for welding?
A When the patch is in place, plate edges and patch piece must be beveled to form a vee which is then filled in with the welding material. If any stays or rivets pass through the part that is cut away, they must be inserted in the same position in the patch plate.

Q How is a repair made if part of the bottom of the shell of a riveted fire-tube boiler has to be cut out and a patch put in?
A If the shell bottom is directly exposed to intense furnace heat, and welding would probably not be allowed, the patch has to be riveted. Cut out the damaged part to leave an elliptical or diamond-shaped hole with rounded corners. The patch should be cut to the same shape with sufficient overlap to allow for riveting, bent to the same curvature as the boiler shell, fitted in place inside the shell, and riveted to it by rivets of the same diameter as those in the shell joints. The rivet patch should be proportioned to give the patch joint strength equal to or greater than the other shell joints. Edges of plate and patch should be calked.

Q How are cracked ligaments between tubes repaired?
A The old method was to make a patch plate called a *spectacle piece* because of its shape, and pin or rivet this to the tube plate over the crack, but now welding does a much better and neater job. Tubes on either side of the crack are removed, the crack vee'd out by chipping or grinding, and then welded in the usual way. After welding is completed, edges of the weld in the tube holes and the surface of the tube sheet are smoothed by chipping and filing, or grinding, and the tubes replaced.

Q What are the advantages of preheating and annealing work that is being, or has been, welded?
A Advantages of preheating are obvious, especially when large parts are welded. If the metal is preheated to a fairly high temperature, heat from the torch is almost entirely available for melting the filler rod and the metal at the edges of the joint. Preheating also causes the metal to "give" and thus avoids expansion and contraction stresses at the weld or in the surrounding metal. If the metal is cold, on the other hand, heat from the torch is rapidly conducted away by the cold metal, and it is more difficult to keep the metal in the joint at a welding heat and do a satisfactory job. Rapid cooling also tends to make the weld metal extremely hard and difficult to chip, file, or machine.

Annealing a part after it has been welded allows the strains that have been set up by unequal heating to equalize and adjust themselves. There is also less danger of an annealed part breaking at the weld or close to it when again put into use. Annealing is done preferably in a proper annealing furnace where temperature is under close control. If such a furnace is not available, the part may be heated to a dark red and then buried in sand, lime, or ashes so that it cools very slowly.

FIGURING HEATING SURFACE

Since most of the parts that make up boiler-heating surfaces are circular or cylindrical shells, every engineer should know the easiest ways to figure these areas.

For practical engineering purposes, the most convenient formulas are:

Circumference of circle = 3,1416 × diameter
Area of circle = 0.7854 × square of diameter

Note that the second formula is just another way of stating that if you draw a square around a circle, the area of the circle is 78.54 percent of the area of the square. These rules are illustrated in Figs. 14-6 and 14-7.

It is not always necessary to use all known "places" or digits of a factor, piece of data, or result of a computation. How many to use depends on the accuracy required. For most boiler computations, it is close enough to use the factors 3.14 and 0.785, respectively. Using 3.14 instead of 3.1416 involves an error of only 0.05 percent of 1 ft² out of 2000 in boiler heating surface. Any attempt to figure closer than this is generally a waste of time because, first, 1 ft² in 2000 is not important, and second, errors in measurement throw the result off more than 0.05 percent in any case. The same applies to use of 0.785 instead of 0.7854.

The area of a flat rectangular surface is merely the product of its two dimensions. To figure the surface area of a cylinder, such as a tube or drum, imagine the surface unrolled to a flat, rectangular shape. Length doesn't change; width is the circumference of the original circle; area is the product, as shown in Fig. 14-7.

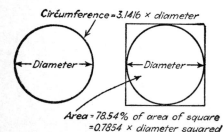

FIG. 14-6. Finding the area and circumference of circle.

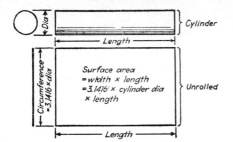

FIG. 14-7. Calculate surface area of cylinder by rolling out flat.

Here are some important points to remember in figuring heating surface:

1. If dimensions are in feet, area comes out in square feet.

2. If dimensions are in inches, area comes out in square inches. Divide square inches by 144 to get square feet.

3. When many tubes are the same length, figure the surface of one tube in square feet. Then multiply this area by the number of such tubes.

4. Boiler-heating surface is always measured on the side exposed to the fire. Thus, the heating surface of *water tubes* is based on *outside tube diameter;* that of *fire tubes,* on *inside diameter.*

5. When tubes enter a sheet or drum, deduct the area of the tube holes from the gross area of the sheet or drum to get net heating surface.

Q   How many square feet of heating surface are there in a boiler furnace flue 3 ft in diameter and 12 ft long?
A   Circumference of flue = 3.14 × 3 = 9.42 ft. Heating surface = area of flue = circumference × length = 9.42 × 12 = 113.04, say 113 ft².

Q   How many square feet of heating surface are there in an 18-ft-long boiler fire tube of 3¹/₄ in. internal diameter?
A   Heating surface of a fire tube must be based on *internal* circumference, which here is 3.25 × 3.14 = 10.205 in. This is 10.205 ÷ 12 = 0.8504 ft. Then, flue surface = 0.8504 × 18 = 15.307 ft², say, 15.31 ft².

COMMENTS ON DECIMAL PLACES

All through the last computation there is the question of how many "places" to carry the various intermediate answers. Many schools still teach that arithmetical problems should be carried a certain number of *decimal places,* but this is not sound practice and often gives foolish results—sometimes needlessly precise, sometimes too rough.

Examples make this point clear. The correct thickness of a certain piece of paper is, say, 0.0055 in. Anyone taught to carry all numbers to three places of *decimals* would write down 0.005 in., which is in error by 10 percent. Asked to express a 1000 in. length in feet, the same person would write 1000 ÷ 12 = 83.333 ft, which is correct to better than 0.005 percent. It may well happen that this degree of precision has no practical meaning. The 83.333 is more than 2000 times as precise as the 0.005, yet both have three decimal places. All this proves that a given number of decimal places doesn't mean a given degree of precision.

What is important is not the number of *decimal places,* but the number of *significant figures,* so called. To determine the number of significant figures, start counting with the first digit *that is not a zero* and count all digits from there on, including zeros.

Thus 4163., 0.4163, and 0.0004163 are all written to four significant figures and all represent the same order of precision in percentages.

Similarly, 3 and 0.00000003 are both written to one significant figure only, and both represent a very *low* order of percentage precision, even though the second number is written to eight decimal places.

Such examples show that the engineer-reader who has been in the habit of counting decimal places should shift over to significant figures to keep out of trouble in computations.

In the problems worked out in the preceding heating surface examples, we used the value 3.14, a constant carried to three significant figures. That choice sets the limit of possible precision for the whole computation at about 0.05 percent. It is a good rule of thumb that the various steps in a computation may carry one or two more significant figures than the data or factors used. Anything beyond that is meaningless. Note that this rule was followed in working out the problems. Most of the steps were carried to four significant figures. The cases where the answer was carried to five figures were justified by the fact that the first digit was a low number. That often warrants using two more significant figures instead of one more. Thus 0.8504 was carried to four figures, but 10.205 to five figures.

HEATING-SURFACE CALCULATION

We follow the above general rule in working out the following problem:

Q An hrt boiler is 5 ft in diameter and 16 ft long. It contains 60 tubes of 3 in. outside diameter and 2.732 in. inside diameter. Find the boiler heating surface. Take the inner surface of tubes, half of the shell surface, and two-thirds of the tube plate area, less the area of the tube holes.

A Circumference of shell = 5 × 3.14 = 15.700 ft
Half circumference of shell = 15.700 × ½ = 7.850 ft
Area of half of shell = 7.850 × 16 = 125.60 ft²

Inner circumference of fire tube = $2.732 \times 3.14 = 8.578$ in.,

or $8.578 \div 12 = 0.7148$ ft

Surface of one tube = $0.7148 \times 16 = 11.437$ ft^2

Surface of 60 tubes = $60 \times 11.437 = 686.2$ ft^2

Total area of one tube sheet = $0.785 \times 5 \times 5 = 19.625$ ft^2

Two-thirds area one tube sheet = $^2/_3 \times 19.625 = 13.083$ ft^2

Area of one tube hole = $0.785 \times 3 \times 3 = 7.065$ in.^{2}

Area of 60 tube holes = $60 \times 7.065 = 423.9$ in.^{2},

or $423.9 \div 144 = 2.94$ ft^2

Net heating surface of one tube sheet = $13.08 - 2.94 = 10.14$ ft^2

Net surface of two tube sheets = $10.14 \times 2 = 20.28$ ft^2

Total net area = $125.60 + 686.2 + 20.28 = 832.08$ ft^2,

say 832 ft^2.   *Ans.*

SUGGESTED READING

Elonka, Stephen M.: *Standard Plant Operators' Manual,* 3d ed., McGraw-Hill Book Company, New York, 1980 (has over 2000 illustrations).

15

INJECTORS AND PUMPS

One of man's oldest aids, the pump, today ranks second only to the electric motor as the most widely used industrial machine. Anything that will flow is pumped, even thick mud and sludge. And to meet the many demands of moving liquids, we have a confusingly large variety of available pumps. They range from tiny adjustable-displacement units to giants handling well over 100,000 gal/min. Here, we cover only the basic types used in boiler and engine rooms. The following information is the minimum that should be known by every operator of these units.

INJECTORS

Q What means are used to force water into a steam boiler against boiler pressure?
A Principal devices to feed water into steam boilers are injectors and pumps, though under certain circumstances an outside source of water supply may be suitable if pressure is high enough and a supply is constantly available. Low-pressure heating boilers, for example, are often fed from city water mains.

Q Sketch a single-tube injector and mark the names of the principal parts.
A Figure 15-1 is a sectional view of a single-tube lifting injector with the principal parts marked.

Q Describe the construction and operation of the single-tube lifting injector.

A This injector uses direct steam pressure to force water into a boiler against the same steam pressure as that which operates the injector. A brass casing contains an expanding steam nozzle, several converging and diverging tubes, which form one practically continuous passage, and a check valve.

When the steam valve is opened, steam enters the injector through the expanding steam nozzle, which converts its pressure energy into velocity. Entering the suction tube, this high-velocity steam jet creates a partial vacuum in the suction pipe. Water is then forced up the suction pipe into the injector by the unbalanced atmospheric pressure on the surface of the water supply. This water mixes with, and condenses, the steam. Some of the steam's velocity is imparted to the combined jet of condensed steam and water as it passes into the tube, and this velocity is again partly converted into pressure in the delivery tube—enough to enable the stream of water flowing through the pipe to lift the check valve and enter the boiler.

Until the jet of water is properly established in the combining and delivery tubes, steam and water escape through the small holes in the sides of the combining tube and pass out through the overflow valve. As soon as the water has gathered enough velocity and pressure to lift the boiler check valve, it passes through the combining and delivery tubes in an unbroken stream, creating a partial vacuum in the overflow chamber, which lifts the sliding washer up against its seat, thus preventing any inrush of air through the overflow opening that might break up the jet. This injector will auto-

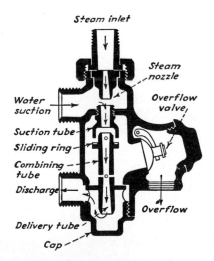

FIG. 15-1. Single-tube lifting injector's principal components.

matically resume work after a momentary interruption in flow without having to be adjusted by the operator.

Various pressure and velocity changes that occur in the injector depend on the design of nozzles and tubes. The diverging steam nozzle allows the entering steam to expand with a consequent fall in pressure and increase in velocity. The converging suction and combining tubes increase the velocity of the water jet, and the expanding delivery tube decreases the velocity of the water jet and increases the pressure. Note that an expanding nozzle *increases steam* velocity but *decreases water* velocity, because steam can expand indefinitely while water cannot expand much.

Q   Sketch a double-tube injector and name the principal parts.
A   Figure 15-2 shows a sectional view of a double-tube injector with names of parts.

Q   Describe the construction and operation of a double-tube injector.
A   It handles hotter water than the single-tube injector and works with a greater lift. It also operates with lower steam pressures and against higher boiler pressures. In operation, one tube lifts the water from the source of supply and delivers it to the other tube, which forces it into the boiler.

Capacity of the injector (Fig. 15-2) can be adjusted by the regulating valve admitting steam to the lifting tube. The lever handle operates the main steam valve, forcer steam valve, and main overflow valve. The main steam valve and forcer steam valve are operated by the same valve stem,

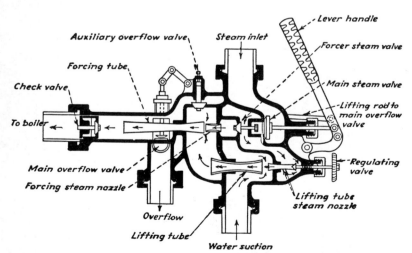

FIG. 15-2.   Double-tube injector uses boiler steam to force feedwater back into boiler against steam pressure.

their construction being such that the main steam valve can be opened a certain amount before the forcer steam valve opens.

To start the injector, assuming the regulating valve is properly adjusted, pull the lever handle to the right, just enough to open the main steam valve. This admits steam to the lifting tube steam nozzle through the regulating valve. The entering high-velocity steam jet creates a partial vacuum in the suction line and lifts water from the supply source. The combined steam-and-water jet issuing from the lifting tube fills the injector body and escapes through the auxiliary and main overflow valves.

Pulling the lever handle farther to the right opens the forcer steam valve and closes the main overflow valve. Steam from the forcing steam nozzle then forces the water being supplied by the lifting tube into the forcing tube, from which it discharges with sufficient velocity to lift the boiler check valve and enter the boiler. As the main overflow valve closes, the pressure equalizes on both sides of the auxiliary overflow valve, which closes automatically.

To stop the injector, pull the lever handle to the left, closing both the main steam valve and the forcer steam valve.

Q What is meant by *lift* of an injector?
A It is the vertical distance from the injector to the surface of the water supply, when the injector is located *above* the supply source. Practically all injectors are of the lifting type, but it is best to place an injector as close to the water-supply surface as possible because a high-suction lift cuts down the capacity. The limit of practical suction lift at sea level is not much over 20 ft, and it is less at higher altitudes because of the lower atmospheric pressure.

Q Explain why an injector can lift water from a lower level and why it will not lift very hot water.
A An injector lifts water because the high-velocity steam jet entering the injector exhausts air from the suction pipe and creates a partial vacuum in the pipe and injector. Then the unbalanced atmospheric pressure on the surface of the water supply forces the water up the pipe into the injector.

If the feedwater is very hot, the following things happen:

1. Entering steam momentarily creates a partial vacuum by exhausting the air from the suction pipe, but continuance of this partial vacuum depends upon the incoming feedwater's condensing the steam, and this does not occur at all if the feedwater is too hot.

2. The injector becomes *steambound*. Creation of a partial vacuum in the suction pipe may cause water in the pipe to evaporate into steam, as lowering the pressure also lowers the boiling point. If this happens, the suction pipe fills with steam and no water enters the injector.

INJECTOR PROBLEMS

Q What are the advantages and disadvantages of an injector as a boiler-feeding device?

A Advantages are:

1. It is small and very compact.
2. Its first cost is low.
3. Upkeep and repair costs are low.
4. It has practically no moving parts.
5. Of very simple construction, it does not readily get out of order.
6. It requires no lubrication.
7. It is easily operated.
8. It has high *thermal* efficiency. This means that practically all heat in the steam used to operate the injector is also used in heating the feedwater. The only heat loss is the small amount of heat radiated from the body of injector. (But note that this "efficient" heating of feedwater by live steam makes it impossible to take full advantage of otherwise wasted exhaust steam to heat the feedwater.)

Disadvantages are:

1. It cannot handle very hot water.
2. Because of limited delivery range, its operation is usually intermittent unless the boiler is carrying a steady load and injector delivery can be regulated to keep a steady water level.
3. With varying load, the injector must be started and stopped frequently, and there is danger that water level may drop below the safe minimum level when the injector is not in operation.
4. It does not operate efficiently on superheated steam.
5. It has low *mechanical* efficiency compared with a pump, its steam consumption being greater for the same amount of work done in forcing water into the boiler.

Q What are some of the common reasons for failure of an injector to work properly and efficiently?

A Failure or inefficient operation may be due to (1) feedwater too hot; (2) suction lift too high; (3) leaks in suction pipe destroying vacuum; (4) steam pressure too low; (5) wet steam preventing proper condensation of steam jet; (6) nozzles and tubes scaled up or plugged with dirt; (7) suction hose collapsed internally, if hose is used; (8) obstruction in suction or discharge line; (9) worn-out injector parts; (10) leaky boiler check valve allowing steam to blow back into injector discharge line.

Q If an injector does not work properly, what should you do to find and remedy the trouble?

A Repair leaks in suction pipe or run a new suction line. Take injector apart and examine nozzles and tubes to make sure that passages are clear and parts not badly worn. If tubes are scaled, they may be cleaned by soaking for several hours in a dilute solution of muriatic acid, about 1 part acid to 10 parts water. Remove any iron or steel parts before placing the injector in the acid. If any parts are worn out, replace them with new parts.

Make sure that suction and delivery are free from obstruction, and that the boiler check valve is in good working order. Wet steam may be caused by the steam connection to the injector being taken off at a point where considerable water comes over with the steam, in which case it may be remedied by changing the connection to a more suitable location. If the water is too hot, use colder water. If the lift is too high, lower the injector or raise the level of the water supply.

Q Sketch an ejector and explain its operation.

A An ejector operates on the same principle as an injector, but its construction is much simpler, and it does not discharge against a very high pressure. It is not suited to boiler feeding. It can, however, be used for many purposes where space is limited and only small quantities of water have to be handled. The type in Fig. 15-3 operates on either steam or compressed air, though steam is preferable. Steam consumption is quite high; so it should be used only where it operates intermittently and for short periods.

In operation, steam or compressed air issues at a very high velocity from the expansion nozzle and creates a partial vacuum in the suction line. Water rises in the suction pipe, and when it meets the steam or air jet, it is forced into the discharge line.

Q Has a pump any advantages over an injector as a means of feeding boilers?

A A pump has several advantages over an injector. Most important are:

1. It can be adjusted to feed continuously at steady or varying rates and does not have to start and stop at frequent intervals.

2. It uses much less steam.

3. It handles hotter water.

4. It has a wider range capacity.

5. It permits the economical use of exhaust steam or bled steam for feedwater heating.

Because of these advantages, pumps are installed in all large steam-boiler

FIG. 15-3. Ejector is simple, used for lifting water.

plants and in most small plants as well, injectors being pretty well confined to very small stationary plants and portable and locomotive boilers.

PUMPS CLASSIFIED

Q How are pumps classified?

A A general classification is:

1. Reciprocating pumps (piston and plunger): *power driven* (simplex, duplex, and triplex); *direct acting, steam driven* (simplex and duplex).

2. Centrifugal pumps (steam turbine, electric motor, or belt driven): *single-stage* or *multistage,* and liquid piston pump.

3. Rotary pumps (turbine, motor, or belt driven): with impellers in the form of gears, lobes that mesh together like gears, or sliding or swinging vanes.

Q Describe a power-driven pump.

A Strictly speaking, all pumps are power-driven in some fashion, but the term *power pump* is usually limited to a pump of the reciprocating type that is driven by a belt or gears from an engine, motor, or line shaft. The pump may have one, two, three, or more cylinders, and the cylinders may be single or double acting. The *triplex* is a very common form of power pump. It has three cylinders set side by side, with their plungers connected to a three-throw crankshaft. The cranks are set 120° apart to ensure a fairly steady flow of water.

Plungers are usually *single acting;* that is, the upward stroke is a suction stroke and the downward stroke is a discharge stroke, the upper end of the cylinder being open.

Figure 15-4 shows a vertical gear-driven triplex power pump, often used as a boiler-feed pump.

RECIPROCATING PUMPS

Q Explain a direct-acting duplex steam pump.

A A sectional sketch of one side of a direct-acting duplex steam pump is shown in Fig. 15-5. It is called a *duplex* pump because it has two steam and two water cylinders, placed side by side.

Q Describe the action in the water end of a duplex direct-acting steam pump.

A For a pump set above the surface of the water supply, the action is as follows: When the pump is started, the movement of the pistons to and fro exhausts the air from the suction pipe and water cylinders. This creates a

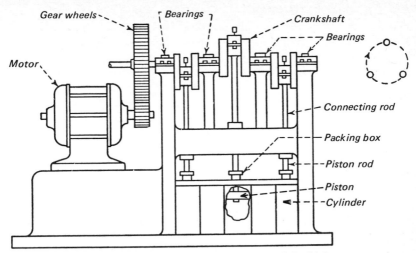

FIG. 15-4.   Vertical, triplex reciprocating pump is used for high pressures.

partial vacuum in the suction line, and the unbalanced atmospheric pressure on the surface of the water supply forces water up the suction pipe into the pump cylinders. Water passes up through the suction valves into the cylinder on one stroke, and is forced up through the discharge valves on the return stroke, as the pressure of the piston on the water closes the suction valves and opens the discharge valves.

In Fig. 15-5, atmospheric pressure is forcing the water into the left-hand side of the cylinder. The piston is forcing the water on the right to flow out through the discharge valves. Since the pump is *double acting*, this process reverses on the return stroke.

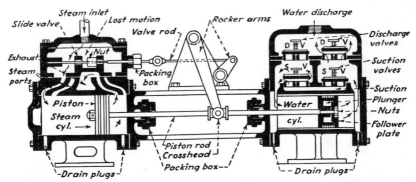

FIG. 15-5.   Duplex direct-acting steam pump is simple and very reliable.

Q What is the difference between a plunger and a piston?
A Though both serve the same purpose, they differ in construction. A plunger is a long cylinder, closed at one or both ends, depending upon the type of pump. In some single-acting power pumps, the outside end of the plunger is open, and the plunger is directly connected to the crank shaft by a connecting rod. In other types of power-plunger pumps, the plunger is closed at both ends and is driven by a short rod connected to a crosshead. In direct-connected steam-driven plunger pumps, the plunger is connected to the same rod as the steam piston.

In all types of plunger pumps, the plunger does not fit closely in the cylinder but is kept airtight and watertight by passing through an *outside-packed* packing box, as shown in Fig. 15-6*b*.

A water piston is a short piston similar in construction to a steam piston. It fits closely inside the pump cylinder and is *inside packed;* that is, it is made watertight by means of rings of square packing fitted into a groove around the circumference of the piston and held in place by a follower plate and a nut on the end of the piston rod. The packing commonly used is made from layers of cotton fabric cemented together with rubber. Sometimes harder materials are used to give longer wear or when the pumps are to be used for pumping liquids that would eat away and destroy softer packing materials. A water piston packed with square rings is shown in Fig. 15-6*a*.

Q Explain the action in the steam end of a direct-acting duplex pump (two steam and two liquid cylinders).
A In Fig. 15-5, two simple steam-engine cylinders are set side by side, the slide valve in each cylinder steam chest being operated by the crosshead on the piston rod of the opposite cylinder, through an arrangement of rods and rocker arms.

Steam valves are the simple *D* type. They slide to and fro over the steam ports, admitting steam alternately to each end of the steam cylinder and exhausting it to the atmosphere when the piston stroke is almost com-

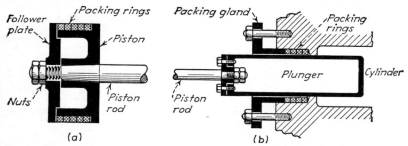

FIG. 15-6. (*a*) Inside-packed liquid piston; (*b*) outside-packed liquid piston.

pleted. There are two steam ports for each end of the cylinder. Steam enters the cylinder through the outer port and passes to exhaust through the inner port, after it has done work on the piston. These separate ports cushion the piston at the end of its stroke and prevent its striking the cylinder heads. As the piston nears the end of its stroke it covers the exhaust port, preventing further escape of steam to exhaust. The steam thus trapped between piston and cylinder head brings the piston to rest and reverses its motion without shock.

Some pumps have hand-operated cushion valves that open or close small passages connecting the steam and exhaust passages at each end of the cylinder, and so regulate the degree of cushioning. These valves should be closed at light loads to give the maximum cushioning effect and shortest possible piston stroke. At heavy loads, the cushion valves should be wide open to give minimum cushioning effect and longest possible piston stroke.

Because each steam valve is operated by the action of the piston rod belonging to the other steam cylinder, the pistons move in opposite directions most of the time. When one piston is momentarily stopped at the end of its stroke, the other piston is in motion, thus ensuring a fairly continuous water flow in the discharge line. This is one point where the duplex pump has an advantage over the simplex type of direct-acting pump. Another is its comparative simplicity of construction and operation.

Q   Describe a simplex direct-acting steam pump (one steam and one liquid cylinder).
A   The action in the water end of this pump (Fig. 15-7) is similar to the action in the water end of the duplex pump, but the water valves are placed at the side instead of on top of the water cylinder, as in the duplex pump. This makes the valves readily accessible without having to disconnect any pipes or remove more than one cover.

The main difference between the simplex and duplex steam pumps lies in the method of operating the steam valves. The entire valve mechanism is contained within the pump itself, and there are no outside moving parts such as valve rods and rocker arms. This pump cannot *short-stroke,* because the main piston must complete its travel before the main steam valve can reverse.

Q   Describe the steam-valve operation in the simplex pump of Fig. 15-7.
A   The main steam piston is driven by steam admitted alternately to each side of the piston by a main slide valve that also allows steam to pass to exhaust when it has done its work in moving the piston. This main slide valve is moved back and forth by a plunger or close-fitting piston valve with hollow ends. These ends are open to the steam chest and are always filled with live steam, which flows through tiny holes marked x in the ends, thus filling the spaces between valve ends and valve-chest covers with live steam also.

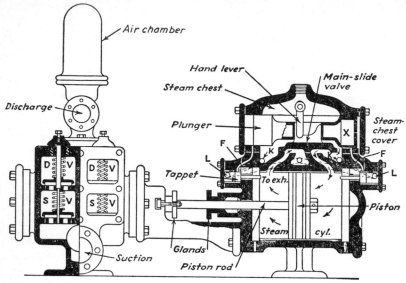

FIG. 15-7.   Simplex direct-acting steam pump complete with an air chamber.

Since pressure on the piston-valve ends is equal, it is ordinarily balanced and motionless. When the main piston travels to the end of its stroke, it strikes the projecting spindle of a small tappet and pushes the tappet back. This places passage F in communication with the exhaust through the passage shown dotted and allows live steam in the space at the end of the piston valve to escape to exhaust. Pressure of the live steam on that end being now relieved, the piston valve is forced over by the live-steam pressure on the other end. As it moves it carries the main slide valve with it, thus reversing the motion of the main piston by admitting steam to the opposite end of the cylinder.

As the piston valve moves over, it shuts off port F and cushions itself on the steam trapped between end of valve and steam-chest cover. Except when pushed in by the main piston at the end of its travel, the tappets are kept closed by constant live-steam pressure on their larger ends, conveyed through the passages KL from the steam chest.

The short lever shown in the center of the large piston valve can be operated by a similar lever on the outside of the steam chest. It is used to start the pump if it happens to stop with the piston valve and slide valve exactly on center.

These valves do not have to be set, since they are operated entirely by steam pressure and are not attached to any moving pump part. They must, however, be kept well lubricated because of their large area of rubbing surface.

Q Sketch some type of vertical steam-driven reciprocating steam pump, and describe the main points in its construction.

A Figure 15-8 is a sketch of a vertical simplex steam pump, showing the steam and water cylinders in section and a section through one set of water-suction and discharge valves, which are connected to the upper part of the water cylinder. Single suction and discharge valves are shown in the sketch, but some pumps have a number of small valves instead of one large valve. These valves are in a separate chamber attached to the water cylinder, and are readily accessible for inspection or repair without having to take any other sections of the pump apart. The area of the delivery valve, or valves, is usually less than the area of the suction valve, or valves, so as to increase the velocity of flow in the discharge line. This also has a tendency to improve the operation of the pump when it is pumping very hot water. Like most other simplex pumps, the steam valves are so constructed that the piston must take a full stroke before it can reverse.

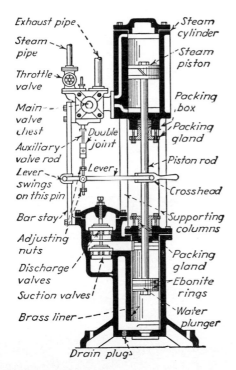

FIG. 15-8. Vertical simplex steam pump is popular boiler-feed pump.

The vertical type of reciprocating pump takes up less floor space than the horizontal type, and there is less wear on the cylinders and pistons, as the weight of the pistons does not rest on the cylinder walls.

Q Explain the valve action in the steam end of the vertical simplex pump shown in Fig. 15-8.

A The steam chest containing the main and auxiliary steam valves is placed with its axis at right angles to the centerline of the pump cylinders, so that the main steam valve moves to and fro horizontally. This arrangement is to prevent the possibility of the weight of the valve causing it to drop down and make the pump stroke short or uneven, as it might do if the travel were vertical instead of horizontal. The main valve is cylindrical in shape with a flat machined on the back to form a seat for the auxiliary valve. The auxiliary valve is a small flat slide valve which slides vertically up and down on the back of the main valve and is operated by a crosshead on the pump-piston rod.

Q How is a steam reciprocating pump controlled?

A Figure 15-9 shows a hookup for a simple manually adjusted regulator. While it controls pressure of steam to the steam chest of the pump, it is not sensitive to change in demand. If a reciprocating pump is operated with reasonably constant capacity or load, it will maintain constant discharge pressure, but only if the steam pressure to the pump's valve chest is fairly constant.

To assure constant steam pressure, the regulator is hooked up as shown in Fig. 15-9. It is an internal-pilot-type piston-operated pressure-reducing

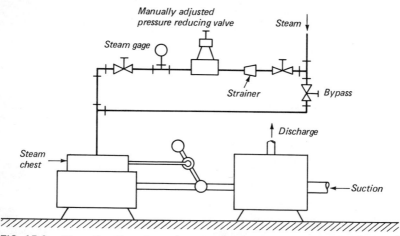

FIG. 15-9.   Manually adjusted regulator controls only pressure of steam to pump.

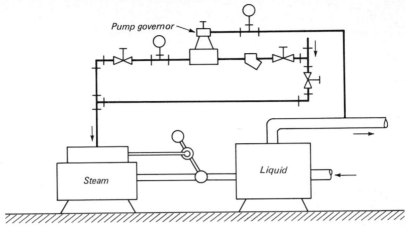

FIG. 15-10. Automatic regulation against changes in both discharge pressure of liquid and steam supply to pump keeps unit running to hold constant pressure.

valve. A tiny pilot in the valve's bonnet provides accuracy, while the large piston in the valve's body provides high lift for quick action.

Q Describe a method of controlling a steam reciprocating pump more accurately than in the previous question.

A Figure 15-10 shows the hookup for regulating the reciprocating pump against changes in the output discharge pressure and also in the steam pressure. The pump governor used is the dial-diaphragm-type regulator which is sensitive to changes in (1) steam-chest pressure and/or (2) pump-discharge pressure, to both of which it is connected. Thus a change in either pressure serves to activate the governor valve in the steam line that controls the pump's chest pressure.

SETTING PUMP VALVES

Q How would you set the steam valves on a duplex pump?

A The first step is to place the pistons in their central positions. They should be at midtravel when rocker arms are vertical, but simply moving the pistons until the arms are vertical cannot be depended upon, since the crossheads may have moved on the piston rods from their original setting. The correct procedure is as follows:

1. Push one of the pistons back until it strikes the cylinder head (Fig. 15-11).

2. Scribe mark *A* on the piston rod at the face of the packing gland.

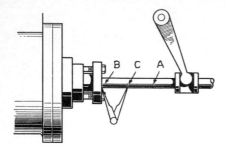

FIG. 15-11. Settling valves of duplex reciprocating steam pump.

3. Pull the same rod out until the piston strikes the opposite cylinder head.

4. Scribe another mark *B* on the piston rod at the face of the same packing gland.

5. Find the center of the distance between these two points and mark this central point *C*.

6. Move the rod back until the central mark is at the face of the gland, and the piston will be exactly at mid-stroke. Repeat this process for the other side of the pump.

7. Set the slide valves exactly in the center of their travel. In this position they will just cover the steam ports with no overlap.

8. Divide the lost motion evenly on each side of the nut or nuts on the valve rod that moves the valves.

9. The setting is now complete, but before replacing the valve-chest cover, be sure to move (with a finger) each valve as far as it will go, in opposite directions, so that the steam ports are uncovered; otherwise the pump will not start when steam is turned on.

To summarize, valves are properly set if they just cover the steam ports, with lost motion equally divided, when the pistons are at mid travel. The term *lost motion* refers to the space between the lugs on the back of the valve and the nuts on the valve rod that moves the valve. This slack is necessary to let the piston almost complete its stroke before the crossheads, rods, and rocker arms move the valve. If no lost motion were provided, the pistons would either remain stationary or make a very short stroke.

Q   What means are used to regulate boiler-feed-pump discharge automatically to suit varying boiler loads?

A   *Pump governors* are used to regulate the speed of boiler-feed pumps of the reciprocating type. These act to throttle the steam supply to the pump when pressure builds up in the discharge line through the partial closing of the boiler-feed valves.

Centrifugal boiler-feed pumps usually run at constant speed. When pressure builds up in the discharge line, it opens a relief valve through which the water flows back to the supply tank.

Q Describe a pump governor.

A Figure 15-12 shows a sectional view of a diaphragm-actuated pump governor. The steam supply to the pump steam cylinder passes through a double-seated valve so proportioned that the upward pressure on the upper valve disk balances the downward pressure on the lower valve disk. The space above the diaphragm is connected to the feed-pump water-discharge line, and any increase in the discharge pressure forces the diaphragm downward. This in turn pushes the valve stem downward and closes the steam valve. When the water-discharge pressure falls, the upward pressure of the heavy coil springs forces the diaphragm and valve stem upward and opens the steam valve. The amount of spring compression and the working limits of the governor can be regulated by adjusting nuts threaded on the valve stem.

Q What is an *air chamber,* and what duty does it perform?

A An *air chamber* (Fig. 15-7) is a hollow casting, closed at the upper end and having a flange at the bottom end for attaching it to the pump or pipeline. Air chambers are placed in the discharge line near the pump or directly on top of the discharge-valve chamber.

The air chamber steadies the flow of water, which is more or less intermittent in any reciprocating pump. The air within the air chamber is com-

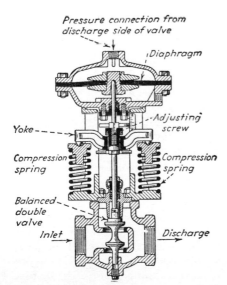

FIG. 15-12. Diaphragm-type governor used on steam pumps.

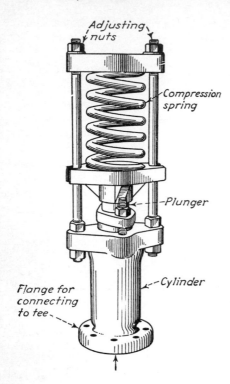

FIG. 15-13.   Alleviator exhausts air to atmosphere, acts like a shock absorber on high-pressure lines.

pressed when the discharge pressure is greatest, and it expands and keeps the water in motion when the discharge pressure is lowest. This also permits the valves to seat with less shock than would be the case if no air chamber were used. Air chambers are sometimes placed on suction as well as discharge lines to give a steadier flow of water in the suction line.

If an air chamber becomes filled with water, it should be drained, since it has no beneficial effect on the operation of the pump when in this state. Most air chambers have small drain valves to drain off water when necessary, and large air chambers usually have a gage glass to show height of water and presence of air in the chamber.

Q What is an alleviator?

A Air chambers are of little use where water pressures are very high, since the air is forced through the pores of the iron or is absorbed by the water, but a device called an *alleviator* (Fig. 15-13) can be used for the same purpose. It consists of a cylinder, flanged at one end for attachment to the pipeline, and having a packing gland at the other end through which a

close-fitting plunger can slide freely. The upper end of the plunger is attached to a heavy coil spring with means for adjusting the compression of
this spring as desired. Excessive pressure in the line forces the plunger upward against the pressure of the spring, and so relieves the pressure in the
pipeline. As the pipeline pressure falls, the spring forces the plunger back
down into the cylinder.

LIQUID-END VALVES

Q Sketch and describe a pump water-valve assembly as commonly used in
reciprocating pumps.

A Figure 15-14 shows a complete valve assembly. The valve is a rubber
disk, soft rubber being used for pumping cold water and hard rubber for
hot water. The valve fits closely on a brass seat which is screwed or pressed
into the valve-deck casting. The valve is held down on its seat by a brass
plate and spring, and these in turn are held together by a stud screwed into
the seat. The valve is opened by the water pressure on the underside lifting
it against the pressure of the spring. Pump valves are also made from cast
and pressed metal and various plastic materials.

Q What is meant by *priming* a pump?

A *Priming* a pump means filling the suction line and water cylinders or
pump casing with water before starting the pump. The pistons or plungers
of a reciprocating pump may not be able to exhaust air from, and create a
partial vacuum in, the suction line if the pump is set very high above the
surface of the water supply and the valves and pistons are badly worn. In
that case it may be possible to start the pump and keep it in operation by

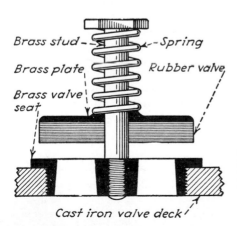

Brass stud

Spring

Brass plate

Rubber valve

Brass valve
seat

Cast iron valve deck

FIG. 15-14.   Valve assembly for
liquid   end   of   reciprocating
pump.

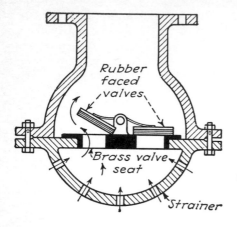

FIG. 15-15. Foot valve keeps suction filled for easier starting.

priming, as the water forms a seal. However, this should not be necessary with a reciprocating pump in good shape and with a moderate suction lift.

The impellers of centrifugal pumps do not have an airtight fit in the pump casing, and their rotation cannot reduce the pressure in the suction line much below atmospheric pressure. It is therefore always advisable to place a centrifugal pump below, or as close as possible to, the surface of the water supply. When the pump is actually lower than the water supply, it will start without any trouble, as the water will flow into the pump by gravity, but when the pump has to lift the water, it must be primed before it will start.

If a check valve is placed on the end of the suction pipe, the pump may be primed by filling the suction line and pump casing with water from some other source of water supply, such as city water mains or the pump discharge line if it should happen to be standing full of water above the pump. When the priming water is turned on, the cocks on top of the pump casing should be opened to let the air escape to indicate when the casing is full of water. A steam- or air-driven ejector may also be used to exhaust air from the pump and suction line and cause the water to rise up into the pump. Various automatic arrangements are used to prime and start centrifugal pumps that operate intermittently and with a suction lift. Most of these use float-controlled switches and electrically driven auxiliary priming pumps or valves on water-pressure lines.

Q Sketch and describe a *foot valve*.

A This is a check valve that will open and allow water to pass freely up the suction pipe to the pump but will close and hold the suction line full of water when the pump stops, thus making it much easier for the pump to pick up its water when it is again started. Figure 15-15 shows a sectional view of a foot valve fitted with a strainer to prevent any large pieces of debris, such as chips of wood, from passing up the suction pipe into the pump. The valves are brass disks, faced with leather or rubber.

CENTRIFUGAL PUMPS

Q Explain the operation principle of a centrifugal pump.

A In a centrifugal pump, one or more revolving wheels, called *impellers,* are attached firmly to a central shaft and surrounded by a stationary casing. As these impellers revolve, water enters at the center from the suction line and is thrown outward by centrifugal force. Water leaves the impeller rim at high velocity, and this velocity is converted into pressure either in the surrounding casing or by diffusion rings surrounding the impeller. In a single-suction pump, water enters one side of the impeller only. In a double-suction pump, water enters the impeller on both sides.

If a centrifugal pump contains only one impeller, it is a *single-stage* pump. If it contains two impellers, it is a *two-stage pump,* and so on. Multistage pumps are constructed with as many as 10 stages. Single-stage pumps are used mainly for pumping against low heads and moderate pressures, although improvements in design have greatly increased the range of this pump. In the multistage pump, delivery from the first stage passes to the suction of the second stage and is delivered from the second-stage impeller at an increased pressure. In this way pressure is boosted in succeeding stages so that the more stages used, the higher will be the final pressure. In practice, the maximum number of stages that can be used is limited by mechanical difficulties of construction.

Q Describe construction of the impellers in centrifugal pumps.

A Figure 15-16 shows two common types, the *closed* and the *open* impeller. The closed impeller has a plate or *shroud* on each side of the vanes, whereas

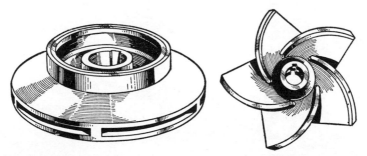

FIG. 15-16.   Closed pump impeller (left) and open impeller (right).

the vanes of the open impeller are not in any way closed in. Side plates of the closed impeller direct the flow and thus make it more efficient than the open impeller for pumping clean water. But the open unit is better for handling thicker liquids such as heavy oil, sewage, or water containing a great deal of dirt or sand.

Q   Explain how the turbine and volute principles are used in centrifugal-pump design.

A   In the turbine centrifugal pump, the impeller is surrounded by a stationary diffusion ring containing passages of gradually increasing cross-sectional area. Water leaves the impeller rim at a high velocity which is converted into pressure as the water passes through the diffusion ring. The casing surrounding the diffusion ring is usually circular in shape and of constant cross-sectional area.

In the volute pump no diffusion rings are used, but the casing is volute in shape—that is, of gradually increasing cross-sectional area—so that as the water flows in the casing its velocity is reduced and its pressure increased.

Q   Describe a centrifugal pump using the turbine principle.

A   Figure 15-17 is a sectional view of a three-stage centrifugal pump having two turbine stages and one volute stage. The liquid from the last impeller flows directly to the discharge, and the use of a final volute stage simplifies construction to quite an extent.

As the impellers are all single-suction, the end thrust is considerable. It is opposed by water pressure acting upon the balance drum in the last stage, and any excess end thrust acting in either direction is taken care of by a center-collar-type Kingsbury thrust bearing. The other bearing is the ring-oiled type.

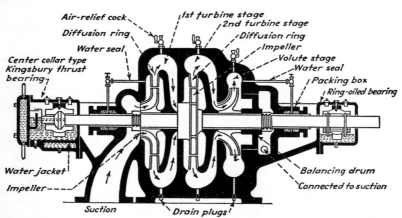

FIG. 15-17.   Centrifugal pump with two turbine stages, one volute stage.

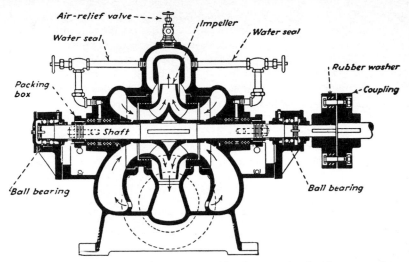

FIG. 15-18.   Single-stage volute-type centrifugal pump piped with water seals.

Q Describe a centrifugal pump using the volute principle.
A Figure 15-18 shows a sectional view of a single-stage double-suction volute pump. In this pump no diffusion rings are used, but the velocity of the water leaving the impeller is converted into pressure in the volute casing. The hydraulic balance is practically perfect in this pump because of the double suction and direct flow to the discharge; so end thrust is negligible. The shaft is supported by ball bearings, one being fixed and the other free to move a little endwise and thus take care of any slight changes in shaft length that result from expansion or contraction.

Use of volute pumps was formerly confined very much to low-pressure work, but volute-type pumps are now made to work against high pressures, in both single-stage and multistage patterns. The volute pump is of simpler construction than the turbine pump and has fewer parts.

Q What methods are used to balance end thrust in centrifugal pumps?
A End thrust may be taken care of by balance plates, ball bearings, multicollar thrust bearings, and special forms of single-collar thrust bearings. The balance plate or drum, as shown in Fig. 15-17, is keyed firmly to the shaft, and its area is such that the pressure of water acting upon it balances the end thrust in the opposite direction.

A ball-thrust bearing, suitable for light or medium end-thrust loads, is shown in Fig. 15-19a. For heavier loads, a multicollar bearing may be used (Fig. 15-19b). A number of collars turned from the solid shaft fit into recesses in the top and bottom halves of the bearing. The multicollar bearing has been largely superseded by single-collar bearings where the thrust is taken up by a number of segments that are flat on the side next to the thrust collar but pivoted at the back so that they can tilt in any direction. A constant stream of oil is supplied to the bearing, and when the collar presses against the bearing surface of the segments, they assume the position shown in Fig. 15-19c, allowing a wedge-shaped stream of oil to pass between the bearing surface of each segment and the collar. The thrust is thus actually supported by a film of oil, and there is little or no metallic contact between the bearing surfaces. If the bearing is meant to support end thrust in either direction, the collar is placed in the center of the bearing with a ring of bearing segments on each side. This type of thrust bearing can support a much greater load per square inch of bearing surface than the multicollar type of thrust bearing.

Q   What valves are necessary in the suction and discharge line of a centrifugal pump?

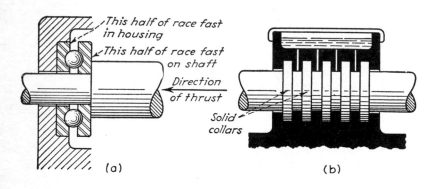

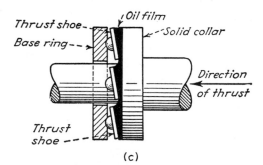

FIG. 15-19.   Three methods of taking end thrust; (a) ball bearing, (b) multicollar thrust bearing, and (c) Kingsbury thrust bearing, all designed for a specific service condition.

A If the pump is placed above the surface of the water supply, there should be a foot valve on the end of the suction line and some means of priming the pump before starting. If the pump is placed lower than the surface of the water supply, no foot valve is needed on the end of the suction line, but a stop valve, preferably of a straightway type, must be placed in the suction line at the pump to enable the flow of water through the suction to be cut off when necessary. In both cases, a check valve should be placed in the discharge line close to the pump, and beyond that, a straightway stop valve.

Q How do you start and stop a centrifugal pump?
A If the pump is above the water-supply surface, prime it first, bring it up to speed, then open the discharge valve gradually to put the load on by degrees. If the pump is below the surface of the water supply, open the suction valve, start the pump, bring it up to speed, then open the discharge valve gradually.

To stop a centrifugal pump, close the discharge valve to take off the load. Stop the pump. Close the suction valve if there is a stop valve in the suction line.

If the pump has to be primed, there should be a foot valve on the end of the suction line.

ROTARY PUMPS

Q What is a *rotary* pump?
A This is a pump in which the liquid is pumped by means of rotating elements of various shapes, contained within a closely fitted casing. Because of the practically airtight fit of the impellers in the casing, their rotation creates a partial vacuum in the suction line when the pump is started, and thus permits the pump to be set with a suction lift. Unlike the centrifugal pump, the displacement is *positive* and entirely independent of velocity of flow and centrifugal force.

Q Describe some common types of rotary pump.
A Figure 15-20a shows the operating principle of the *gear* type of rotary pump. The pump body contains two spiral gears that mesh closely with each other and have a minimum amount of clearance between the points of the gear teeth and the casing wall. As the gears revolve, liquid is drawn in at the suction opening, carried around between the gear teeth and the pump casing, then forced into the discharge line.

The pump shown in Fig. 15-20b is somewhat similar in principle to the one in Fig. 15-20a, but the impellers have large *lobes* instead of teeth. Meshing gears on the ends of the impeller shafts drive the impellers and keep

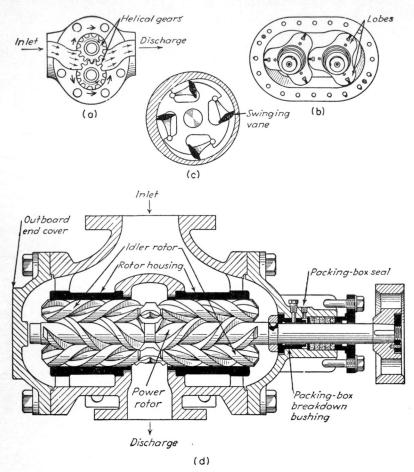

FIG. 15-20.   Rotary pumps come in many designs. Here are four popular ones: (*a*) gear type; (*b*) lobe type; (*c*) swinging-vane type; (*d*) screw type.

the lobes spaced so that they do not actually touch each other. Close contact between the points of the vanes and between the sides of the casing and the lobes is secured by flat metal strips pressed outward by springs.

In Fig. 15-20*c* the drive shaft and rotor are eccentric with the casing bore, and the pumping is done by swinging vanes that push back into recesses in the rotor during one part of the revolution and swing outward during the rest of the revolution to carry the liquid from the solution opening to the discharge.

Figure 15-20*d* shows a *screw* pump in which the liquid is drawn in at the center and forced outward to the discharge along the spiral grooves in the rotating impellers.

Reciprocating and centrifugal types of pumps are usually preferred for water pumping, but compact form, simplicity of construction, and positive displacement make the rotary pump very suitable for pumping heavy oils and other thick or viscous liquids possessing some lubricating property.

Q What is an *air lift* and how does it operate?

A An *air lift* is a device for raising water from wells by compressed air, no moving mechanical parts being used. As shown in Fig. 15-21, a large pipe is placed in the well with the lower end submerged. Compressed air is led into the bottom of the large pipe through a small air line, run either inside the larger pipe, as shown in sketch, or on the outside and curved up into the larger pipe at the bottom.

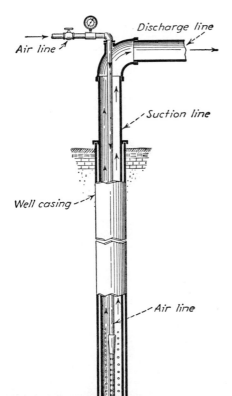

FIG. 15-21. Air lift is simple, works on compressed air.

As the compressed air escapes into the water at the bottom of the suction pipe, it produces a mixture of air bubbles and water which is lighter than the well water on the outside of the suction pipe. This column of air bubbles and water is forced upward by the pressure of the heavier column of water outside. A supply of compressed air is necessary to operate the air lift.

Rate of discharge depends upon *lift* and *submergence*. Lift is the vertical distance from the surface of the water in the well to the point of discharge. Submergence is the vertical distance from the point at which the compressed air enters the suction pipe to the surface of the water in the well. The depth of water in the well is, of course, lowered to some extent when the air lift is in operation; so the running submergence varies in practice between 35 and 75 percent of the total vertical distance from point of air entry to point of water discharge.

Q What is an *air pump*?

A An *air pump* (also called a *vacuum pump*) is a pump used to exhaust air from an enclosed vessel or system such as a steam-turbine condenser or a heating system. Reciprocating and centrifugal pumps of special design are used for this purpose, and also steam jets.

Figure 15-22 is a cross section of a centrifugal displacement liquid-piston pump, widely used for wet-vacuum service on steam condensers. Primarily an air-removal pump, it depends for its operation on a water seal main-

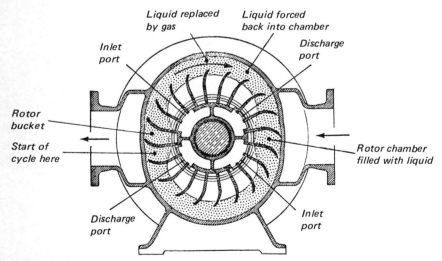

FIG. 15-22.   Liquid-piston-type vacuum (air) pump widely used for condenser service.   (*Courtesy of Nash Engineering Co.*)

tained at constant level from a source of supply. When the rotor is turning at operating speed, this water, revolving with the rotor, is thrown to the outer periphery of the elliptical casing by centrifugal force. Moisture and noncondensable gases are admitted to the rotor chambers by ports and are entrained between the walls of the chambers and the water.

As a rotor chamber rotates from the narrow part of the elliptical casing to the wide part, the water, following the contour of the casing, leaves a gap at the inner portion of the chamber, and into this gap, moisture and gases entering the chamber through slots in the rotor at the bottom of each chamber are drawn from an inlet port in the fixed central cone in the rotor hub. As a chamber passes the wide part of the casing and again nears the narrow portion of the casing at the opposite side, the entrapped air is forced through a discharge port in the cone. The air-removal capacity of single-stage pumps is fairly constant up to a vacuum of about 22 in. of mercury.

SUGGESTED READING

Elonka, Stephen M.: *Standard Plant Operators' Manual,* 3d ed., McGraw-Hill Book Company, New York, 1980 (has over 2000 illustrations).

16

PUMPING THEORY
AND CALCULATIONS

Humans live on the floor of an ocean of air which weighs 14.7 psi. And because air has weight, they use it to lift liquids by reducing the pressure between the liquid level and the suction side of a pump. There are many things that must be learned about pumps in order to solve each problem that comes up when working with them. Here, we round out the theory begun in the preceding chapter and cover some calculations the operating engineer should know how to perform.

PUMPING THEORY

Q What is meant by *static suction lift* in connection with pumps and injectors?

A *Static suction lift* is the vertical distance in feet (Fig. 16-1) from the surface of the water supply to the intake of the injector or centerline of the pump suction, when the pump or injector is placed above the source of water supply. Always slope lone suction lines upward toward the pump as illustrated, to avoid air pockets forming at the high point and thus reducing pump capacity.

Q Where a suction line loop for a centrifugal pump is unavoidable when going through a wall, or some object is in the way, illustrate how you would solve the problem.

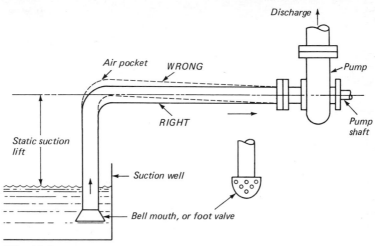

FIG. 16-1.   Right and wrong methods of installing long suction line to pump.

A   Figure 16-2 shows the *right* method, dropping instead of raising the loop. Even when pumping from a pressure tank as in the figure, the pump would not deliver the expected flow with an upward loop, in which air and other noncondensable gases liberated from the water will cause a centrifugal pump to lose capacity.

Q   What is *dynamic suction lift?*
A   This is static suction lift plus friction and velocity losses in the suction line.

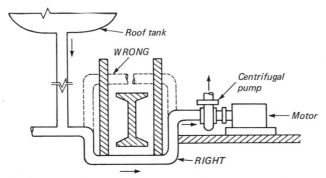

FIG. 16-2.   Right and wrong methods of looping pump suction line past obstructions in its path.

Q What is the maximum theoretical static suction lift of a pump or injector?

A The maximum theoretical static suction lift depends upon the atmospheric pressure. At sea level (where the atmospheric pressure is approximately 14.7 psia), if a long length of pipe were plugged at one end, filled with water, inverted without spilling any of the water, and the open end submerged in a tank of water, the atmospheric pressure on the surface of the water in the tank would support a column of water in the pipe 34 ft high. This, then, would be the maximum theoretical static suction lift of a pump or injector at sea level, if the steam jet or pump plungers could create a perfect vacuum in the suction pipe.

Atmospheric pressure decreases as we ascend above sea level, and so, also, does the maximum theoretical static suction lift. Dividing the atmospheric pressure at sea level by the height of the column of water that can be supported by the atmospheric pressure (14.7 ÷ 34) gives the constant number 0.434. Then to find the maximum theoretical static suction lift for any given atmospheric pressure, divide the atmospheric pressure in psia by 0.434, and the answer will be the maximum theoretical static suction lift in feet.

Q How is atmospheric pressure usually measured?

A It is usually measured by the height in inches of a column of mercury in a glass tube. A mercury barometer for indicating atmospheric pressure can be made by taking a long glass tube closed at one end, filling it with mercury, inverting the tube without spilling any mercury, and immersing the open end in a dish containing mercury. The atmospheric pressure on the surface of the mercury in the open dish holds the column of mercury up in the tube just as a column of water is held in the suction line of a pump or injector by atmospheric pressure, but as mercury is heavier than water, the column is not so high. The height of the mercury column in a mercury barometer is around 30 in. at sea level.

Q How can atmospheric pressure as measured in inches of mercury be converted into pounds pressure per square inch?

A Since 1 in.^{3} of mercury weighs 0.49 lb, we multiply the reading in inches of mercury by 0.49 in order to convert it into pounds pressure per square inch.

Q What would be the maximum theoretical static suction lift of a pump when the mercury barometer read 26.4 in.?

A Multiply inches of mercury by 0.49 to get pounds pressure per square inch, then divide by 0.434 to get lift in feet, thus:

26.4 × 0.49 = 12.936 psi
Suction lift = 12.936 ÷ 0.434 = 29.8 ft   *Ans.*

Q Why is the practical suction lift of a pump or injector always less than the theoretical suction lift?
A The maximum theoretical suction lift cannot be attained with the ordinary pump or injector because of the frictional resistance to the flow of water offered by the piping and fittings, the pressure used up in imparting velocity to the water, and the practical impossibility of securing a perfect vacuum in the suction line.

The temperature of the water also has a bearing on the practical suction lift. If the water is very hot, it may vaporize under the reduced suction pressure, filling the pump and suction line with steam and thus preventing water from rising in the suction line by destroying the partial vacuum. Even moderately warm water reduces the possible suction lift by the head equivalent to the vapor pressure (see steam tables in Chap. 7, Tables 7-3 and 7-4). Thus for water at 102°F abs, steam or vapor pressure is 1 psi. This will reduce head by $1 \div 0.434 = 2.3$ ft. When pumping cold water, the practical suction lift is never more than about 80 percent of the maximum theoretical suction lift.

Q What is meant by *head* in connection with pumping?
A The term *head* is commonly used to express difference of elevation in feet or the equivalent in psi, but in pumping there are other factors to be considered. The vertical distance in feet from the centerline of the pump to the point of free discharge (or the free surface of the liquid in the discharge well) is the *static discharge head*. The static discharge head plus the equivalent head required to overcome friction and impart velocity to the water in the discharge line is the *dynamic discharge head*.

The *total dynamic head* against which a pump operates is the sum of the dynamic suction lift and the dynamic discharge head.

Q What is the relationship between head in feet and pounds pressure per square inch?
A 1 lb pressure per in.² = $1 \div 0.434 = 2.3$ ft of head
   1 ft of head = $1 \div 2.3 = 0.434$ psi   *Ans.*

Q Tabulate some common data used in pumping calculations.
A Some of the principal constants used in pumping calculations and their equivalent values in common units of weight and volume are as follows:
   1 lb pressure per in.² = 2.3 ft of head
   1 ft of head = 0.434 psi
   1 ft³ of water weighs 62.4 lb
   1 ft³ of water contains 7.48 U.S. gal
   1 ft³ of water contains 6.24 imperial (imp.) gal
   1 U.S. gal weighs 8.33 lb
   1 imp. gal weighs 10 lb

1 U.S. gal contains 231 in.³
1 imp. gal contains 277 in.³
1 ft³ of hot water weighs approximately 60 lb
1 boiler hp is equivalent to the evaporation of 34¹/₂ lb of water from water at 212°F into saturated steam at 212°F.

1 ft · lb is the mechanical work done when a weight of 1 lb is lifted through a height of 1 ft, or work done when a force of 1 lb is exerted through a distance of 1 ft. It is the unit of mechanical work.

In addition to the above quantities, the following formulas relating to circles and cylinders are used in many pumping calculations:

The area of a circle is 78.5 percent of the area of the square it fits into. That is the diameter squared and then multiplied by 0.785, or

$$A = D^2 \times 0.785$$

The diameter of a circle is equal to the square root of the quotient of the area divided by 0.785, or

$$D = \sqrt{A \div 0.785}$$

The volume of a cylinder is equal to the area of the end multiplied by the length, or

$$V = A \times L$$

The values given have been carried to three significant figures only. Practically all pumping calculations are based on very approximate data and observations. Using more exact values would therefore mean a great deal more arithmetical work with no real advantage in the accuracy of the final result. For the same reason it is a mere waste of time to take approximate measurements and work an answer out to some such figure as 24,689.764 when all we can be reasonably sure of is that the quantity is somewhere around 24,700.

CALCULATIONS

Q How do we calculate the *useful work* done by a pump?
A *Useful work* done is the product of the water pumped, in pounds, multiplied by the height in feet to which the water is raised. This gives foot-pounds of work done.

Q How do we calculate the *useful horsepower* of a pump?
A The *useful horsepower* is found by dividing the useful work, in foot-pounds, done *in one minute* by the unit of horsepower (33,000 ft · lb of work done in 1 min).

Q What is the *mechanical efficiency* of a steam pump, and how is it calculated?

A The mechanical efficiency of a steam pump is the percentage of the power developed in the steam cylinder that is delivered as useful work in raising water against a certain head or pressure. In the case of any other type of pump, it is the percentage of the power put into the pump, which is delivered as useful work. Useful power delivered is always less than power supplied to the pump, because some power is used up in overcoming friction of the pump parts.

Stated as a formula,

$$\text{Mechanical efficiency} = \frac{\text{power output of pump}}{\text{power input of pump}}$$

Q What is the horsepower developed in a simple steam-pump cylinder of 8-in. diameter and 12-in. stroke, with the pump making 50 working strokes per minute, and with steam pressure 100 psi gage?

A Use the common formula for horsepower of steam engines; thus,

$$\text{ihp} = \frac{P \times L \times A \times N}{33,000}$$

where ihp = indicated horsepower
P = mean effective pressure of steam, psi, as found by a steam-engine indicator. In this case the mean effective pressure (average steam pressure throughout the stroke) will be 100 psi, since the inlet of steam is not cut off before the end of the stroke in most simple steam pumps
L = length of stroke, ft
A = area of steam piston, in.^{2}
N = number of working strokes per min

Applying this formula,

$$\text{ihp} = \frac{100 \times 1 \times 50.24 \times 50}{33,000} = 7.6 \text{ ihp} \qquad \textit{Ans.}$$

Q If the pump in the previous question is delivering 330 gal of water per min to a height of 60 ft, what useful horsepower is being exerted?

A Useful hp $= \dfrac{330 \times 8.33 \times 60}{33,000} = 5.0$ hp   *Ans.*

Q What is the mechanical efficiency of the above pump?

A Mechanical efficiency $= \dfrac{\text{hp output}}{\text{hp input}} = \dfrac{5}{7.6}$

$$= 0.658 \text{ or } 65.8 \text{ percent} \qquad \textit{Ans.}$$

Q What is *piston speed?*

A *Piston speed* is the distance traveled by a pump piston in feet per minute. In a simplex pump the piston speed and the total piston travel per minute are the same. In a duplex pump, piston speed is the speed of one piston in feet per minute or one-half of the total piston travel per minute.

Q What are the piston speed and total piston travel in a duplex pump having an 8-in. stroke and making a total of 120 strokes per minute?

A Piston speed $= \dfrac{8 \times 60}{12} = 40$ ft/min   *Ans.*

Piston travel $= 40 \times 2 = 80$ ft/min   *Ans.*

Q What is the most suitable piston speed for a reciprocating pump?

A This depends entirely upon the size of the pump and the purpose for which it is used. Speeds should vary for different kinds of work and should be low enough, in the first place, to allow for wear and slip at this speed while supplying all wants. Then in an emergency, the capacity can be greatly increased by speeding up the pump. Table 16-1 gives the piston speeds recommended by a well-known pump manufacturer for various strokes and types of service.

Q What is the capacity in cubic inches of the water cylinder of a reciprocating pump if the cylinder diameter is 6 in. and the stroke is 8 in.?

A First find the area of the pump piston in square inches, using the common formula, area $= 0.785 \times D^2$. Then

Area of piston $= 0.785 \times 6 \times 6 = 28.26$ in.^{2}
Volume $=$ area in in.$^2 \times$ length of stroke
$= 28.26 \times 8 = 226.08$ in.^{3} *Ans.*

TABLE 16-1 **Average Piston Speeds, in Feet per Minute, for Direct-Acting Pumps**

Stroke, in.	Boiler feed	General service	Low service
3	18	28	25
4	22	34	30
5	24	38	34
6	26	42	36
8	30	48	44
10	38	58	52
12	40	60	54
15	50	75	68
18	60	90	80

Q What is the capacity of this pump in gallons per minute if the pump makes 90 strokes per minute?
A If there is no slip,

$$\text{Capacity} = \frac{226.08 \times 90}{231} = 88 \text{ gal/min} \qquad \textit{Ans.}$$

Q What is the actual capacity of this pump if the "slip," or leakage of water past the valves and pistons, amounts to 8 percent of the theoretical capacity?
A If the slip is 8 percent, then the actual capacity is $100 - 8 = 92$ percent of the theoretical capacity and

$$\text{Actual capacity} = \frac{88 \times 92}{100} = 80.96, \text{ say } 81 \text{ gal/min} \qquad \textit{Ans.}$$

Q The discharge pipe from a pump at the bottom of a mine shaft, 460 ft deep, has an internal diameter of 4 in. What will be the static pressure on the pump piston in psi when the pipe is standing full of water, and what will be the total weight of water in the pipe?
A Static pressure on pump piston = 460×0.434
$$= 199.64, \text{ say } 200 \text{ psi} \qquad \textit{Ans.}$$
Area of pipe = $4 \times 4 \times 0.785 = 12.56$ in.^{2}
Volume of water – $(12.56 \times 460) \div 144 = 40.12$ ft^3
Weight of water = $40.12 \times 62.4 = 2503$ lb $\qquad$ *Ans.*

Q How much water will be discharged in gallons per minute through a pipe of $^1/_2$-in. internal diameter if the velocity of the water in the pipe is 4 ft/sec?
A 4 ft/sec = $4 \times 60 - 240$ ft/min
Area of discharge = $0.5 \times 0.5 \times 0.785 = 0.196$ in.^{2}
Volume of discharge = $(0.196 \times 240) \div 144 = 0.326$ ft^3
Volume in gallons = $0.326 \times 7.48 = 2.44$ gal/min $\qquad$ *Ans.*

Q How much water will be discharged in gallons per minute by a centrifugal pump with a 5-in. discharge and velocity at the discharge opening of 8 ft/sec?
A 8 ft/sec = $8 \times 60 = 480$ ft/min
Area of discharge = $5 \times 5 \times 0.785 = 19.6$ in.^{2}
Volume of discharge = $(19.6 \times 480) \div 144 = 65.3$ ft^3
Volume in gallons = $65.3 \times 7.48 = 488$ gal/min $\qquad$ *Ans.*

Q How is the size of a reciprocating steam pump usually given?
A In giving the size we state the diameter of the steam cylinder, the diameter of the water cylinder, and the length of the stroke, all in inches and in the order mentioned. Thus, a pump having a 10-in.-diameter steam cylin-

der, an 8-in.-diameter water cylinder, and a 12-in. stroke would be a 10 × 8 × 12 in. pump.

Q An 8 × 4½ × 10 in. duplex pump makes a total of 80 strokes per minute. How many imperial gallons per minute does it discharge if slip is 10 percent?

A If the pump makes a total of 80 strokes per minute, each side makes 40 strokes per minute, and if slip is 10 percent of total capacity, the volumetric efficiency of the water end of the pump is 90 percent. Then

Water delivered in gal/min
= area of piston in ft² × total piston travel in ft × gal in 1 ft³
  × volumetric efficiency of pump

$$= \frac{4.5 \times 4.5 \times 0.785}{144} \times \frac{10 \times 80}{12} \times \frac{6.24}{1} \times \frac{90}{100}$$

$$= \frac{4.5 \times 4.5 \times 0.785 \times 10 \times 80 \times 6.24 \times 90}{144 \times 12 \times 100}$$

$$= 41.33 \text{ gal/min} \qquad Ans.$$

ESTIMATING PUMP CAPACITY

Q How would you calculate the size of a boiler-feed pump for any given conditions?

A In the case of steam-driven boiler-feed pumps of the reciprocating type, the steam end must be considerably larger than the water end to give a large reserve power capacity and enable the pump to work against high pressures. As boiler-feed pumps usually handle hot water, it should be noted that hot water weighs less than cold water, and for pumping calculation purposes, its weight may be taken as, say, 60 lb/ft³ instead of 62.4. Piston speeds are also lower as a rule on boiler-feed pumps than on general-service or low-service pumps.

The best procedure when calculating the size of a boiler-feed pump for a given boiler installation is to assume a fairly low piston speed for normal operation, and then calculate the size of pump needed to deliver the required amount of water at this speed, making a generous allowance for wear and slip. If the capacity of the boilers is given in boiler horsepower, it should be remembered that 1 boiler hp is equivalent to the evaporation of 34.5 lb of water, from water at 212°F into steam at 212°F.

Q What will be the dimensions of an inside-packed double-acting duplex steam-driven pump to supply a battery of six 150-hp fire-tube boilers?

A In order to use the table of piston speeds (Table 16-1), we must assume a length of stroke. This is, of course, a matter of guesswork, but if we find at the end of our calculation that the stroke is not in proportion to the rest of the pump dimensions, we can easily change it to some more suitable length and check the problem over to find if any other changes in dimensions are necessary.

Assume that one side of the pump will supply half of the water and that the pump stroke will be 10 in., which gives us a suitable piston speed of 38 ft/min from the table. Also, allow 20 percent for wear and slip, which means that we are figuring on only 80 percent of the theoretical capacity being actually delivered by the pump. Then

Quantity of water to be supplied per hour by one side of the pump

$$= \frac{6 \times 150 \times 34.5 \times 100}{2 \times 80} = 19,406 \text{ lb}$$

Quantity of water required per minute $= 19,406 \div 60 = 323.4$ lb

Quantity of water required per stroke $= 323.43 \div \dfrac{38 \times 12}{10}$

$$= 7.093 \text{ lb}$$

Assume that the pump will be handling hot water weighing approximately 60 lb/ft³. Then

Volume of water cylinder $= 7.093 \div 60 = 0.1182$ ft³ $= 204.2$ in.³
Area of water cylinder $= 204.2 \div 10 = 20.42$ in.²

Diameter of water cylinder $= \sqrt{\dfrac{20.4}{0.785}} = \sqrt{26.0} = 5.1$ in., say $5\frac{1}{2}$ in.

Assume that the area of the steam cylinder is twice the area of the water cylinder, to give a good margin of power. Then

Diameter of steam cylinder $- \sqrt{\dfrac{2 \times 5.5 \times 5.5 \times 0.785}{0.785}}$

$$= \sqrt{60.5} = 7.78, \text{ say } 8 \text{ in.}$$

Size of pump is $8 \times 5\frac{1}{2} \times 10$ in.   *Ans.*

Q Calculate the size of simplex pump that would be needed under the same conditions as in the previous question.
A In this case we have only one cylinder; so it must be larger than the cylinders of the duplex pump and the stroke should also be longer. Assume a 12-in. stroke, which is given a piston speed of 40 ft/min in Table 16-1, and again allow 20 percent for slip. Then

Quantity of water required per hour $= \dfrac{6 \times 150 \times 34.5 \times 100}{80}$

$$= 38.812 \text{ lb}$$

Quantity of water required per minute = 38,812 ÷ 60 = 647 lb
Quantity of water required per stroke = 647 ÷ 40 = 16.2 lb.

Assume the pump handles hot water weighing 60 lb/ft³. Then

Volume of water cylinder = 16.2 ÷ 60 = 0.270 ft³ = 466 in.³
Area of water piston = 466 ÷ 12 = 38.83 in.²

$$\text{Diameter of water piston} = \sqrt{\frac{38.83}{0.785}} = \sqrt{49.46} = 7.03, \text{ say } 7 \text{ in.}$$

Assume the area of the steam piston is twice the area of the water piston. Then

$$\text{Diameter of steam piston} = \sqrt{\frac{2 \times 7 \times 7 \times 0.785}{0.785}} = 9.9, \text{ say } 10 \text{ in.}$$

Size of pump is 10 × 7 × 12 in.   *Ans.*

Q What is meant by the *duty of a pump?*
A The *duty of a pump* is the foot-pounds of work done per 1,000,000 Btu supplied by the steam. It is used as a basis of comparison of pump performance.

Q During a pumping test, a duplex pump raised 18,400,000 lb of water against a head of 60 ft. The total heat supplied to the pump in the steam during the test was 30,600,000 Btu. What was the duty of the pump?
A Duty of pump $= \dfrac{18,400,000 \times 60}{30.6}$

$\qquad\qquad$ = 36,078,000 ft · lb per 1,000,000 Btu   *Ans.*

Q What would be the thermal efficiency of the pump in the previous question?
A Dividing foot-pounds of work done per million Btu by 778 converts foot-pounds to Btu, and this divided by 1,000,000 Btu gives the thermal efficiency. Thus,

$$\text{Thermal efficiency} = \frac{\text{work done per 1,000,000 Btu} \div 778}{\text{heat equivalent of 1,000,000 Btu}}$$

$$= \frac{36,078,000 \div 778}{1,000,000}$$

$$= \frac{36,078,000}{778,000,000} = 0.0464 \text{ or } 4.64 \text{ percent}\qquad \textit{Ans.}$$

Q What is the difference between a "low" duty and a "high" duty pump?
A Large pumping engines which have compound cylinders and which use steam expansively have a fairly high thermal efficiency, say around 10

percent, and are classed as high-duty pumps. Small pumps that do not use steam expansively have a low thermal efficiency, around 2 to 3 percent, and are classed as low-duty pumps.

Q Calculate the horsepower that is expended by the water end of a reciprocating steam pump in forcing 450 gal of water per min to a height of 230 ft, assuming 10 percent of the power is used up in overcoming friction in the pump and pipelines. What will be the required horsepower of the steam end if the mechanical efficiency of the pump is 85 percent?

A hp of water end $= \dfrac{\text{ft} \cdot \text{lb of work done per min}}{33,000 \times 0.9}$

$\qquad\qquad = \dfrac{\text{lb of water per min} \times \text{height raised in ft}}{33,000 \times 0.9}$

$\qquad\qquad = \dfrac{450 \times 8.33 \times 230}{33,000 \times 0.9} = 29 \text{ hp} \qquad Ans.$

$\qquad$ hp of steam end $= \dfrac{29 \times 100}{85} = 34 \text{ hp} \qquad Ans.$

Q The capacity of a pump is given as 4860 U.S. gal/min; how much is this in imperial gallons?

A $\dfrac{4860 \times 231}{277} = 4053 \text{ imp gal} \qquad Ans.$

Q In the steam cylinder of a pump, the mean effective or average steam pressure is 60 psi, the length of the stroke is 14 in., the diameter of the cylinder is 12 in., and total number of strokes per minute is 80. Find the indicated horsepower.

A $\text{ihp} = \dfrac{PLAN}{33,000} = \dfrac{60 \times 14 \times 12^2 \times 0.785 \times 80}{12 \times 33,000} = 19.2 \text{ hp} \qquad Ans.$

Q Find the diameter of a double-acting simplex pump to discharge 120 gal of water per min at a piston speed of 40 ft/min, allowing 10 percent extra for slip.

A Capacity $= 120 \times 10 \text{ percent} = 132 \text{ gal/min}$

$\qquad \dfrac{132}{7.48} = 17.65 \text{ ft}^3/\text{min}$

$\qquad$ Area of piston in in.$^2 = \dfrac{\text{volume in ft}^3}{\text{piston speed in ft}} \times 144$

$\qquad\qquad\qquad = \dfrac{17.65}{40} \times 144 = 63.5 \text{ in.}^2$

$\qquad$ Diameter of piston $= \sqrt{\dfrac{63.5}{0.785}} = \sqrt{81} = 9 \text{ in.} \qquad Ans.$

Q How long would it take the pump in the preceding question to empty a sump 20 ft long, 10 ft wide, and 10 ft deep, standing full of water?
A Volume of water = 20 × 10 × 10 = 2000 ft³

2000 × 7.48 = 14,960 gal

$$\text{Time required} = \frac{14{,}960}{120} = 124.7 \text{ min, or 2 hr 4.7 min} \qquad Ans.$$

Q How long would it take to empty this sump if a feeder of 50 gal/min is flowing into it?
A The amount by which water is actually lowered will be 120 − 50 or 70 gal/min. Then

$$\text{Time required} = \frac{14{,}960}{70} = 213.7 \text{ min or 3 hr 33.7 min} \qquad Ans.$$

CHARACTERISTIC PUMP CURVES

Q Why are characteristic curves used for centrifugal pumps and what do they mean?
A Figure 16-3 shows characteristic curves of a centrifugal pump at constant speed. These curves are used because the relations between head, discharge, and pump speed are best determined by test rather than by calculation—unlike the same relations for positive-displacement pumps (rotary and reciprocating). Horsepower and efficiency curves are included in Fig. 16-3 with the head-discharge curve for a pump running at constant speed. The head and capacity at which the pump is rated at this speed are those values obtained when the efficiency is a maximum. The value of the

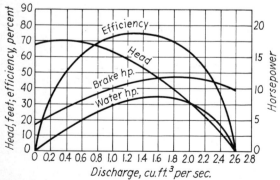

FIG. 16-3. Characteristic curves of a centrifugal pump at constant speed are necessary information for operators.

head at the point of maximum efficiency is usually less than at shutoff, but may be higher in exceptional cases.

The values used in plotting such curves are those computed for a given speed of the pump, usually that at which the pump operates at the best efficiency, or that at which it was designed to operate. For centrifugal pumps, the relation between the various quantities is largely controlled by the angles and curvatures of the impeller blades and the shape of the volute, or the arrangement and design of the diffusion vanes. If the vanes are radial or inclined forward in the direction of rotation, the head will increase with increased capacity; if the vanes are curved backward sufficiently, the head will remain constant or decrease as the capacity increases, as shown in the curve.

In general, for a given centrifugal pump, the quantity of water delivered varies directly with the speed, the head with the square of the speed, and the power with the cube of the speed. Thus, doubling the speed of a pump impeller doubles the quantity of water pumped, which produces a head 4 times as great and requires 8 times as much power to drive the pump.

Because manufacturers furnish curves for their specific output, and as pump-performance curves vary within wide limits, just remember that general curves are for general approximations only. In Fig. 16-3, the highest point on the total head curve is 70 ft; thus the pump could not be used for a greater head, since at the existing constant speed no water would be delivered.

SUGGESTED READING

Elonka, Stephen M.: *Standard Plant Operators' Manual,* 3d ed., McGraw-Hill Book Company, New York, 1980.

17

PIPING, ACCESSORIES, AND CALCULATIONS

In the broad field of energy-systems engineering, pipes are arteries interconnecting machines. Without piping, today's energy systems—steam, compressed gas, air conditioning, refrigeration, liquid handling—could not exist. Piping and piping standards have been with us a long time. In 1820 the first standards appeared for cast-iron pipe in Britain; they were later adopted in America.

Here we cover piping details and calculations for stationary engineers who, although they do not design piping systems on a full-time basis, may find such information valuable for an occasional job. Before installing boiler-pipe connections, or any piping that will be subjected to severe stresses, always consult the ASME and other codes.

HOW PIPE IS MADE

Q What materials are used for making pipe?
A Pipe is made from clay, cement, concrete, lead, brass, copper, cast iron, wrought iron, wrought steel, alloys of different metals, and various plastic materials.

Q For what particular purpose is each kind of pipe used?
A Cement, concrete, and clay (tile) pipes are used mainly for water supply and drainage purposes. Lead pipe has a high resistance to corrosion and can be bent readily to any desired shape, but it cannot withstand high tem-

perature or much internal pressure. Copper and brass pipes resist corrosion better than steel or iron but are more expensive and not so suitable for very high pressures. Copper pipe is very flexible and easily formed into coils or short-radius bends.

Cast-iron pipe is used largely for low-pressure and low-temperature purposes, such as water supply and drainage. It is weaker than wrought-iron or wrought-steel pipe, but less subject to corrosion. The addition of small amounts of nickel or chromium to cast iron increases its strength and resistance to corrosion considerably. Pure wrought-iron pipe resists corrosion better than wrought-steel pipe, and hence is preferred for purposes where this quality is important. However, most of the pipe used for high-pressure lines and general purposes is made from wrought steel. Where corrosive liquids have to be conveyed, steel pipes may be lined with lead or cement.

Pipe with high corrosion-resisting properties and able to withstand considerable internal pressure is made from a number of nonmetallic plastic compounds.

Q How are wrought-iron and wrought-steel pipes made?
A There are three common methods (Fig. 17-1): (1) Butt weld, formed by drawing a hot skelp through the dies. The edges aren't beveled; drawing produces parallel edges at the weld. (2) Lap weld, formed by drawing a beveled skelp through the dies. The bevel gives overlap at the joint; drawing produces a tight weld. (3) Fusion weld, rolled from standard plate. The gap is fill-welded to seal the joint. This method is used in sizes up to 12 in.

Q What are the various grades of wrought-iron and wrought-steel pipe?
A Wrought-iron and wrought-steel pipe were formerly divided into three grades known as Standard, Extra Strong, and Double Extra Strong, but this classification has been superseded by one published by the American National Standards Institute, formerly the American Standards Association. The ANSI specification consists of 10 different grades or schedules numbered as follows: 10, 20, 30, 40, 60, 80, 100, 120, 140, and 160. These grades are based on pressure-stress ratios which relate the pipe dimensions directly to the maximum pressure it is intended to carry. The outside diameter is the same for each size in all schedules, but the wall thickness varies,

Butt weld   Lap weld   Fusion weld

FIG. 17-1. Three types of pipe weld: butt, lap, and fusion.

increasing gradually from the lightest grade, schedule 10, up to the heaviest grade, schedule 160. Pipe sizes go by the nominal inside diameter up to and including 12 in. Above 12 in. the outside diameter is taken, hence the term OD used for the larger sizes. ANSI dimensions for pipe sizes from $1/8$ in. to 12 in. inclusive, in schedules 40 and 80, are given in Table 17-1.

The new grading makes a wider range available in the larger pipe sizes, but there is very little change in the dimensions of the smaller sizes. Up to and including 8 in., schedules 40 and 80 correspond to the old Standard and Extra Strong grades. There is no size corresponding to the old Double Extra Strong, and pipe in the heaviest grade, schedule 160, is still much lighter in weight and larger in bore than the Double Extra Strong.

Latest editions of standards may be purchased from American Standards Association, 70 East 45th St., New York, N.Y. 10017.

TABLE 17-1   Welded and Seamless Pipe Dimensions

Nominal size, in.	External	Actual diameter, in. Internal Schedule 40	Schedule 80	No. of threads per in.	Depth of thread, in.	Length of effective thread, in.	Size of tap drill for pipe taps
$1/8$	0.405	0.269	0.215	27	0.029	0.264	$21/64$
$1/4$	0.540	0.364	0.302	18	0.044	0.402	$29/64$
$3/8$	0.675	0.493	0.423	18	0.044	0.408	$19/32$
$1/2$	0.840	0.622	0.546	14	0.057	0.534	$23/32$
$3/4$	1.050	0.824	0.742	14	0.057	0.546	$15/16$
1	1.315	1.049	0.957	$11\frac{1}{2}$	0.069	0.683	$1\,3/16$
$1\frac{1}{4}$	1.660	1.380	1.278	$11\frac{1}{2}$	0.069	0.707	$1\,15/32$
$1\frac{1}{2}$	1.900	1.610	1.500	$11\frac{1}{2}$	0.069	0.723	$1\,23/32$
2	2.375	2.067	1.939	$11\frac{1}{2}$	0.069	0.756	$2\,3/16$
$2\frac{1}{2}$	2.875	2.469	2.323	8	0.100	1.137	$2\,11/16$
3	3.500	3.068	2.900	8	0.100	1.200	$3\,5/16$
$3\frac{1}{2}$	4.000	3.548	3.364	8	0.100	1.250	$3\,13/16$
4	4.500	4.026	3.826	8	0.100	1.300	$4\,3/16$
5	5.563	5.047	4.813	8	0.100	1.406	$5\,1/4$
6	6.625	6.065	5.761	8	0.100	1.512	$6\,5/16$
8	8.625	7.981	7.625	8	0.100	1.712	
10	10.750	10.020	9.564	8	0.100	1.925	
12	12.750	11.938	11.376	8	0.100	2.125	

Q How would you determine which grade of pipe to use for any specific purpose?

A The proper grade to use will depend upon the service for which it is intended. Pipe in schedule 40 (old Standard grade) is most commonly used for general purposes, but pipe in schedule 80 (old Extra Strong), or heavier if necessary, should be used in high-pressure work.

PIPE THREADS

Q What is meant by *depth of thread, lead,* and *pitch,* in connection with pipe threads?

A *Depth of thread* is the depth to which the thread is cut, measured at right angles to the axis of the pipe. *Lead* is the distance that the screw advances in one turn. *Pitch* is the distance from center to center of adjacent threads. In a single-threaded screw, lead and pitch are the same. Pipe threads are all single-threaded. Double- and treble-threaded screws are used as transmission screws on machine tools and other machinery.

Q What form of thread is used on screwed pipe?

A The form of thread commonly used is known as the American Standard taper pipe thread. It is V-shaped, with a 60° included angle, but the sharp point of the V is flattened at the top to a depth equal to 0.033 of the pitch, and flattened at the root of the thread to a similar amount. This makes the depth of the thread equal to 0.8 of the pitch.

The Whitworth form of thread is used in Great Britain and other countries. It is also a V-shaped thread with a 55° included angle, and with the points of the vee rounded off at the point and root of the thread. The depth of the Whitworth thread is 0.64 of the pitch.

The plain V thread is simpler than either of the foregoing, but it is difficult to keep sharp V points on taps and dies. Flattening or rounding the points of the threads increases the life of the threads and thread-cutting tools without weakening the thread to any appreciable extent.

Q Explain how pipe threads make tight connections when screwed together.

A Pipe threads are cut with a taper of $3/4$ in./ft, or $1/16$ in./in., in order that they may make mechanically strong and gastight connections when the taper on the pipe tightens in the taper in the fitting. Because of the taper, a pipe thread is not of perfect form throughout its whole length. The nature of this imperfection is shown in Fig. 17-2, and the length of effective thread is given by the formula

$$L = \frac{1}{n}(0.8D + 6.8)$$

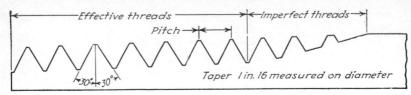

FIG. 17-2.   Standard pipe-thread taper is designed for self-sealing at this end.

where D = actual external diameter of the pipe
$\quad\quad n$ = number of threads per inch
The remaining threaded part has a few shallow threads that have little mechanical or jointing value.

Q   What is the length of effective thread on a 1-in. pipe?
A   From the formula,

$$\text{Length} = \frac{1}{11^{1}/_{2}}\,(0.8 \times 1.315 + 6.8) = {}^{2}/_{23}(1.052 + 6.8)$$

$$= {}^{2}/_{23} \times 7.852 = 0.68 \text{ in., or a little more than } {}^{5}/_{8} \text{ in.} \qquad Ans.$$

Q   What precautions should be observed when cutting and threading pipe?
A   Pipe should be cut with a hacksaw, parting tool, or power saw in preference to a wheel cutter, since the last tool leaves a burr on the inside of the pipe and some wheel cutters leave a burr on the outside as well. A burr on the outside makes threading more difficult unless it is ground or filed off. A burr on the inside reduces the effective area of the pipe unless it is reamed out. Quite a clean cut can be made by a cutting torch in the hands of a skilled operator.

When threading a pipe, care should be taken to cut the thread the proper length and depth. Threading machines and most hand dies are provided with guide marks and stops, and the threads are automatically cut to proper dimensions if the dies are in good shape and properly set. Blunt dies require much more power to turn than sharp dies, and cut a poor thread that will give trouble sooner or later. If threads are not cut deep enough or are too short, the pipe will enter only a few threads into the fitting, making a weak joint that will be almost sure to leak or break in a very short time. If threads are cut too deep or too long, the pipe will go too far into the fitting and tighten only on the last few threads. It may also block other openings into the fitting or butt against a valve seat and damage it.

After a pipe is threaded, it should be tapped sharply to remove cuttings from the threads, and a standard fitting should be tried on to make sure that the thread is properly cut. If a jointing paste is used, it should be put on the pipe threads and not in the fitting. Paste smeared in a fitting is simply pushed ahead of the pipe when it is screwed in, and washed into the system. Very little paste is required if the threads are good. Its main function is to lubricate the threads when the joint is being tightened and enable it to be broken when necessary, without stripping the threads. Keep dies well oiled and clean when cutting, and store them in a proper place to prevent damage to the cutting edges when they are not in use.

PIPE FITTINGS

Q What materials are used in the manufacture of pipe fittings?

A Pipe fittings are made from cast iron, alloy cast iron, bronze, malleable iron, cast steel, alloy cast steel, and forged steel. Chromium, nickel, and molybdenum are the usual alloying metals, and they are added to increase strength and resistance to corrosion. Fittings are made from all these materials in a variety of weights to suit different purposes, but in general, cast iron and bronze are used for low and medium pressures and temperatures, alloy cast iron and malleable iron for medium pressures and temperatures, cast steel and alloy cast steel for fairly high pressures and temperatures, and forged steel for very high pressures and temperatures. Note that when choosing fittings for any particular purpose, the temperature to which the fittings will be exposed must be taken into account, as well as pressure.

Q Sketch some common forms of pipe fittings.

A Some of the commonly used types of screwed pipe fittings are shown in Fig. 17-3.

Q Sketch some common forms of welding pipe fittings.

A Some types of fittings for welded connections are shown in Fig. 17-4. These fittings are seamless, presenting a smooth inner and outer surface. The ends are beveled as shown for V welding.

Q Sketch and describe some common forms of flanged pipe joints.

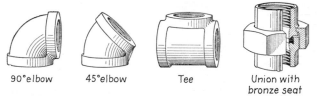

90°elbow   45°elbow   Tee   Union with bronze seat

FIG. 17-3. Angle, and designs of other standard threaded pipe fittings.

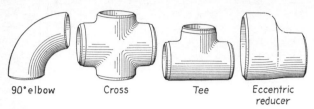

90° elbow   Cross   Tee   Eccentric
                                               reducer

FIG. 17-4.   Pipe fittings for welded joints come in many
shapes.

A   Sectional views of three typical flanged joints are shown in Fig. 17-5.

In the *screwed flange* joint the pipe is threaded in the usual way, and the flanges are bored out and threaded with a taper so that they will be a tight fit on the pipe thread. The flanges are screwed on until the ends of the pipe are flush with the face of the flange or projecting slightly. The projecting ends are smoothed off by grinding or chipping and filing, and sometimes the ends are expanded a little in the flange by peening—that is, hammering the inside of the pipe lightly with a ball-peen hammer. Peening should not be necessary if the threads are good and fit properly.

The *slip-on welded flange* is a plain flange, slipped over the end of the pipe and welded inside and outside as shown.

The *Van Stone joint* is made by boring the flange out to slip easily over the end of the pipe, then flanging the end of the pipe and facing off the flanged end thus providing a smooth surface for the gasket. Being loose, the flanges can be turned to any position to match bolt holes. In a modification of this joint the edges of the flanged ends of the pipes are seal-welded after the bolts are tightened, thus preventing any possibility of leakage. The screwed and slip-on welded flanges are suitable for low and moderate pressures, and the Van Stone and welding neck flanges (not shown) are preferable for very high pressures.

Q   How are pipes connected by welding alone?

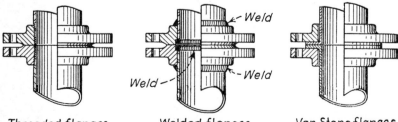

Threaded flanges   *Welded flanges*   *Van Stone flanges*

FIG. 17-5.   Pipe flanges of these designs are used where welding isn't practical.

A   The ends of the pipes are beveled to an angle of 30°, then butted together, and the V joint filled by a suitable welding rod and gas or electric
welding apparatus. The use of welding rings, as shown in Fig. 17-6, keeps
the ends of the pipes in proper alignment and strengthens the joint.

Q   How are flanged joints made steam-, gas-, or watertight?
A   This is done by placing a gasket or ring between the joint faces and
tightening the flange bolts, taking care to pull them up evenly all around so
that there is an equal pressure on the ring or gasket at all points. Gaskets
may be made from sheet rubber, rubber with wire insertion, sheet asbestos,
corrugated soft-metal rings, or various other materials, depending upon
the purpose of the pipeline and the pressure to be carried. Metal rings of
oval or octagonal cross section that fit into grooves in the faces of the
flanges are often used on high-pressure lines, and sometimes flanges are
made with a projecting ring on one flange which fits into a similar recess in
the companion flange, a gasket being placed in the recess to make the joint
tight. No gasket is needed when the flanges are seal-welded.

Q   Has a welded joint any advantages over a screwed joint?
A   When properly made, welded joints are more permanent than screwed
joints and less likely to give trouble in course of time, but the making of
welded joints requires more skill and judgment than the joining together of
screwed fittings. Factory-welded jobs such as large pipe headers are probably more reliable than screwed joint headers, but for general erection and
repair in the field, welding alone is not particularly safe unless it is done by
thoroughly competent and experienced welders. People can be trained easily and quickly to make up screwed joints, but good welders are not trained
in a day.

   In general, welded joints have a number of advantages over screwed
joints. Joints can be built up at any angle without the use of special fittings.
Pipes do not have to be in exact alignment, which may save considerable
time when laying large pipes. Connections can be made to existing lines

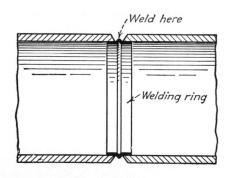

FIG. 17-6. Welding ring used
in piping prevents weld metal inside.

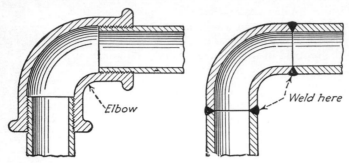

FIG. 17-7.   Comparison of older screwed with welded pipe joints.

without having to disconnect them. If the welding is properly done, there should be no projection on the inside of the pipe or fitting and therefore less resistance to the smooth flow of gas or liquid than that offered by the screwed type of fitting. This is shown clearly in Fig. 17-7. There is a saving of space occupied and metal used in places where a great number of joints have to be made, if welded joints can be used instead of screwed joints.

Q What other forms of pipe joints are used in addition to screwed and welded joints?
A Four other common forms are (1) calked bell-and-spigot joint, (2) screwed bell-and-spigot joint, (3) Dresser joint, and (4) soldered joint.

The *calked joint* (Fig. 17-8) is used mainly for drainpipe and low-pressure water supply. The pipes are of cast iron and the joint is made by inserting a spigot end into a bell end, calking a few rounds of oakum in the neck of the bell to stop the lead from running into the pipe, plastering clay or putty around the mouth of the bell, leaving a small opening on top, and pouring molten lead into the joint through this opening. After the lead has cooled, it is calked tight with a calking tool. The joint may also be made with lead wool instead of molten lead. The lead wool method is very useful where pipes have to be calked under water or with water running through them.

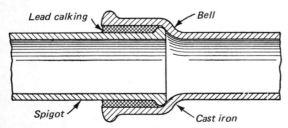

FIG. 17-8.   Calked bell used in spigot joint of cast iron.

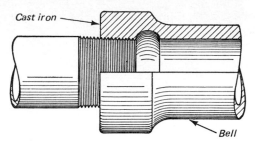

FIG. 17-9.   Threaded cast-iron bell of spigot joint.

The *threaded bell-and-spigot joint* is also used on cast-iron pipe. The pipe joint is not calked in this case but screwed together, as shown in Fig. 17-9.

In the *Dresser joint* (Fig. 17-10) a packing-ring assembly is used, consisting of two iron rings with a ring of soft packing between. This is slipped over the plain end of the pipe before it is inserted in the bell of the other pipe. Lugs in the outer iron ring pass through grooves in the flange of the bell, and a slight turn causes the lugs to enter a groove in the interior of the bell flange, thus locking the entire assembly in place. Tightening the long cap screws, which pass through the packing ring and screw into the inner iron ring, brings the two iron rings closer together and squeezes the soft packing ring so that it presses hard against the spigot and bell, making a tight joint that permits considerable movement without leakage.

Soldered joints are used on copper pipe in plumbing work. The ends of the pipes and the inside of the fittings are tinned, and the joint is made by

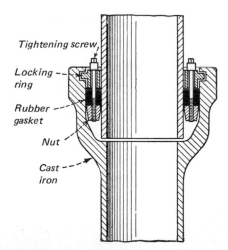

FIG. 17-10.   Dresser pipe joint allows   movement,   prevents straining.

slipping the fitting over the end of the pipe and heating it with a blowtorch to cause the solder to flow.

BENDS AND SUPPORTS

Q How are pipe bends made?
A Small-diameter pipe can often be bent cold by a roller-bending machine, or slight bends may be made by bending the pipe over some curved object, but large-diameter pipe must be heated before it can bend properly. If apparatus for heating and bending the pipe in one operation is not available, the usual practice is to heat a small portion of the pipe at a time and bend it a little at the heated part, continuing the process until the desired bend is secured. Care must be taken not to bend the pipe too much at one time when using this method, or the walls may be squeezed out of shape. Filling the pipe with dry sand and plugging the ends helps to keep it in its proper shape while it is being bent, but all important large-sized bends should be ordered from a pipe manufacturer or engineering works equipped to do this work.

Q How are pipelines supported and anchored?
A Pipelines are supported by steel or timber trestles, concrete pillars, cast-iron or wrought-steel wall brackets, or steel hangers of various designs. They are anchored—that is, held rigidly at one or more points—by clamps attached to supporting columns or brackets.

Figure 17-11 shows adjustable and nonadjustable hangers, single- and double-roller bracket supports, and a constant-support type. The first hanger shown has no adjustment for length, but the second hanger is provided with a turnbuckle. This turnbuckle has right- and left-hand threads so that it can be lengthened or shortened for proper length adjustment and even distribution of weight.

The rollers on the bracket supports permit the pipe to move freely as it expands and contracts. The upper roller on the double-roller support

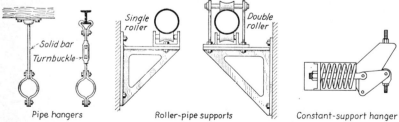

FIG. 17-11. Many hanger and support designs are available for unrestricted movements, as all piping expands and contracts with temperature and pressure.

holds the pipe down on the lower roller while still allowing free lengthwise movement.

In the last type of pipe hanger, the load is carried by a heavy coil spring. Means are provided for adjusting the compression of the spring, and an indicator on the side of the case enclosing the spring shows the load that is being supported by the hanger, in pounds, and the extension of the spring in inches.

EXPANSION

Q   What provision must be made for taking care of expansion of pipelines, especially those carrying high-temperature fluids and gases?

A   Expansion is taken care of by *expansion bends* and *joints*. The pipeline is anchored at certain points where excessive movement due to expansion is undesirable, and bends or expansion joints are placed in the line between these anchored points. As a rule, pipelines are anchored close to points where they connect to boilers or engines, and the devices for the absorption of expansion are inserted in the long runs. Supports at points other than anchors are designed to permit free movement of the pipes. If adequate provision is not made for expansion, there is danger of rupture of pipes, valves, or fittings, with disastrous consequences.

Q   Sketch a good type of expansion pipe bend.

A   A good form of expansion bend is shown in Fig. 17-12. If the bends are made to a large enough radius, this form of expansion pipe bend will take care of considerable expansion in a pipeline, without throwing an excessive stress on the flanges or other fittings.

Q   Sketch and describe several forms of expansion joints for pipelines.

A   The expansion joint shown in Fig. 17-13 consists of a smooth sleeve which slides in or out through a packing box as the pipeline expands or contracts in length. The packing box is packed with soft packing, held in place by a gland and studs. Tie rods guard against the possibility of the two

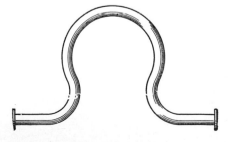

FIG. 17-12.   Large expansion pipe bend is simple but takes up space.

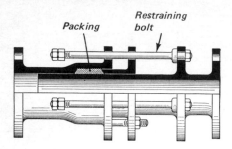

FIG. 17-13. Low-pressure expansion joint is linear, saves space.

halves of the joint pulling apart, and flanges are provided for connecting the joint into the pipeline.

The expansion joint shown in Fig. 17-14 also consists of a smooth sleeve sliding through a packing box, but it has internal traverse stops to guard against pulling apart instead of outside tie rods, and metallic packing instead of soft packing. It is more suitable for high pressures than the joint shown in Fig. 17-13.

Corrugated pipes and metal bellows are also used in the construction of expansion joints. An expansion joint with a welded metal bellows is shown in Fig. 17-15. Since the bellows is welded to the two halves of the joint, it is perfectly steamtight, and no packing box is necessary.

Q How can the probable amount of expansion in a pipeline be calculated?
A The expansion can be calculated fairly closely for moderate temperature ranges if we know the coefficient of expansion of the metal from which the pipes are made. The coefficient of expansion is the amount of increase of length per unit of length, per degree Fahrenheit rise in temperature, or

Expansion in in. = original length in in. × rise in temp. in °F
× coefficient of expansion

Q What would be the amount of expansion in a steam pipeline 340 ft long

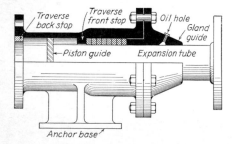

FIG. 17-14. High-pressure expansion joint has metallic packing.

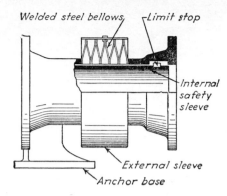

FIG. 17-15. Metal-bellows expansion joint does not need packing.

if the outside temperature is 60°F and the pipe is carrying steam at 150 psig?

A A coefficient of expansion of 0.0000068 is given by various authorities for the metals from which steel and iron pipe are manufactured. Using this value, and taking temperature of steam as 366°F,

Expansion = 340 × 12 × (366 − 60) × 0.0000068
        = 340 × 12 × 306 × 0.0000068
        = 8.49 or approximately 8½ in.   *Ans.*

STEAM SEPARATORS

Q What is a steam separator?

A A steam separator is an apparatus for separating out moisture that may be carried in suspension by steam flowing in pipelines and for preventing this moisture from reaching and perhaps damaging engines, pumps, or other machinery that may be driven by the steam. The separation of the moisture from the steam is accomplished by either giving the steam a whirling motion or causing it to strike baffles that change the direction of its flow.

Q Sketch and describe two types of steam separator.

A A *centrifugal* steam separator is shown at the left in Fig. 17-16. In passing through this separator, the steam strikes a spiral-shaped plate which gives it a whirling motion. The moisture is thrown outward by centrifugal force, and it drains down the walls to the bottom of the separator while the steam passes up the central pipe and out to the line.

In the *baffle type* of steam separator shown at the right, the direction of flow of the steam is changed by suitable baffles. The moisture adheres to

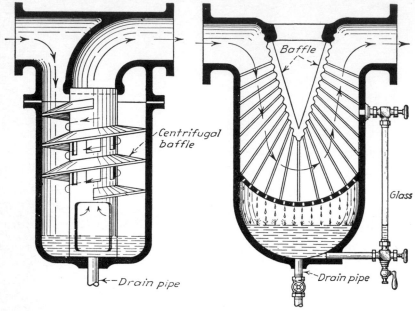

FIG. 17-16.　Both types of steam separators throw out droplets of condensation.

the baffle walls and drains down to the bottom of the separator. In all types of separators the water is drained away through a steam trap.

STEAM TRAPS

Q　Name some of the common types of steam traps and explain how they operate.

A　Traps may be roughly divided into six classes: bucket traps, float traps, tilting traps, expansion traps, thermostatic traps, and impulse traps.

In the *bucket* trap, the movement of an open bucket as it fills with condensate causes the discharge valve to open, and the steam pressure then forces the condensate out of the trap. As the bucket empties, it returns to its original position and closes the discharge valve.

Float traps contain a float, usually a hollow metal ball. As the trap fills with condensate, the float rises and opens the discharge valve. As the steam pressure empties the trap, the float falls and closes the discharge valve.

In the *tilting* trap the condensate drains into a counterbalanced cylindrical- or spherical-shaped vessel, and the weight of water causes the vessel to tilt and in doing so to open a discharge valve. When the steam pressure has forced out the water, the counterbalance brings the tilting bowl back to the closed position.

Expansion traps may be actuated by the expansion and contraction of some metal element in the trap, by oil in a sealed cartridge acting upon a sealed plunger, or by a volatile liquid in a sealed container, usually a metal bellows to facilitate expansion.

The *thermostatic* trap is a form of expansion trap. The discharge valve is opened or closed by the contraction and expansion of a metal bellows containing a highly volatile liquid with a low boiling point. When the trap is cold, the valve is open and water discharges freely. When steam enters the trap, the liquid in the bellows vaporizes. This causes the bellows to expand, since the vapor exerts a greater pressure than the liquid, and the expansion of the bellows closes the discharge valve.

The bucket trap is perhaps the most widely used trap for general purposes. Expansion traps, thermostatic traps, and combination float and thermostatic traps are extensively used in heating systems. The impulse trap is used on high-pressure steam lines.

Q What are *return* and *nonreturn* traps?

A *Return* traps are designed to return water of condensation to the boiler, against the boiler pressure. Tilting traps and some forms of float traps are commonly used for this purpose. They are set a few feet above the boiler and a live steam line is run from the boiler to the trap. Tilting of the bowl or movement of the float opens the discharge valve and at the same time opens a valve admitting live steam to the trap. The steam pressure from the live steam line, together with the few feet of gravity head from the trap to the boiler, causes the condensate to flow from the trap into the boiler.

The term *nonreturn* applies to traps of any type that have no live-steam connection direct from the boiler and that simply discharge the condensate to waste or to a hot-water receiver against little or no back pressure.

Q Sketch and describe two common types of bucket steam traps and explain how they operate.

A A common type of open-bucket trap is shown in section in Fig. 17-17. When water enters the trap, the bucket floats and presses upward on the discharge valve, keeping it closed so that no steam escapes. When the water rises above the level of the rim of the bucket, it fills the bucket, causing it to sink and open the discharge valve. The steam pressure acting upon the surface of the water in the bucket forces it up through the discharge valve. As the bucket empties, it again floats upward and closes the discharge valve.

The trap shown in Fig. 17-18 is also of the bucket type but with the bucket inverted. When the trap is filled with water, the bucket sinks to the bottom and opens the discharge valve, allowing the water to escape freely, but as soon as steam begins to enter the trap, the bucket fills with steam and

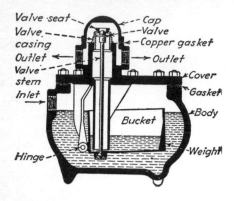

FIG. 17-17.   Open-bucket
steam trap is reliable workhorse.

air that may be carried with the steam, and floats upward, closing the discharge valve. As fresh condensate enters the trap, the steam and air are
forced out of the small vent hole in the top of the bucket by the rising water
level; the bucket loses its buoyancy and sinks to the bottom again, opening
the discharge valve.

Q   Sketch and describe some form of thermostatic trap.

A   A form of thermostatic trap that is in very common use on steam-heating systems is shown in Fig. 17-19. The discharge valve of this trap is at-

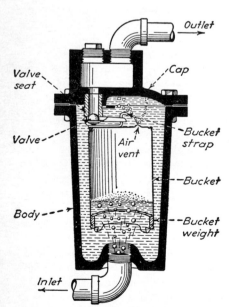

FIG. 17-18.   Inverted-bucket
steam trap is also rugged.

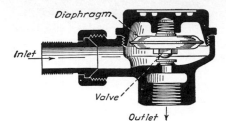

FIG. 17-19. Thermostatic trap
is very simple and reliable.

tached to a thin metal bellows or other thin-walled metal device containing
a highly volatile liquid that will evaporate into a gas at a comparatively low
temperature. When the trap fills with condensate, the bellows or diaphragm contracts and opens the discharge valve so that the water may flow
freely out of the trap, but when steam enters the trap, the volatile liquid in
the bellows or other sealed container vaporizes, and the resulting expansion of the bellows or container diaphragm closes the discharge valve and
prevents the escape of steam.

Q Explain the impulse steam trap.

A The impulse steam trap (Fig. 17-20) operates on the principle that hot
water under pressure tends to flash into vapor when the pressure is re-

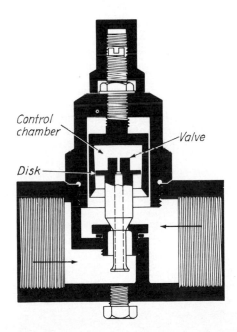

FIG. 17-20. Thermodynamic
steam trap works on the impulse
principle. (*Courtesy of Yarway
Corp.*)

duced. When hot water is flowing to the trap the pressure in the control chamber is reduced, causing the valve to rise from its seat and discharge water. As steam enters the trap, the pressure in the chamber increases, causing the valve to close, thus reducing the flow of steam to that which can pass through the small center orifice in the valve. It requires 3 to 5 percent of full condensate capacity to prevent steam from being discharged through this small orifice. Impulse traps are available for all pressures, they are compact, and the few parts subject to wear can be easily replaced. It is necessary to install a strainer, since small particles of foreign matter will interfere with the operation of the valve.

PIPING LAYOUTS

Q How are piping layouts usually represented in drawings, and how are measurements indicated on the drawings?
A There are two methods of representing pipelines and fittings in drawings, *single-line* drawing and *double-line* drawing. The first method is commonly used in rough sketches, and the second is used in finished drawings made to scale.

Pipe measurements are taken from center to center of pipes and fittings, and the radius of a pipe bend is always measured to the center of the pipe. Examples of dimensioning are shown in Figs. 17-21 and 17-22.

Q Show how dimensions are inserted in double-line pipe drawings.
A Figures 17-21 and 17-22 are reproductions of scale drawings, reduced from the original size. Figure 17-21 shows a short run of straight pipe with a valve and two elbows, and Fig. 17-22 shows a gooseneck bend.

Q Show some of the symbols used to represent valves and fittings in single-line pipe sketches, and give an example of a single-line pipe sketch.
A Figure 17-23 shows a few of the commonly used single-line symbols, and Fig. 17-24 shows their application.

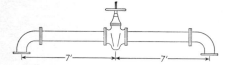

FIG. 17-21. Double-line drawing of pipe with valve and fittings showing points of measurements.

Q Draw the piping layout for a battery of two dry-back fire-tube boilers.
A Figure 17-25 shows the header supported by wall brackets and anchored midway between the boiler connections. Provision is also made for expansion by curving the main steam line, and a steam trap is installed under the header for drainage.

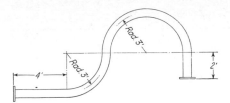

FIG. 17-22. Double-line layout of pipe bends with flanged ends.

Q What general precautions should be observed in the erection of pipelines?

A The proper size and grade of pipe, valves, and fittings should be selected for the service for which the line is intended. Joints, whether welded, screwed, or flanged, should be carefully made, with pipes in exact alignment. Threads on screwed pipe should be clean and smooth, and a good jointing compound should be used to preserve the threads, lubricate the

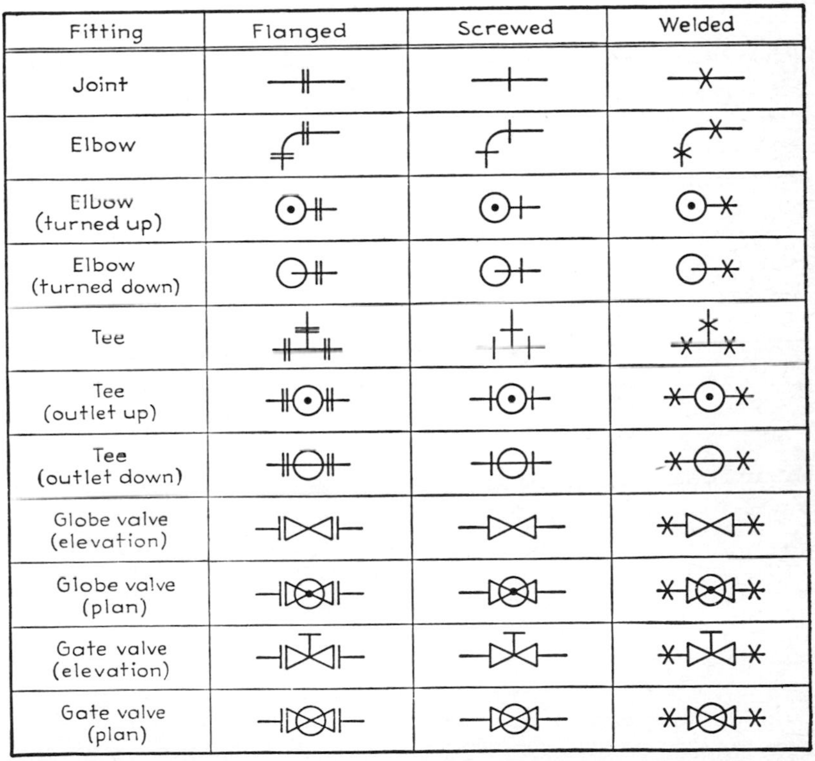

Fitting	Flanged	Screwed	Welded
Joint	⟶‖⟶	⟶+⟶	⟶X⟶
Elbow			
Elbow (turned up)	⊙‖	⊙+	⊙＊
Elbow (turned down)	◯‖	◯+	◯＊
Tee			
Tee (outlet up)	‖⊙‖	+⊙+	＊⊙＊
Tee (outlet down)	‖◯‖	+◯+	＊◯＊
Globe valve (elevation)	◁▷	◁▷	＊◁▷＊
Globe valve (plan)			
Gate valve (elevation)			
Gate valve (plan)			

FIG. 17-23. Standard single-line symbols used for pipe fittings and valves.

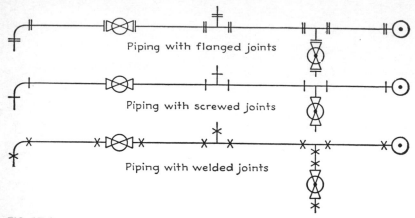

FIG. 17-24.   How standard piping symbols are used with single-line pipeline drawings.

joint while it is being tightened, and facilitate the breaking of the joint if it ever has to be taken apart. Faces of flanges should be clean and smooth, and bolt holes should be in line. Gasket material should be suitable for the pipeline service. Sheet rubber, for example, may be excellent for a low-pressure water line, but it would be entirely unsuitable for a high-pressure steam line. Flange bolts should be tightened up evenly all around so as to keep flange faces exactly parallel. Pipelines should be well supported, with an even distribution of weight on the hangers or other supports, and anchored at suitable points and well drained by steam traps. There should be no low points where water can collect, but if this is unavoidable, traps should be placed at these places. Steam lines should be covered with insulating material to prevent loss of heat by radiation. The welding of pipe joints should be done only by experienced welders; so always check the ASME Welding Code.

Q Why are schemes for pipeline identification desirable, and what methods of identification are used?
A If pipelines are clearly marked in some manner that indicates the kind of material being conveyed through the line, they can be easily traced by persons not acquainted with the piping layout, and valves that will shut off the flow through any accidental leak or break can be quickly located and closed. Newcomers to the operating staff can also get the run of the plant in less time and be less likely to make costly mistakes if some system of marking pipelines is used.

In some identification systems, the entire line is painted a distinctive color, different colors being used to represent steam, water, gas, etc. In other schemes wide color bands are painted at intervals. The colors used usually have some relation to the nature or use of the substance being carried through the line; thus, red is invariably chosen for fire-protection lines. Orange and yellow are common choices for lines carrying dangerous or poisonous materials, as they are easily distinguished even in a poor light.

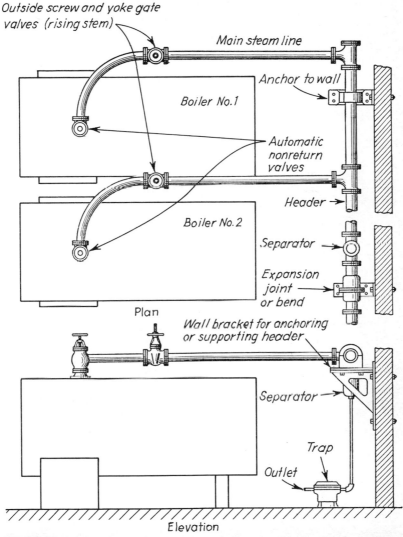

FIG. 17-25.   Plan and elevation piping layout for battery of two or more boilers.

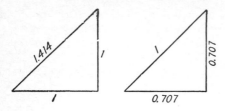

FIG. 17-26. Forty-five-degree right-angle triangle ratios for calculating proper piping lengths.

For safe materials, green or other less conspicuous colors may be used. Where pipe-identification schemes are in use, explanatory charts should be posted in places where they may be readily consulted in case of emergency.

CALCULATIONS

Q What are some common cases where exact lengths of pipe have to be calculated?
A Probably the most common cases requiring some arithmetical calculation are (1) finding length of a 45° offset, (2) finding length of a pipe bend, (3) finding length of a spiral coil, and (4) finding length of a helical coil.

Q How would you find the length of pipe required to make a 45° offset?
A This is calculated from the ratios between the lengths of the sides of a right-angled triangle. As shown graphically in Fig. 17-26, these ratios may be stated in two ways, $1:1:1.414$ and $0.707:0.707:1$.

Q What is the length of the 45° offset shown in Fig. 17-27, measured from center to center of fittings?
A Length of offset = $16 \times 1.414 = 22.62$, say $22^5/_8$ in.   *Ans.*

Q How would you find the length of a pipe bend?
A Bends are usually made in the form of an arc of a circle. A circle contains 360°. If its radius is known, the complete circumference can be found by multiplying the radius by $2\pi = 2 \times 3.14$.

   In order to calculate the length of pipe required to make a pipe bend, we must know the angle of the sector of which the bend is an arc and the

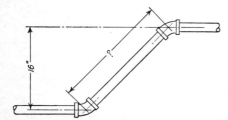

FIG. 17-27. Calculating pipe length in 45° angle with 16-in. offset.

radius to which the pipe is to be bent. Then, by simple proportion, the length of the bend is to the circumference of the circle as the angle of the sector is to 360, or, stated as a formula,

$$\text{Length of bend} = \frac{\text{circumference of circle} \times \text{angle of sector}}{360}$$

Q What length of pipe is required to make the bend shown in Fig. 17-28?

A $\text{Length} = \dfrac{\text{circumference of circle} \times \text{angle of sector}}{360} = \dfrac{2\pi R \times 60}{360}$

$$= \frac{2 \times 3.14 \times 42 \times 60}{360} = 44 \text{ in.} \qquad \textit{Ans.}$$

Q How would you find the length of a spiral coil?

A Add the inner and outer diameters together and divide by 2. This gives the mean or average diameter. Multiply the mean diameter by π to get the average circumference, and then multiply the average circumference by the number of coils to get the total length of pipe in the coil.

This rule can be used to find the lengths of rolls of material such as belting and sheet lead, without having to unroll and actually measure the length. Diameters are measured from center to center of coils or rolls.

Q What length of 1-in. pipe will be required to make a spiral containing 10 coils, outside diameter of 36 in. and inside diameter of 8 in.?

A $\text{Mean diameter} = \dfrac{36 + 8}{2} = 22 \text{ in.}$

Mean circumference $= 22 \times 3.14$
Total length $= 22 \times 3.14 \times 10 = 691 \text{ in.} = 57 \text{ ft } 7 \text{ in.}$ \qquad *Ans.*

Q How would you find the length of a helical pipe coil?

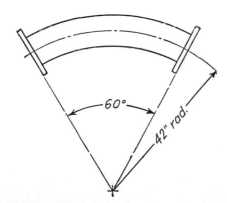

FIG. 17-28. Sixty-degree bend.

A If the coils are close together, each coil is very nearly a true circle, and the error will be very slight if we assume the length of one coil in this case to be π (3.14) times the mean diameter. If the coils are not close together, the length of one coil will be a little more than $3.14 \times D$, and we must use the right-angled triangle rule as well as the formula for circumference in order to get the true length. The right-angled triangle rule is: "The square on the hypotenuse (longest side) of a right-angled triangle is equal to the sum of the squares on the other two sides."

First find the mean circumference by multiplying mean diameter by π. Then measure the *pitch* of the spiral, that is, the distance from center to center of adjacent coils. The pitch is one leg of the right angle and the mean circumference is the other, and

$$\text{Length of one coil}^2 = \text{mean circumference}^2 + \text{pitch}^2$$

or

$$\text{Length of one coil} = \sqrt{\text{mean circumference}^2 + \text{pitch}^2}$$

Q Find the length of one turn in the helical pipe coil shown in Fig. 17-29.
A Mean circumference of coil = mean diameter $\times \pi$
$$= 12 \times 3.14 = 37.7 \text{ in.}$$
Length of one coil $= \sqrt{\text{mean circumference}^2 + \text{pitch}^2}$
$$= \sqrt{37.7^2 + 4^2} = \sqrt{1,437}$$
$$= 37.9 \text{ in., say 38 in.} \qquad Ans.$$

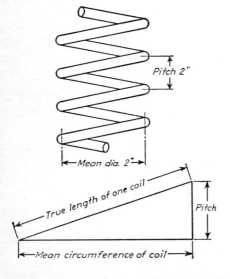

FIG. 17-29. Method of finding length of any helical pipe coil.

Q How can we find the size of a steam pipe, given the weight of steam to be carried per hour in pounds, the pressure in pounds per square inch, and the velocity of flow in feet per minute?

A If the length of the pipe is so short that pressure drop may be neglected, the required pipe area in square inches may be found by multiplying the volume of steam flowing in cubic feet per minute by 144 and dividing this by the velocity of steam flow in feet per minute. The pipe diameter can then be found by the usual formula for finding diameter from area. The volume of 1 lb of steam at the given pressure is found in the steam tables (Tables 7-3 and 7-4). Stated as a formula,

$$\text{Area of pipe in in.}^2 = \frac{\text{volume of steam flowing in ft}^3/\text{min} \times 144}{\text{velocity of steam flow in ft/min}}$$

$$\text{Diameter of pipe in in.} = \sqrt{\frac{\text{volume of steam in ft}^3/\text{min} \times 144}{\text{volume of flow in ft/min} \times 0.785}}$$

Q What size of steam pipe would be required to handle a flow of 16,000 lb of steam per hr, at a pressure of 250 psia and a velocity of 8000 ft/min?

A Area of pipe $= \dfrac{16,000 \times \text{ft}^3/\text{lb} \times 144}{60 \times 8000}$

Diameter of pipe $= \sqrt{\dfrac{16,000 \times 1.84 \times 144}{60 \times 8000 \times 0.785}} = \sqrt{11.25}$

$\qquad\qquad\qquad$ = 3.36 in., which is the internal diameter
$\qquad\qquad\qquad$ of a schedule 80, 3½-in. pipe.   *Ans.*

Q How are sizes of pump suction and discharge pipes calculated?

A Sizes of pipes will depend upon velocity of flow; 500 ft/min for the discharge and 200 ft/min for the suction are fairly common values. Then, if

D = diameter of pipe, in.
A = cross-sectional area of pipe, ft²
G = volume of water pumped per min, gal
V = velocity of flow, ft/min

$$G = A \times V \times 7.48 = \left(\frac{D}{12}\right)^2 \times 0.785 \times V \times 7.48$$

$$= \frac{D^2 \times 0.785 \times V \times 7.48}{144}$$

and

$$D = \sqrt{\frac{G \times 144}{0.785 \times V \times 7.48}} = 4.95 \times \sqrt{\frac{G}{V}}$$

Q What should be the sizes of the suction and discharge pipes of a direct-acting steam pump delivering 80 gal/min?

A Diameter of suction $= 4.95\sqrt{\dfrac{80}{200}} = 4.95 \times 0.63 = 3.12$ in.   *Ans.*

(A schedule 40, 3-in. pipe will do.)

Diameter of discharge $= 4.95\sqrt{\dfrac{80}{500}} = 4.95 \times 0.4 = 1.98$ in.   *Ans.*

(A schedule 40, 2-in. pipe will do.)

THERMAL INSULATION (LAGGING)

Q Why should bare surfaces be insulated?

A By retarding the flow of heat, thermal insulation will: (1) reduce heat gain or loss from piping, equipment, and structures, thus decreasing the amount of heating or refrigeration capacity needed; (2) control surface temperature for personnel protection and comfort; (3) prevent icing or water vapor condensation on cold surfaces; (4) make it easier to control space and process temperature. Insulation can also (1) add structural strength to floors, walls, or ceilings; (2) prevent or retard the spread of fire; (3) reduce noise level; (4) absorb vibration. It is also a good support for surface finish and helps keep out water vapor if the joints are properly sealed.

Figure 17-30 shows a cross section of an insulated metal surface, which might be a steam drum, steam piping, or any hot surface. Temperature drop through metal is negligible; thus heat loss through bare metal is very great until it strikes the insulation, where it drops sharply.

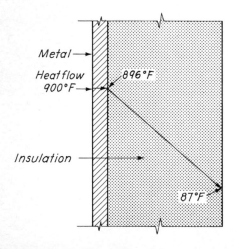

Metal →
Heat flow
900°F →
896°F
Insulation →
87°F

FIG. 17-30. Temperature drop through insulation is very sharp.

Q What are some insulating materials used for high temperatures?
A Asbestos, alumina silica ceramic fiber, fibrous potassium titanate, diatomaceous silica, mineral fiber, expanded silica (perlite), felted glass fiber, glass (cellular), and 85 percent magnesia. The maximum temperature for alumina silica ceramic fiber is 2300°F, and for 85 percent magnesia is 600°F. The others fall in between these limits.

Today reflective insulation of foil of aluminum, stainless, or Inconel is used for temperatures to 1400°F. Uses run from building insulation to nuclear reactors and space craft.

SUGGESTED READING

Elonka, Stephen M.: *Standard Plant Operators' Manual,* 3d ed., McGraw-Hill Book Company, New York, 1981.

18

FEEDWATER TREATMENT

Natural waters contain impurities, depending largely on the source. Wells and springs are classed as groundwater, rivers and lakes as surface water. Groundwater picks up impurities as it seeps through the rock strata, dissolving some part of almost everything it contacts. But the natural filtering effect of rock and sand usually keeps it free of suspended matter.

Surface water often contains organic matter such as leaf mold and insoluble matter such as sand and silt. Pollution from industrial waste and sewage is frequently present. Stream velocity and amount and location of rainfall can rapidly change the character of the water.

In general, impure water tends to cause trouble in industry by forming deposits on surfaces, thus impairing heat transfer (as from fire side to water side of boiler) or product quality. Also, it may damage metal, wood, or other construction materials used to store, transport, or otherwise contain water. Thus, the broad aims of treatment are to prevent deposits and control corrosion.

For industry, undesirable water impurities fall into these main groupings: (1) dissolved mineral and organic matter, (2) dissolved gases, such as oxygen and carbon dioxide, (3) turbidity, (4) color, (5) taste and odor, and (6) microorganisms.

IMPURE FEEDWATER

Q What are the effects of impure feedwater in boilers?

A Effects are (1) foaming and priming; (2) mud deposits on heating surfaces with consequent danger of burning or bagging plates; (3) scale deposits on heating surfaces which retard transmission of heat through metal, thus wasting heat and causing danger of plates' overheating and burning or bagging; (4) corrosion of plates and other metal surfaces.

Q   What are the common impurities in boiler feedwater?
A   Impurities in boiler feedwater may be divided into:

1. Scale-forming substances: The principal ones are the carbonates and sulfates of lime and magnesium.

2. Scum-forming substances (may be mineral or organic): (*a*) Mineral scum-forming or foaming impurities usually contain soda in the form of a carbonate, chloride, or sulfate. (*b*) Organic impurities that form scum or cause foaming are usually found in water containing sewage.

3. Sludge-forming substances: These are usually solid mineral or organic particles, carried in suspension.

4. Corrosive substances: These may be chemical compounds, chloride of magnesium, free acids, or gases such as oxygen and carbon dioxide.

Q   What are the chemical impurities commonly found in boiler feedwaters?
A   Principal impurities, with their technical names, chemical symbols, common names, and effects, are given in the table below.

Name	Symbol	Common name	Effect
Calcium carbonate	$CaCO_3$	Chalk, limestone	Soft scale
Calcium bicarbonate	$Ca(HCO_3)_2$		Soft scale
Calcium sulfate	$CaSO_4$	Gypsum, plaster of Paris	Hard scale
Magnesium carbonate	$MgCO_3$	Magnesite	Soft scale
Magnesium sulfate	$MgSO_4$	Epsom salts	Corrosion
Silicon dioxide	SiO_2	Silica	Hard scale
Calcium chloride	$CaCl_2$		Corrosion
Magnesium bicarbonate	$Mg(HCO_3)_2$		Scale, corrosion
Magnesium chloride	$MgCl_2$		Corrosion
Sodium chloride	$NaCl$	Common salt	Electrolysis
Sodium carbonate	Na_2CO_3	Washing soda or soda ash	Alkalinity
Sodium bicarbonate	$NaHCO_3$	Baking soda	Priming, foaming
Sodium hydroxide	$NaOH$	Caustic soda	Alkalinity embrittlement
Sodium sulfate	Na_2SO_4	Glauber's salts	Alkalinity

Q   What is meant by *hard* and *soft* water?
A   *Hard* water contains an excess of scale-forming impurities, while *soft* water contains little or no scale-forming substances. Hardness is mainly a

result of the presence of the carbonates of calcium and magnesium and can easily be recognized by its effect on soap. Much more soap is required to make a lather with hard water.

Q How is hardness expressed?
A In United States practice, it is expressed either in parts of calcium carbonate per million parts of water (ppm) or in grains of calcium carbonate per gallon.

Q What is the relation between ppm and grains per gallon?
A 1 ppm = 0.058 grain per U.S. gal
   1 grain per U.S. gal = 17.1 ppm

Q What are the principal scale-forming impurities in boiler feedwater, and what kind of scale does each deposit?
A Principal scale-forming impurities are the sulfates and chlorides of lime and magnesium, which form a hard scale, and the carbonates of lime and magnesium, which form a soft scale. Silica also forms a dense hard scale and has a hardening influence on the soft-scale-forming substances.

ROUTINE WATER TESTS

Q Complete water analysis requires an experienced chemist, but what routine water tests should every plant operator know how to make?
A While there are many analytical methods for each impurity, titration is one of the more commonly used. Here a standard solution from a burette or calibrated container is added to the water sample (Fig. 18-1). When the end point is reached, usually signaled by a color change in the sample, the burette solution level is noted. It's best to use an automatic burette which starts each titration at the zero mark, thus eliminating the need for subtraction to get the amount of standard solution used. This reading is proportional to the amount of substance present.

Q Describe the colorimetric method, which is widely used because it is simple.
A This test is based on developing a color in the sample proportional to the amount of substance present. Then the concentration of the substance sample is found by comparing its color with the color standards of known concentration. Some comparators have color standards sealed in glass ampules; others use standard colored-glass disks.

Q How is water hardness found?
A Methods include (1) gravimetric analysis for calcium and magnesium, (2) colorimetric titration, (3) the soap test, (4) palmitate titration, and (5) the soda reagent procedure. Methods 4 and 5 are rarely used. Gravimetric

analysis is restricted to the laboratory for precise work. Colorimetric titration and the soap test are widely used.

Q   Describe the soap test for hardness.
A   Figure 18-1 has details. Fill a burette with standard soap solution (APHA), add to 50-mL sample of water in a bottle, and shake until a permament lather forms and lasts about 5 min. Subtract 0.30-mL lather factor from total soap used; then multiply the remainder by 20 to find the hardness of the sample expressed in ppm as calcium carbonate.

Q   Describe the chloride test.
A   This calls for a burette filled with standard silver nitrate solution. Take a 50-mL water sample and neutralize it by adding 4 or 5 drops of phenolphthalein indicator plus enough sulfuric acid to make the solution colorless. Then add 5 drops of potassium chromate indicator. This turns the solution a bright yellow. Slowly add silver nitrate from the burette until one drop produces a permanent reddish color. Subtract 0.2 mL from the amount of nitrate used and multiply the remainder by 20 to find the amount of chloride present in ppm as Cl.

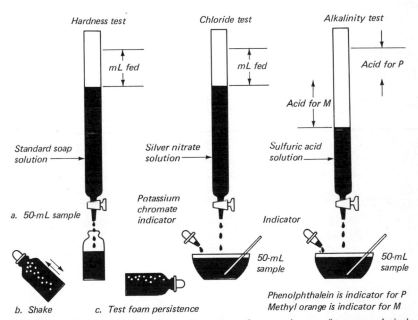

FIG. 18-1.   Titration with ordinary laboratory glassware is one of many analytical methods available for analyzing water. It's commonly used for hardness, chloride, and alkalinity tests.   (*Courtesy of* Power.)

NOTE: Two other tests you should make are alkalinity titration and hydrogen-ion concentration. They are described, along with complete instructions on all phases of water testing, in *Manual on Industrial Water*, published by ASTM, and *Standard Methods for the Examination of Water, Sewerage and Industrial Wastes*, published jointly by the American Public Health Association and the American Water Works Association.

Q Explain what is meant by *temporary* and *permanent* hardness and the treatment necessary to reduce each.

A Water containing only the carbonates of lime and magnesium is said to have *temporary* or *carbonate* hardness because these impurities can be caused to precipitate as a soft sludge if water is heated to a temperature close to boiling point at atmospheric pressure.

Water containing the sulfates of lime and magnesium is said to have *permanent* or *noncarbonate* hardness because these impurities do not precipitate as solids until the water is heated to a temperature over 300°F. Thus, where water passes directly through an open heater to the boiler, heater temperature is not high enough to precipitate the impurities causing permanent hardness, and they therefore deposit in the boiler.

PURIFYING WATER

Q What methods are commonly used to purify feedwater or neutralize the bad effects of impure feedwater?

A There are three general methods of treatment: (1) mechanical (Fig. 18-2), using filters and settling tanks; (2) chemical, using chemical treatment before or after feeding the water into the boiler; (3) thermal, using feedwater heaters.

Q What are the mechanical methods of feedwater treatment?

A Water containing solid matter in suspension may be purified (1) by chemical means alone; (2) by settling tanks or filters; or (3) by both. Settling tanks are large tanks into which the feedwater is pumped or run by gravity and allowed to stand for a time before being pumped into the boiler. This permits most of the suspended solids to settle to the bottom of the tank whence they are washed out through a drain at intervals.

Filters are tanks containing charcoal, coke, excelsior, or other finely divided filling through which the feedwater passes before it reaches the boiler-feed pumps.

Q What are the chemical methods of treating boiler feedwater?

A There are two general methods:

1. Water is pretreated in a suitable purifying apparatus by adding chemicals to neutralize any acids present and precipitate the scale-forming

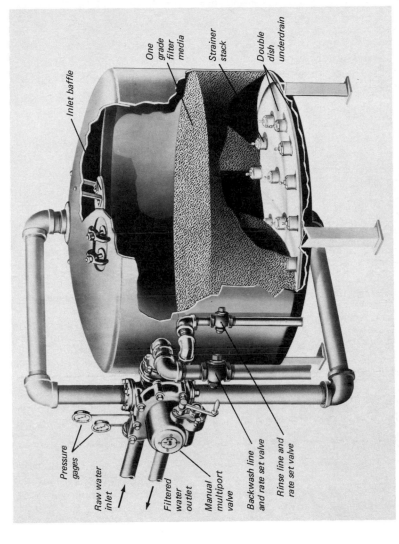

Inlet baffle

One grade filter media

Strainer stack

Double dish underdrain

Pressure gages

Raw water inlet

Filtered water outlet

Manual multiport valve

Backwash line and rate set valve

Rinse line and rate set valve

FIG. 18-2. Mechanical filters use filter media and also settle out some solids. (*Courtesy of Permutit Co.*)

impurities. The water is then filtered to remove these solids before it passes into the boiler.

2. Chemicals are simply added to the water as it is pumped into the boiler, or they are introduced separately into the boiler. Thus the solids precipitate and accumulate in the boiler itself.

Outside treatment is generally much better. With internal treatment, though the impurities are neutralized and rendered nonscaling and non-corrosive, there is a heavy concentration of chemicals and possible solids in the boiler itself.

Q What should be done before adopting any form of chemical treatment for boiler feedwater?

A Before starting to use any treatment, send samples of the feedwater to a competent analyst, who will report on the nature of the impurities and recommend the proper chemical treatment. Follow the analyst's advice. See Fig. 20-2 in Chap. 20.

Q Explain *coagulation.*

A This is partly a mechanical and partly a chemical process of water puri-fication used in connection with filtering. Much of the impurity in surface waters is in a finely divided state and takes a long time to settle, or it may even be fine enough to pass through voids in the filtering material. By add-ing certain chemicals to the water, gelatinous or jellied substances are formed which cause the small particles of mud and silt to coalesce into groups too large to pass through the filtering material.

Q What are boiler compounds?

A They are commercial preparations in powdered, solid, or liquid form, made to treat impure boiler feedwater. Carbonate of soda, caustic soda, phosphate of soda, and tannin are common ingredients of boiler com-pounds. It is possible, though unlikely, that a compound may prove quite beneficial even if no tests have been made to determine its suitability, but the blind buying of a chemical compound guaranteed to cure all boiler ills is a mistake. No single mixture of chemicals can possibly be effective in neu-tralizing the impurities in all feedwaters.

If you use a boiler compound, buy it from a reputable company that makes a specialty of analyzing feedwater and making up a compound to suit each case, or submit samples of the feedwater and the proposed com-pound to a qualified analyst, for test and advice on the compound's suitabil-ity for treating the feedwater. Further tests should be carried out at fairly close intervals, as the nature of the impurities in the feedwater may vary greatly from time to time.

Q What are the commonest chemicals used in feedwater treatment?

A Slaked lime [$Ca(OH)_2$] and soda ash ($NaCO_3$) are the commonest.

Slaked lime reduces temporary or carbonate hardness caused by the carbonates of lime and magnesium. Soda ash reduces permanent or noncarbonate hardness caused by the sulfates of lime and magnesium.

Q Describe the lime-soda treatment for boiler feedwater.
A In this process lime is added to the water to reduce the carbonate hardness. The carbonates of calcium and magnesium, which are the main impurities producing temporary hardness, are usually combined with dissolved carbonic acid (CO_2), forming bicarbonates which are soluble in water. The lime combines with the carbonic acid; so the bicarbonates are then reduced to calcium carbonate and magnesium hydroxide, which are insoluble and precipitate as a sludge.

Noncarbonate, or permanent, hardness resulting from the sulfates of lime and magnesium is reduced by adding soda ash. This decomposes the soluble sulfates into carbonates, which precipitate as solids, and soluble salts, which do not form scale in the boiler.

The water can be treated cold, but the chemical reactions are much quicker if it is heated almost to the boiling point. Use settling tanks and filters to remove the sediment.

Most feedwater treatments do not produce "pure" water. They may really increase the impurities, as chemicals are added in most cases, but the original objectionable impurities are now changed to a less harmful form.

Q Describe the hot-process feedwater-treating plant (Fig. 18-3).
A In Fig. 18-3, raw water enters the top of an open heater where its temperature is raised by direct contact with exhaust steam. This heating eliminates some of the dissolved gases carried by the water, and vents them to the atmosphere through the vent condenser. The treating chemicals are added to the water as it passes into the sedimentation tank beneath, and the sediment deposited by the combined heating and chemical action is washed out at intervals through a drain in the bottom of the tank. Treated water is drawn off through the inverted funnel and, if necessary, passed through a filter before it reaches the boiler-feed pump. The water takes 1 hr or more to pass through the system.

Q Describe the zeolite process of softening water.
A This process uses a sandlike substance called *zeolite* (natural or artificial) arranged in a filter bed. This substance has the remarkable property of *base exchange*. When hard water passes through a bed of zeolite, the calcium and magnesium in the hardness compounds pass *into* the zeolite and are replaced by sodium *from* the zeolite. Thus calcium bicarbonate becomes sodium bicarbonate, and magnesium sulfate becomes sodium sulfate. Since none of the sodium compounds are scale-forming, this exchange frees the water from hardness.

As this process continues, the sodium supply of the zeolite bed is gradually depleted until it no longer properly softens the water. The zeolite is then restored to its original condition by the action of a strong brine of common salt. This puts sodium back into the zeolite and removes calcium and magnesium.

Zeolite-treated waters show *zero hardness* by the soap test but are, of course, laden with soluble sodium salts. To avoid foaming and priming, the concentration of these salts in the boiler must be kept within limits by suitable blowdown.

Because of the danger of embrittlement under certain conditions, a high percentage of sodium carbonate is undesirable in boiler water. Thus, zeolite softeners are better suited for waters in which sulfate hardness predominates. Sometimes zeolite softening is preceded by hot-process lime treatment to remove most of the carbonate hardness.

Figure 18-4 shows a modern zeolite softener with automatic control. During the softening part of the cycle, water flows downward through the zeolite bed. When the water-softening capacity of the zeolite approaches exhaustion, the automatic valves cut off the downward flow and backwash upward to loosen the material and remove deposited dirt. The third step is admission of a measured quantity of common salt brine at the top of the bed. After a suitable interval, a stream of rinsing water removes excess salt and cleans the zeolite, which is now ready for another softening cycle.

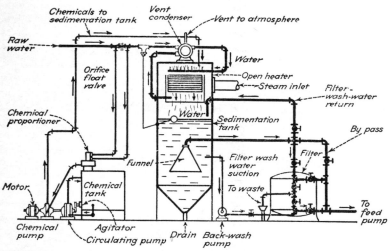

FIG. 18-3.   Hot-process feedwater treating plant mixes steam with raw water and chemicals to produce the required degree of water purity for large steam generators.

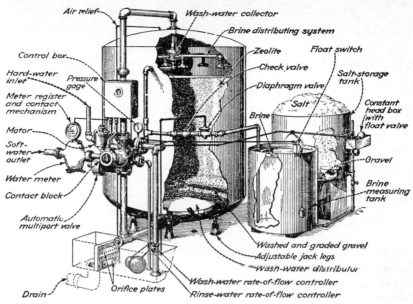

FIG. 18-4. Zeolite feedwater-treating plant is old standby in many boiler rooms.

The unit pictured makes the various valve changes automatically at the right time, as determined by the water meters.

Q Explain the zeolite softener monitoring alarm system (Fig. 18-5).

A This water-hardness monitor keeps a close check on the water coming out of the zeolite softeners. As soon as the hardness rises to the crucial

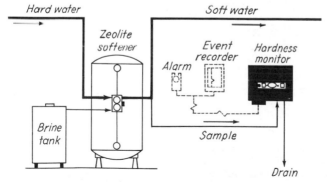

FIG. 18-5. Water-hardness alarm, event recorder, and hardness monitor for zeolite softener. (*Courtesy of Colgon Corp.*)

point, the monitor signals an alarm. The operator knows when softened water is flowing even with fluctuating hardness levels in the makeup water. It is one way to make sure that only soft water is going to the boilers.

Q   What are the thermal (heating) methods of feedwater treatment?

A   In thermal purification the water is heated before it enters the boiler to a temperature where the chemical impurities in solution will solidify and deposit in the heater. Feedwater heaters are of two main types: *open* and *closed*. Open heaters are commonly used both to heat and to purify water containing carbonate hardness. Closed heaters also reduce noncarbonate hardness, but the difficulty of cleaning the tubes of a closed heater is a strong objection to its use as a water purifier.

The open heater is usually open to the atmosphere. The essential feature of an open heater is that it contains no heat-transfer surfaces, the steam and water being in direct contact. Usually open heaters operate at atmospheric pressure, or very slightly higher. Hence it is impossible to raise the temperature of the water in the heater much above the boiling point at atmospheric pressure (212°F at sea level). In practice, the temperature is often kept a few degrees below the boiling point because of the difficulty of pumping very hot water. The boiler-feed pumps are placed between the open heater and the boiler so that they draw water from the heater and discharge it to the boilers. The pumps must be placed considerably lower than the heater so that the water will run into the pump suction by gravity. Water close to 212°F cannot be raised by pump suction. Open heaters are generally placed on a high platform in the boiler room.

A closed feedwater heater is simply a steel shell containing rows of tubes. Live or exhaust steam enters this shell and passes around the tubes (ordinarily). The feedwater circulates through the tubes and is warmed by the heat conducted through the walls of the tubes from the steam. The closed heater is usually placed between the feed pump and the boiler; therefore the water in the heater is always under boiler pressure, or higher, and its temperature thus can be raised much higher than that of the water in an open heater.

HEATERS

Q   How is water purified in an open heater?

A   Heating water almost to the boiling point causes the carbonates of lime and magnesium to solidify and deposit in the heater. These are the impurities that cause temporary, or carbonate, hardness. Water containing these impurities would deposit a soft porous scale in the boiler. Gases such as oxygen and carbon dioxide, which would cause corrosion in the boiler, are also driven off in an open heater and vented to the atmosphere.

Open heaters usually contain filtering compartments to retain the precipitated impurities, and also any other solid matter that may be in the feedwater.

Q What is the action in a closed heater?
A It is sometimes possible to raise the temperature high enough in a closed heater to cause the most troublesome impurities, the sulfates of lime and magnesium, to solidify and precipitate, as well as the carbonates of lime and magnesium which deposit at a much lower temperature. Sulfates remain in solution up to 300°F, at which temperature they solidify and settle. Because these impurities may form a hard scale on the heater tubes, it is not particularly profitable to use a closed heater as a water purifier. Water containing these impurities should be chemically treated before it passes through the heater.

Q Describe the open feedwater heater shown in Fig. 18-6.
A This is a common type of direct-contact open heater, using exhaust steam to heat the feedwater. Entering steam passes the oil baffle on the side and mingles directly with the water, which enters at the top and trickles down over the trays. The purpose of the trays is to break the water up into a great number of fine streams and thus allow it to mix more intimately with the steam. The heated water passes through a coke filter before entering the pump suction. An overflow pipe is provided to prevent flooding of the heater. A float-controlled valve holds the water level constant.

The feed pump must be placed at a lower level than the heater, as the pressure in the pump suction is always less than atmospheric. If the temperature is carried too close to the boiling point in the open heater, the

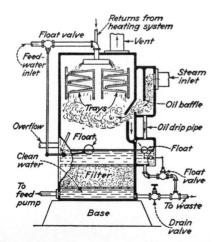

FIG. 18-6. Open feedwater heater removes oil, vents off noncondensable gases with heat from exhaust steam.

water will flash into steam in the suction line and pump cylinders, unless the heater level is many feet higher than pump suction.

Q Describe a closed feedwater heater.

A Figure 18-7 shows a common type. This heater consists of a steel shell containing a large number of small tubes through which water passes while steam circulates on the outside. The water and steam do not come into actual contact as they do in the open heater. Either exhaust or live steam can be used for heating the water, but the use of live steam for this purpose does not save any fuel.

The closed feedwater heater is usually placed between the feed pump and the boiler. Hence the water passing through the heater is under a pressure higher than the boiler pressure, and its temperature can be raised much higher than in an open heater. With this arrangement, water is heated after leaving the feed pump, so the pump does not have to handle the heated water.

Many plants use an open heater before the pump, and a closed heater (or several closed heaters in series) after the feed pump.

Q What is an *evaporator*?

A It is an apparatus for securing pure feedwater by distillation. A bent-tube evaporator, operating on much the same principle as a fire-tube boiler, but using steam instead of hot gas as the heating medium, is shown in Fig. 18-8. Steam enters at the top of the manifold, passes through the rows of bent tubes, returns to the bottom of the manifold, and drains away

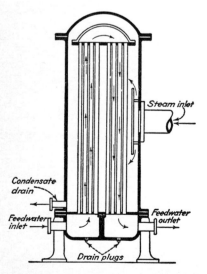

FIG. 18-7. Closed feedwater heater is two-pass unit, does not filter.

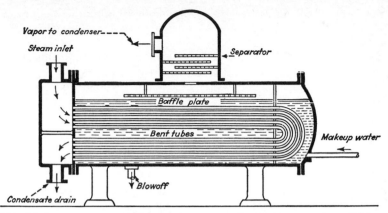

FIG. 18-8. Evaporator distills water, removes dissolved gas, retains solid impurities requiring regular removal by mechanical or chemical means.

as condensate. Water in contact with the outer surfaces of the tubes evaporates and passes out as vapor at the top of the shell, either to a closed condenser, such as a closed feedwater heater, or to an open feedwater heater, where it is condensed.

Distillation of the water in this way removes dissolved gas, but all solid impurities are left in the evaporator, which must be cleaned regularly to remove scale from the outside surface of the tubes. One method of scale removal is to crack it off quickly by varying the steam and water temperatures, causing the tubes to expand and contract. Sometimes a water softener is used ahead of the evaporator to prevent scale formation, in which case the impurities are deposited as a soft mud and can be removed by blowdown. Evaporators are commonly used in condensing power plants to purify the raw makeup water required for boiler feed. In most industrial plants the large percentage of makeup would make evaporators too large and costly.

OPERATING PROBLEMS

Q What are the cause and cure of *foaming* and *priming?*
A *Foaming* is said to occur when the steam space in a boiler is partially filled with unbroken steam bubbles. *Priming* is the carrying over of liquid water from the boiler by the steam.

Foaming is caused by impurities in the water preventing the free escape of steam as it rises to the surface, or by an oily scum on the surface of the water. This surface scum may be caused by oil, vegetable matter, or sewage in the water. The remedies are treatment and filtration of the water to re-move solid impurities, plus enough blowdown to avoid excessive concentra-tion of salts.

An exceedingly dangerous operating condition, foaming indicates a pressing need for boiler cleaning, feedwater treatment, or both. It is practi-cally impossible to tell the true water level in a badly foaming boiler, and therefore it may be difficult to prevent damage due to low water.

Priming, carrying over of water slugs with the steam, may be caused by the same conditions that cause foaming, but it may occur with perfectly pure feedwater if the steam space is too small, if the boiler is forced way above its normal capacity, or if the water level is carried too high. Priming is highly dangerous because of the possibility of suddenly lowering the water level below the danger line, and also because the water slugs carried over may do great damage to pipelines, valves, fittings, pump, and engine cylin-ders.

To avoid priming and probable water-hammer damage through sudden opening of large steam valves, all main stop valves on boilers and steam lines, and throttle valves on engines, should be opened very slowly and water in the boiler should not be carried above its normal level. In plants with superheaters any liquid water going to the superheater is there eva-porated, so troublesome deposits are formed in the superheater. Liquids carried over may damage steam turbines.

Q What are the causes of *corrosion,* and how do you prevent it?

A *Corrosion* is the wasting away of the metal surfaces of boiler parts. It may be external or internal. External corrosion is a rusting process caused by water leaks or drips on the outside surface of the boiler, or at places hid-den by insulation or brickwork. It may also occur in fire tubes and flues from the action of sulfuric acid formed from moist soot and the sulfur dioxide in combustion gases. Internal corrosion is caused by acids, oxygen (from air) or other gases in the feedwater, or by electrolytic action.

Corrosion evenly distributed over the metal surface is called *uniform* cor-rosion; it may be in the form of little holes or pits, called *pitting,* or it may occur as the eating away of a narrow groove along the edge of a riveted joint, called *grooving.* The latter is induced by the slight bending action of the metal along the riveted joints as the boiler expands and contracts. This movement keeps a narrow strip of metal surface clean and free from scale, thus offering a clean surface for corrosive elements to attack.

External corrosion may be prevented by keeping external surfaces, also tubes and flues, clean and dry. Internal corrosion may be prevented by suitable water treatment and the removal of dissolved gases, particularly oxygen.

Q What is *caustic embrittlement,* and how do you prevent it?

A So-called *caustic embrittlement* is a hardening and crystallizing process, sometimes called *intercrystalline corrosion,* which causes fine cracks in metal, particularly in riveted joints and around rivet holes. Its exact cause has not been fully determined, but it commonly occurs where feedwater contains a large amount of free bicarbonate of soda or has been treated with excessive amounts of soda ash, caustic soda, or compounds containing these chemicals. This is one of the dangers of using a boiler compound without knowing its exact composition and the feedwater's impurities.

Sodium bicarbonate and sodium carbonate *hydrolyze* under heat and pressure into sodium hydroxide (caustic soda). This caustic soda tends to concentrate in the riveted joints where its contact with the steel is believed by some authorities to liberate free hydrogen gas, which penetrates and embrittles the steel.

Results of extensive tests in laboratories and in boilers under ordinary working conditions seem to indicate that embrittlement can be prevented if all alkalies that might yield sodium hydroxide are eliminated and if sodium phosphate is used to maintain the correct degree of alkalinity in the boiler. Tannin and lignin derivatives have proved to be effective inhibitors of embrittlement over a wide range of boiler pressures.

All welded boilers offer no cracks for concentration of caustic soda and hence should be free from embrittlement trouble. The same holds true of boilers with riveted joints if they are calked inside and not outside, so that any parts that are not tightly calked can be checked by external signs of leakage through the joint.

Q For what are *deconcentrators* and *continuous blowdown* systems used?
A When there is high concentration of soluble or suspended solids in boiler feedwater, some will undoubtedly precipitate as sludge and bake into scale on tubes and plates. This condition can and should be remedied by proper treatment and filtering to remove troublesome solids before the feedwater enters the boiler. Where further filtering is desirable, some of the boiler water can be removed continuously, passed through a deconcentrator, which is simply a pressure filter, and returned to the boiler.

Strong concentrations of soluble impurities, while they may not be scale-forming, may cause foaming, priming, and even corrosion. Blowing down the boilers at intervals will reduce this concentration. Figure 18-9 shows the principle of continuous blowdown. It keeps concentration within reasonable limits and at the same time prevents loss of heat if the blowdown water is passed through a *heat exchanger* where its heat is given up to incoming feedwater. The heat exchanger is similar to a closed feedwater heater or a shell-and-tube condenser. Hot blowdown water passes through the tubes contained within the shell, and the cold water circulates around the outside of the tubes, or vice versa. A control valve placed between boiler and heat exchanger controls blowdown flow rate.

Q What is meant by the pH value of boiler feedwater?

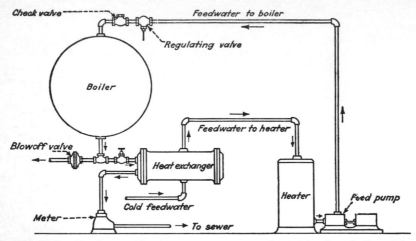

FIG. 18-9. Continuous blowdown system removes impurities from boiler while saving much of the heat by raising feedwater temperature.

A   All water contains free hydrogen (H^+ ions) and hydroxide (OH^- ions). H^+ ions give acid reactions, and OH^- ions give alkaline. In pure water that is neither acid nor alkaline these ions balance each other, and the water is neutral. If the balance is disturbed so that H^+ ions predominate, the water will be acid; if OH^- ions are in excess, it will be alkaline.

In feedwater treatment it is important to know the degree of acidity or alkalinity, and the pH system of notation has been devised for this purpose. It is a simple number scale based on the hydrogen-ion concentration. Number 7 on the pH scale is the neutral point. Below 7, acidity increases, and the lower the number the greater the acidity. Above 7, alkalinity increases, and the greater the number the greater the alkalinity.

Feedwater should be alkaline rather than acid to prevent corrosion, but too high a degree of alkalinity may induce caustic embrittlement. Correct pH value can be determined only by careful tests in each case.

A simple test for finding pH value can be made by adding a definite amount of a prepared dye, called an *indicator,* to a measured sample of water, allowing it to stand for a few minutes, then matching the color of the mixture with a set of numbered standard color tubes. Dye indicators and color standards can be procured from any chemical supply house.

REMOVING IMPURITIES

Q   What dangers may arise from oil passing into the boiler with feedwater?

A Oil will coat the heating surfaces and cause them to overheat, keeping the water from making direct contact with plates and tubes. Oil will also form a scum on the surface of the water and prevent steam bubbles from breaking away readily from the surface, thus causing foaming. Oil can be removed from the feedwater by proper filtration.

Q Why is it desirable that air and other dissolved gases be removed from feedwater before it enters the boiler?
A Some of these gases, particularly oxygen, have a very corrosive action on metal. Carbon dioxide may also produce corrosion or combine with other impurities to produce scale.

Q Describe an apparatus for removing air and other dissolved gases from boiler feedwater.
A If the water is heated close to the boiling point in an open heater, most of the dissolved gases can be separated out and vented to the atmosphere. For more complete oxygen removal, use a deaerator or deaerating heater.

An improved form of open heater which also acts as a very effective deaerator is shown in Fig. 18-10. Feedwater passes first through the tubes of a small vent condenser, then discharges into the upper part of the

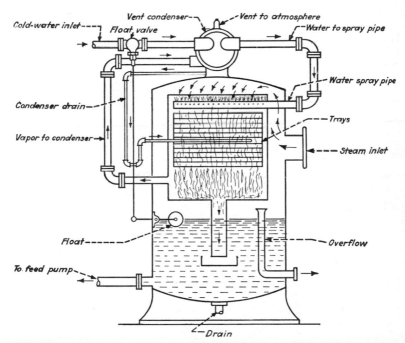

FIG. 18-10. Deaerating feedwater heater vents off noncondensable gases.

deaerator in the form of a great number of fine sprays. Entering steam mingles with these finely divided streams of water, heating the water and liberating the air and other dissolved gases. This process continues as the water trickles downward over banks of small trays, which still keep it in a finely divided state.

Most of the steam is condensed by contact with the water, but any remaining vapor passes with the separated air and gases to the vent condenser, where the steam is condensed by contact with the tubes containing the incoming water supply. The condensed steam drains back to the lower compartment of the deaerator, and the air and gases are vented to the atmosphere if the deaerator is being operated at a pressure above atmospheric pressure. If the pressure in the deaerator is below atmospheric pressure, a vacuum pump is used to draw off air and other gases from the vent condenser.

SUGGESTED READING

Elonka, Stephen M.: *Standard Plant Operators' Manual,* 3d ed., McGraw-Hill Book Company, New York, 1980.

19

OPERATION, INSPECTION, AND REPAIRS

Standing a boiler-room or engine-room watch (shift) properly, lighting off boilers, making internal inspections, cleaning heating surfaces, replacing tubes and making other repairs, laying out and pouring foundations for boilers, and maintaining refractory all make up the interesting, frustrating, or rewarding day of the operating engineer, depending on how informed he or she may be.

And how is a heating boiler protected when it is laid up for the summer season or for any long period? Is it enough just to shut off the fuel, close all the valves, and let it wait for the next start-up? Here we cover many of these vital subjects, all based on experience, to help keep energy flowing.

TAKING OVER THE SHIFT

Q What certificates (licenses) should be displayed in the power plant?
A Certificates of the engineers operating the plant, boiler and pressure vessel inspection certificates, and any others that may be required by local authorities.

Q What are the objects of the various boiler acts and regulations?
A The main objects are public safety and efficiency. The principal means of securing these are:

1. Setting up high standards for materials, design, and workmanship in the construction of steam boilers and pressure vessels to ensure strength, durability, and safety.

2. Provision for periodic inspection of boilers and pressure vessels by qualified inspectors appointed for that purpose.

3. Certification of properly qualified engineers to have charge of and operate steam power, process, and heating plants.

Q What should be your first duty on taking over a shift?
A Check the water level in the boilers or have the firemen or water tenders check it under your supervision, by blowing down the water gage glasses and the water columns if the boilers are fitted with water columns. Note if the water returns quickly to its proper level in the glass when the drain valve is closed, thus showing the passages to be free from obstruction, and check the level in each glass with the gage cocks.

Q After the water level in the boilers has been checked, what should be the next procedure?
A Check the steam pressure, then examine the logbook for anything out of the ordinary routine on the previous shift and for any instructions left for the oncoming shift. After water levels have been checked, a complete inspection should be made of the entire plant. Examine all lubricators and oil pumps to see that they are full and working properly. Feel accessible bearings for signs of excessive heating. Read all recording gages, meters, and switchboard instruments and note any deviation from normal. Ascertain the cause of any unusual noise, and remove or remedy it. Make this plant inspection immediately on taking over, as the time to find trouble left over from the previous shift is at the start of a shift and not some hours later.

Q What should be done if the feedwater supply to the boilers is suddenly shut off through injector, pump, or feed-line trouble?
A If the cause of the trouble is not immediately apparent, or is such that it will take some time to remedy it, firing must be stopped at once, before the water can reach a dangerously low level.

If gas, oil, or pulverized fuel is being used, shut off the fuel supply and reduce the draft. If boilers are hand-fired with stationary grates, smother the fire with ashes, sand, earth, or fresh coal and reduce draft. Even with dumping grates it may be preferable to smother the fire rather than dump it in the ashpit and perhaps warp the grates. If boilers are stoker-fired, increase the stoker speed and feed to run out the live fire, cover the grates with green coal, and reduce the draft.

Q What action would you take on discovering a dangerous defect in any boiler or engine under your charge?
A Shut down the boiler without delay and immediately notify the local boiler inspecting authority and your employer.

If you discover a dangerous defect in an engine under your charge, immediately shut down the engine and take steps to replace or repair the defective part.

LIGHTING OFF THE FURNACE SAFELY

Q How should stoker-fed coal-burning boilers be lit off?
A Run the stoker until the grates are partially covered with coal. Then put a quantity of light wood and some oily rags or paper on top and ignite under a light draft. As the coal catches on fire, increase feed and draft to produce the desired burning rate. Heat output is largely a matter of the amount of air supplied. Regulate coal feed to maintain a fuel bed thick enough to prevent blowing holes in the fire, yet thin enough to burn out by the time it reaches the discharge end of the stoker.

Q Explain steps needed to light off an oil burner safely.
A 1. Make sure oil is at the right temperature and pressure for the burner.

2. Make sure the damper is wide open. It should be so weighted that if holding screws let go, it will swing open.

3. Purge the furnace for a few minutes before applying the torch. If a blower is used, start it before oil is turned on.

4. Make sure that the oil burner is clean and in working order. If in doubt, insert a clean burner.

5. Insert a lighted torch. Place in the direct path of the oil; then turn the oil on.

6. Regulate the draft so the flame of the torch will not blow out. Keep the torch in front of the burner until the oil flame will maintain itself.

7. As the furnace temperature comes up, check burner and oil flame. Adjust oil flow and draft as necessary.

TWO IMPORTANT RULES: (1) Any increase in oil flow to burners *must be preceded* by increased airflow. (2) Any decrease in oil flow *must be followed by* decrease in airflow.

Q Should oil burners be fired off hot brickwork?
A *Never.* Numerous furnace explosions and personal injuries have resulted from this dangerous practice. *Always* use a torch.

Q Why, how, and how often are heating surfaces of a steaming boiler cleaned?
A On the fire side, soot and slag deposits not only reduce heat-transfer rates from fuel to water, but buildup of such deposits increases draft loss and causes smoke. As the fire side of a boiler coats with soot, the exhaust stack temperature climbs, sending heat that could not be absorbed by the

dirty tubes up into the atmosphere. Remember that $^1/_8$ in. of soot on a surface has the insulating effect of a $^5/_8$-in. layer of asbestos.

Steam or air is used to blow soot from heating surfaces (Fig. 19-1). Frequency of soot blowing depends on exhaust gas temperature rise, which tells the boiler operator when to blow tubes. When blowing, be sure to keep the boiler steaming at a reasonably high rate. This avoids the possibility of flameout or the explosion of dead pockets filled with unburned combustibles or fuel-rich gas in some portions of the boiler. Always increase draft slightly before blowing soot.

Water-washing the fire side is another method for keeping fire-side surfaces clean. This is done when the boiler is cold and out of service and the brickwork protected by tarpaulin (canvas).

Q   What precautions should you observe when firing a boiler with gas?
A   1.   See that all gas valves are tight and that there are no leaks in pipes or fittings.

2.   Perforate all dampers to ensure sufficient draft, even when dampers are closed, to carry off any gas that may enter the furnace through a leaky valve or from the accidental opening of a gas valve.

3.   Before lighting burners, make sure gas valves are tightly closed, dampers are open, and there is no trace of gas already in the furnace. Use a proper torch for lighting up.

4.   If burners blow out from any cause, allow a long enough interval to clear the furnace of gas before lighting up again. Do not attempt to light burners from hot brickwork; use a torch. Many serious accidents have occurred from neglecting these precautions.

5.   When starting to fire up, warm furnace slowly so as not to damage the setting or the boiler through excessive expansion of some parts while others remain cold.

6.   When forced or induced draft is used, start the draft fan and let it run for a short time before lighting up in order to sweep away any possible

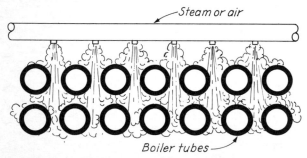

FIG. 19-1.   Stationary soot blowers used for in-line tubes.

accumulation of gas in the furnace. Keep the fan running for some time after shutting off the gas supply, to clear the furnace of unburned gas so there will be no danger of explosion.

Q What precautions should you observe when firing a boiler with oil?

A 1. See that oil pressure and temperature are correct before lighting up and, when steam-atomizing burners are used, that there is sufficient steam pressure. Open dampers.

2. If the oil spray does not ignite at once, or if the burners blow out from any cause, allow enough time to elapse to clear the furnace of oil vapor before lighting up again.

3. If for any reason unburned oil collects on the furnace floor, the gas distilled from this oil must be carried away and the furnace entirely cleared of any explosive gas-and-air mixture before attempts are made to light the burners.

4. Use strainers where necessary in oil lines and clean them regularly.

5. When atomizers are not in use, remove and thoroughly clean them.

Q What causes smoke and how is its formation prevented?

A Visible smoke is composed of particles of soot formed by the incomplete combustion of the hydrocarbon constituents of the fuel. Smoke is produced if:

1. Air supply is insufficient for complete combustion.

2. Air and combustible gases are not thoroughly mixed.

3. Furnace temperature drops below the ignition point of the gases.

4. Combustion space is too small, preventing proper mixing of air and gases, and allowing gases to come into contact with cool boiler surfaces before they are ignited.

To prevent smoke formation, avoid these conditions. The air supply must be sufficient for complete combustion. Air and gases must mix thoroughly. Keep furnace temperature above the ignition temperature of the gases. There must be ample combustion space and proper baffling, where necessary, to mix air and gases and guide them on the proper path.

FIRING NEW BOILERS

Q When putting a new boiler into service, what precautions must be observed with the setting?

A Start a light fire after the boiler has been filled with water. The setting should be allowed to dry out under this fire for several days. This gives the walls a chance to settle and lets any strains equalize. Firing up with a full load on a new setting will almost certainly crack the brickwork and damage the setting badly.

Q   What care does a boiler setting need during operation?
A   When the setting has no outer steel casing, give the outside of the brick-work a heavy coat of paint or cover with a plastic sealing preparation for this purpose, keeping a watchful eye out for cracks or breaks. Cracks should be pointed up and broken brickwork repaired at once, for cold-air leakage into the furnace lowers its efficiency considerably. Leakage of gas out through the setting into the boiler room is also objectionable. Give the same care to the inside of the setting, and repair or reline inner walls when-ever necessary.

Q   What provisions are made for preserving and extending the life of fur-nace walls in large boiler installations where combustion rates are very high?
A   Where furnace temperatures are extremely high, use special air-cooled tiles, or protect the walls by banks of tubes through which water circulates. The latter is called a *water wall*.

Q   How would you get a boiler ready for inspection?
A   Cool the boiler slowly. When it is cool, open the blowoff valve and drain the boiler. At the same time, open some small valve above the water line to permit air to enter the boiler as it drains and so prevent formation of a vac-uum. When boiler is drained, remove manhole and handhole cover plates, wash out the interior, and remove any loose scale and mud. Clean soot from outside of tubes and shell and inside of firebox. Remove all soot and ashes from ashpit and combustion chamber. Have ready all tools, clamps, and fittings that may be needed for valve or gage setting or testing. If the boiler is in a battery with others, make sure that steam, water, and blowoff valves cannot be accidentally opened while workers are in the empty boiler.

Q   How would you assist the boiler inspector during his inspection?
A   Give him all the help he requires. Point out any known defects. Station someone immediately outside the boiler when he is making the internal in-spection and make sure, if the boiler is in a battery with others, that all steam, water, and blowoff valves are locked shut or otherwise secured so that they cannot be opened accidentally. Make provision for the application of a hydrostatic test, if this is required, and in general, assist in every way to make the inspector's examination thorough and complete.

Q   How is a hydrostatic test applied?
A   Fill the boiler until water comes out of the air vent; then close all valves, gag the safety valve so that it cannot open to release the pressure, and apply and hold water pressure of $1^1/_2$ times the maximum allowable working pressure, by means of a boiler-feed pump or a hand pump if other pumps or sources of water pressure are not available. While the water pressure is on, the outside of the boiler is examined for leaks, distortion of plates, or

other defects, and light hammer blows may be applied to test the strength of any parts suspected of weakness.

INTERNAL INSPECTION

Q What defects would you look for when making an internal inspection of a steam boiler?

A After the boiler has been thoroughly cleaned, examine all plates, tubes, rivets, bolts, stays, and internal fittings for signs of corrosion, cracks, breakage, or distortion of shape. Cracked stays or rivets may be detected by tapping the parts lightly with a hand hammer, for there is a decided difference between the ringing sounds given off by a cracked part and by one that is not cracked. Corrosion is indicated by an eating away of the metal surface, and a great number of fine cracks may indicate *embrittlement.*

Q What defects would you look for when making an external inspection of a steam boiler?

A Such inspection should include the setting as well as the boiler proper, unless the boiler is self-contained. If the boiler has an external setting, examine it for cracks or other signs of deterioration in the brickwork, particularly in the refractory lining of the furnace. Carefully examine the supporting columns, bolts, and beams to see whether they are sound. The outside of the shell and all attachments to it should also be examined closely for signs of corrosion, cracks, or breaks.

Carefully check parts subjected to intense heat for signs of burning, blistering, or bagging of plates because of weakness in the metal or overheating by accumulations of oil, scale, or mud on the water side of the plate. When any part of a boiler shell or attachment to the shell is concealed by a brick or other insulating covering, remove enough of this covering to examine plates, riveted joints, and points of attachment of fittings to the shell.

Q What is your duty when you discover any defects in the boiler or boilers under your charge?

A The answer depends largely on such local conditions as nature and size of plant, nature of load, plant personnel, system of boiler inspection, regulations regarding inspection of boilers, and distance from the boiler inspection headquarters.

If the defect is obviously dangerous, the engineer has no alternative. He must shut down the damaged boiler immediately and notify his employer and the boiler inspector. The inspector should then give instructions regarding the repairs or replacements necessary to put the boiler in safe working condition, and he should inspect and approve these changes before the boiler is again put in service.

Where the boiler inspector's services are immediately available, as is usual in thickly populated centers, the engineer should always follow the above procedure and leave matters to the inspector's judgment, although a shutdown may not actually be necessary unless the inspector orders it.

In sparsely settled communities or isolated plants far from any boiler-inspecting center, the stationary engineer, like the marine engineer, may sometimes have to rely upon his own judgment as to carrying on or making minor repairs before the inspector can be notified and inspection made. This is especially true where a shutdown may result in great inconvenience and perhaps human suffering; but under no circumstances should risks be taken that might result in a boiler explosion. Where it is absolutely necessary, in these isolated cases, to carry out minor repairs, notify some competent boiler-inspecting authority at the earliest possible moment.

Q What is *corrosion?* How is it caused and prevented?
A *Corrosion* is an eating away of the surface of plates, tubes, and rivets or stays by acid impurities in the boiler feedwater. Prevent it by changing over to a purer feedwater. If this is impossible, treat the feedwater so as to neutralize the contained acids. Air dissolved in the feedwater is also a cause of corrosion, and a special apparatus called a *deaerator* is sometimes used to remove oxygen and other dissolved gas from the water before it enters the boiler. Some types of open feedwater heaters are also efficient deaerators, particularly those called *deaerating heaters.*

Zinc slabs, securely fastened to some part of the inner boiler surface, are sometimes used to reduce corrosion. The zinc instead of the boiler metal is attacked and eaten away by the galvanic action of the corrosive agents. When zinc plates are used, suspend metal trays beneath them to catch falling debris, or it may pile up on some part of the boiler heating surface, causing this surface to become overheated and burned.

Q What is *pitting?*
A This term is applied to corrosion that is not uniform over the entire metal surface but occurs only in small spots or pits and gives the metal a honeycombed appearance. Pitting is caused by lack of uniformity in the metal. It may be that the parts which are eaten away contain impurities or material more susceptible to corrosion than the rest of the metal. Or the differences may set up local galvanic couples, as when copper and zinc are in contact in an acid solution.

CLEANING HEATING SURFACES

Q How is soot mechanically removed from the inside of fire tubes?
A Soot deposits are usually removed by hand scrapers made to fit closely

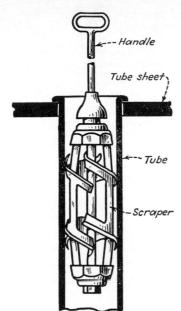

FIG. 19-2. Hand-operated scrapers for removing stubborn soot.

in the tube or by steam blowers operated by hand or built into the boiler setting. A hand scraper is shown in Fig. 19-2. In the permanently installed soot-blowing attachment for a fire-tube boiler in Fig. 19-3 the short vertical pipe carrying the blowing jets can be rotated through a semicircle by the outside handle. When the blower is not in use, the short horizontal pipe is slid back through the steamtight sleeve and the blower pipe placed horizontally in the brickwork recess away from the direct sweep of flames and hot gas.

Blowers for water-tube boilers usually consist of horizontal pipes carrying a number of small steam jets set to blow downward vertically or diago-

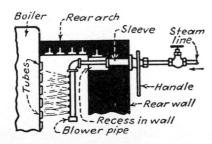

FIG. 19-3. Handy steam-jet soot blower installed in fire-tube boiler.

nally between the tubes. The blower pipes are supported on top of the tubes and can be rotated from the outside of the setting.

Q How can boiler-scale deposits be removed by mechanical means?

A Remove soft-scale deposits on plates and tubes by hand scrapers, using special long-handled tools to reach parts not readily accessible. If the scale is hard it must be chipped off with chisels and hand hammers, or with air hammers if compressed-air tools are available. Remove hard scale deposits on fire tubes by using an air- or steam-driven vibratory hammer. Water tubes are usually cleaned by compressed air, water, or steam-driven cutting tools.

Q Discuss the use of compressed air, steam, and water power for driving water-tube cleaners.

A The water-driven tube cleaner has a small water turbine of conventional design to rotate the cutter head, and the steam- and air-driven cleaners use a small rotary engine cylinder for this purpose. The form of cutting head and outside construction is similar in all three types.

The water turbine requires a large volume of water, and the high temperature of steam presents risk of injury to the operator, making its use inadvisable if other forms of power are available. Compressed air has none of the drawbacks associated with water or steam power, and its use is preferable to either, with water power as a second choice.

Figure 19-4 shows an air-driven water-tube cleaner fitted with a wheel cutter, and a cross-sectional view of the rotary motor used to drive the cutter head. This motor has only two moving parts, the shaft and a blade which fits into a slot in the shaft and can slide freely in the slot as the shaft revolves in the eccentric casing or cylinder. Compressed air enters through the inlet port, acts upon the sliding blade to cause the shaft to rotate, and then escapes through the exhaust port when it is uncovered by the sliding blade.

The cutter head shown in Fig. 19-4 has several arms carrying a number of hardened-steel cutter wheels. These arms swing outward by centrifugal force as the motor revolves the cutter head, and the sharp teeth of the cutter wheels cut away the scale. Other types of cutter head can be fitted to the

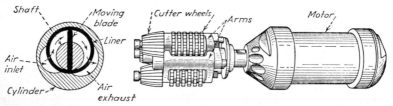

FIG. 19-4.   Air-driven tube cleaner used for scale inside of water tubes.

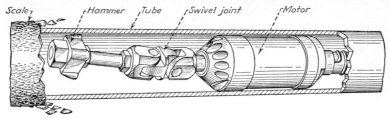

Scale — Hammer — Tube — Swivel joint — Motor

FIG. 19-5. Air-driven tube knocker for scale on fire side of tubes.

motor for such operations as opening up a passage through a tube that is partially blocked by a thick accumulation of mud or scale before using the wheel cutter to complete the cleaning process.

Q Describe a power-driven cleaner for removing scale from fire tubes.
A The use of steam to drive fire-tube cleaners is objectionable because of the risk of injury to the operator and because water is likely to damage the boiler setting; therefore, compressed air is preferable for this purpose.

Figure 19-5 shows an air-driven cleaner in the process of scaling the outside of a fire tube. The hammer is connected by a swivel joint to the same type of rotary motor as that used in the water-tube air-driven cleaner. As the hammer head rotates, it strikes light and rapid blows all around the inner circumference of the tube, causing the outer coating of scale to crack and fall off. The cleaner should be kept constantly in motion so as not to concentrate the hammering action on one particular part of the tube for any length of time, as this might damage the tube, nor should it be operated too close to the tube sheet, since it may tend to loosen the tube in the tube sheet.

Q Before internal inspections, how can the water side of hot boilers be cooled down quickly for entering 6 hr after cutout?
A Figure 19-6 shows methods of cooling down (a) fire-tube, (b) straight, and (c) bent-tube water-tube boilers. Entering hot boilers is a problem. It sometimes helps to change airflow so the heated stack draws fresh air through the boiler, but the best result isn't always obtained with most accesses open. Assume that the fire-tube boiler must be entered 6 hr after cutout, as would be required in a creamery where milk is received 7 days a week and steam is needed until all equipment is washed—usually 4 or 5 P.M. To get in, haul fire and open drafts to cool refractory; then haul ashes and wet down. While it is not good practice, pumping cold water in and blowing down may get steam off within an hour. Do this until you can hold your hand against the blowpipe. Drop in the top manhole cover and start spraying water inside the shell from the outside. Keep your head to one side to avoid the cloud of vapor from the manhole.

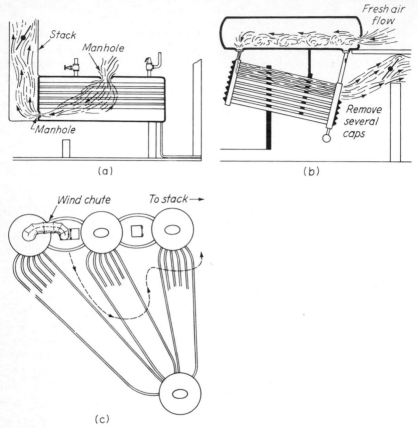

FIG. 19-6. Methods used for cooling three types of boilers before entering water side: (a) fire-tube boiler; (b) inclined water-tube type; and (c) bent-tube boiler.

Refractories lose some heat after a few hours, but the boiler shell remains hot. Continue to spray and lower water in boiler. Drop front manhole cover, shown in Fig. 19-6a, and remove it. Now close front uptake doors. Some heat will be drawn from the furnace and combustion chamber through the tubes and go up the stack. At the same time, fresh air is drawn into the top manhole and goes down and out the front manhole, picking up vapor on the way. Continue spraying. When vapor stops, spray for a few more minutes. The top of tubes under the manhole will dry quickly.

Longitudinal-drum boilers (Fig. 19-6b) must have four or five handhole plates knocked out in the rear header under the drum. Lay burlap over downcomer tubes in the rear saddle. Close ashpit, cleanout, and firing doors. Circulation will be through manholes and length of drums, down through front nipples to front headers, back through tubes, and then through uptake or smoke trunk.

Bent-tube boilers are difficult to ventilate through the drums. You may make a wind chute of canvas and hoops, fit over a manhole, and place the other end at a pass cleanout door (Fig. 19-6c). Bent-tube boilers have to be well cooled before anyone can work inside.

While it is best to allow plenty of time to cool down boilers gradually, even boiler inspectors use these expedient methods when time is short.

REMOVING, REPLACING TUBES

Q   How do you remove a defective tube in a fire-tube boiler if the tube is in the center of the tube sheet?
A   Since the tube is surrounded by other tubes, it must be pulled out through the front tube sheet. The beads are cut off both tube ends and the ends split in three or four places by means of a thin round-nosed chisel or a diamond point. Take great care not to cut or otherwise damage the tube sheet. The split tube ends are then closed in a little with a blunt-nosed tool to loosen them in the tube sheets, and the tube is driven out of the tube sheet, using a piece of pipe or bar of suitable diameter. If the tube is covered with scale, it may be necessary to use a set of blocks to pull it out through the front tube sheet. Figure 19-7 shows the method of ripping the tube end with a diamond-point chisel after the bead is cut off.

Q   How do you remove a defective tube in the bottom row of a fire-tube boiler?
A   You can avoid the laborious job of pulling the tube out through the front tube sheet by cutting the tube off just inside of both front and rear tube sheets, dropping it down, and pulling it out through the front man-hole or handhole. The ends left in the tube holes are then split and knocked out. This is the fastest and easiest way to remove tubes when a boiler is being retubed. An internal wheel cutter which resembles a tube expander in construction and operation may be used to cut off the tube

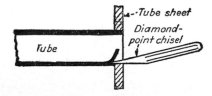

FIG. 19-7.   Ripping end of a tube.

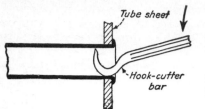

FIG. 19-8.   Using a hook-cutter bar.

ends or, if this is not available, a simple hook chisel bar. The method of using this bar is shown in Fig. 19-8.

Q   Explain the correct procedure for putting a new tube or set of tubes in a fire-tube boiler.

A   After the old tube or tubes are removed, the holes in the tube sheet are cleaned and any ragged or sharp edges rounded off slightly. The tube ends are annealed to make the metal ductile and prevent cracking when the ends are being expanded and beaded. In an hrt boiler, the holes in the front tube sheet will be an easy fit for the tubes, and the holes in the rear tube sheet will be a driving fit unless the holes have been enlarged by excessive tube expanding.

After the holes are properly prepared, the tube ends are cleaned and the tubes driven into place until the ends project about $1/4$ in. beyond the tight tube sheet. The tight ends are now expanded and beaded. Best procedure is to tighten the tube in the tube sheet with the tube expander just enough to keep it from moving. Next splay the tube end out to an angle of 45° with a ball-peen hammer (Fig. 19-9a), bead it back on the tube sheet with a beading tool (Fig. 19-9b), and finally, finish expanding the tube. If it is fully expanded before beading, it will probably be loosened up a little by the beading process and require retightening.

When all the tubes are expanded and beaded in the tight tube sheet, the opposite ends are cut, if necessary, to the proper length for beading, then

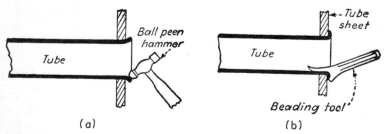

FIG. 19-9.   How to splay end (a) and bead the end (b) of boiler tube.

expanded in the same way. Sometimes beading is dispensed with if the tube ends are not directly exposed to intense heat, as in the smokebox tube sheet of a fire-tube boiler. The main object of beading is to prevent the tube end from burning off. It is the expanding that really holds the tube in the sheet.

The welding of beads to the tube sheet is also a fairly common practice. This cuts down leakage at the tube ends and avoids the necessity for frequent reexpanding which is often characteristic of this type of boiler.

Q Describe some type of tube expander for boiler tubes.

A The self-feeding roller tube expander is most commonly used for this purpose. The body of the roller tube expander, or cage, fits loosely in the tube, and the expanding rolls are pressed outward by a tapered mandrel. When expanding is done by hand, the mandrel is turned by means of a short bar inserted through a hole in the large end of the mandrel. When mechanical power is used, the end of the mandrel is shaped to fit into a socket in the driving tool. The expanding rolls are held loosely in the cage and are free to turn as the mandrel revolves. They are set at a slight angle to the longitudinal axis of the expander, and this tends to feed both the mandrel and the cage inward.

The expander in Fig. 19-10 is power-driven and has a set of flaring rolls in addition to the expanding rolls. These flaring rolls flare out the part of the tube that projects beyond the edge of the tube plate, a common practice where tube ends are not exposed to fire and beading is therefore not necessary. This expander also has a stop collar threaded on the cage extension. This collar can be adjusted to give the expander just the correct amount of inward travel to expand and flare the tube end fully. When the collar comes in contact with the tube sheet or header, the tube should be fully expanded and a few more turns is all that is needed to set the tube and complete the operation. In simpler types of tube expanders with no stop-collar adjustment, the operator has to depend upon the feel of the expander to know when expansion is complete, and this feel can be acquired only by practice. In order to do good work, the expander must be kept well lubricated and both expander and tube end must be clean and free from dirt or scale.

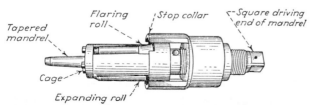

FIG. 19-10. Tube expander of the self-feeding roller type.

Q What should you do when a new tube fits very loosely in the holes in the tube sheets?

A In an old boiler where leaky tubes have perhaps been expanded over and over again, the holes in the tube sheets may become enlarged to such an extent that a new tube will be a very loose fit. Here the new tube should not be expanded out until it tightens in the tube sheet, as this stretches the tube too much and weakens it so that it will not remain tight.

Fill the space between the tube and the tube sheet with a thin metal bushing called an *outside ferrule*. Ready-made seamless copper ferrules can be bought for this purpose. If they are not easily procurable, a thin strip of copper plate, flattened to a taper at the ends and bent in a circular shape so that the tapered ends overlap, does very well. The ferrule is driven into place between tube and sides of tube hole, and the tube is then expanded and beaded in the usual way. Inside ferrules are sometimes driven into the ends of old leaky tubes as a temporary means of stopping leakage. However, they reduce the tube area and interfere with cleaning.

Q How are water tubes fastened in tube sheets and headers?

A Water tubes are secured the same way as fire tubes by expanding the ends with a tube expander, but the projecting ends are usually left untouched or simply splayed out a little without being beaded except when tubes enter the bottom of a drum. Ends of these tubes may be beaded to offer as little obstruction as possible to flow of water into the tubes, and to drain the drum completely when the boiler is emptied for cleaning, inspection, or repair.

Q How are defective water tubes removed and replaced?

A Ends are split with a narrow chisel, then bent inward with a crimping tool, which is a short bar with a narrow slot in the end that fits over the tube. Straight tubes may then be pulled out through one of the tube holes. This is not always possible with bent tubes, whose ends may have to be worked out of each tube hole in turn and the body of the tube removed sideways. If the defective tube in a bent-tube boiler is in an inner row, it may be necessary to remove several tubes to get at the damaged one. This may require replacing almost an entire row instead of only one tube, unless the spacing permits removal of a single tube.

FOUNDATIONS

Q Explain laying out a foundation for a battery of boilers.

A First outline the foundation wall on the ground with reference to some base line, such as the boiler-room wall. Use a measuring tape, preferably steel, and a large square. Corner markers, which are set well back from the

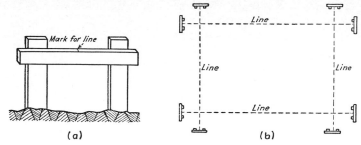

FIG. 19-11. Boiler foundation layout (be sure to leave room between walls and machinery to remove and replace boiler tubes).

excavation edge so that digging the trench won't disturb them, may be wooden stakes. However, *batter boards* (Fig. 19-11a) are better, and their tops should be leveled. Then lines are run to guide the excavation work and provide checking points for height of foundation walls (Fig. 19-11b).

Depth and width of excavation depend on the soil. In solid ground, a shallow trench may be all that is necessary to reach *hardpan*. If really solid footing means going to a great depth, and the ground is fairly firm and well drained, it may be possible to ensure a stable foundation by widening the trench and placing a concrete slab on the bottom. This footing slab, or *mat*, should be at least 18 in. thick and reinforced with steel or iron bars.

When the ground is soft to a considerable depth, and a more suitable location cannot possibly be selected, you may have to drive a large number of piles and then run a wide reinforced-concrete mat over their tops to give a still greater bearing area.

Q What are average bearing capacities, in tons per square foot, of rock, gravel, sand, and clay?
A No exact figures can be given because these materials vary greatly in composition, and their bearing capacity is strongly affected by presence or absence of moisture. Various competent authorities give these average values:

Nature of ground	Safe bearing strength, tons/ft^2
Hard rock	40
Soft rock	8
Hard dry clay	4
Wet clay	1
Cemented gravel (hardpan)	10
Gravel	6
Coarse dry sand	4
Wet sand	2

Q How do you make and set forms for a concrete foundation?
A Make them of common boards, at least $^7/_8$ in. thick, with one or both sides smooth. Nail these boards, smooth side inward, to two-by-fours, making the joints so tight that concrete can't leak through. Place forms in the trenches on the footing slab, so braced to each other and to the sides of the trench that they cannot move when the concrete is poured. The space is filled with concrete, and the top is smoothed off to provide a level footing for the brick walls.

Bottom bolts for columns are embedded in the concrete if steelwork is to support the boilers, and the whole mass is allowed to set a few days before the forms are removed and the walls built. Make sure concrete is set hard before removing the forms.

Carefully check alignment of the forms with the base line and dimensions on the plan before pouring.

GRAVEL AND SAND

Q What are proportions of sand, gravel, and cement for boiler or engine foundation?
A Strength of a concrete mixture depends on nature and proportions of ingredients and on mixing method. Proportions of cement, sand, and gravel again depend upon size and type of machinery to be supported. Less cement is needed if the aggregate (crushed rock or gravel) is of uniform size.

The strongest concrete mixture commonly used runs about $1:1^1/_2:3$, indicating proportions by volume of cement, sand, and gravel in that order. The weakest mixture runs about $1:3:6$. The first mixture would be suitable for foundations for heavy vibrating machinery, the second for light smooth-running machines. Boiler foundations carry heavy weights but are not often subjected to excessive vibration; so a medium mixture, say $1:2:4$, should usually be satisfactory.

For maximum strength, use clean sand of good quality and aggregate composed of gravel or crushed hard rock. Cinders, soft rock (like limestone), or dirty clayey sand produce a weak concrete and may ruin the foundation, causing great trouble and expense later.

Q How do you mix concrete for a boiler or engine foundation?
A If a mechanical mixer is available, the proper amounts of sand, gravel, and cement for one batch are thoroughly mixed while dry. Then only enough water is added to make the concrete sufficiently plastic to fill all spaces when it is poured and tamped into the forms. Too much water tends to float out cement and weaken concrete.

For hand mixing, dump the measured amounts of sand, gravel, and cement on a large watertight board platform and mix thoroughly while dry. Then add enough water to make the mixture into a stiff plastic for further mixing. Main essentials of good concrete are thorough dry and wet mixing, clean sand and gravel, and not too much water.

Q What proportion does the dry material bear to the mixed concrete?
A Separate dry ingredients of concrete always occupy more space than the wet mixture. Shrinkage averages 25 percent. The right amount of water in mixing is important.

Typical proportions of cement, sand, gravel (or crushed rock), and water for a 1:2:3 mixture are

6 sacks cement (approx 6 ft³)
12 ft³ sand
18 ft³ gravel or crushed rock
38 gal water (6¹⁄₃ gal per sack of cement)

Q How do you pour concrete in boiler foundation walls?
A It is common practice in a small job to load the concrete into wheelbarrows by hand shovel, or direct from the mixer if a machine is used, and to arrange plank tracks so that the concrete can be wheeled and dumped directly into the forms. As the concrete is poured, it is tamped with an iron bar to make sure no holes or cavities are left unfilled. This is particularly necessary when the concrete is mixed fairly stiff.

Do not run concrete in freezing weather if it can possibly be avoided. If it must be run in cold weather, heat the sand and gravel, use warm water in mixing, and cover the concrete with some heavy material to prevent freezing before it starts to set.

When the forms are filled to the required height, place the column foundation bolts in position, if supporting steelwork is to be used. Then smooth the top of the concrete to provide a level base for the brickwork. Allow the concrete to set properly before the forms are removed and the walls built.

Q How do you erect boilers after foundation walls are finished?
A When the concrete in the walls has set hard, the forms are removed and the excavation filled in. The ashpit and combustion chamber floor can now be laid and allowed to set. The next step depends on the types of boilers being erected. Haul hrt boilers into position over the foundation, and jack up and support on blocks about where they will finally stand. Small hrt boilers are usually supported by brackets resting on the side walls. Suspend large hrt boilers from steel beams supported on steel columns.

If the boilers are to be supported by brackets resting on the side walls, walls are built up to the finished height and the blocking removed. If steelwork is used, the supporting columns and crossbeams are placed in position and bolted to each other and to the foundation bolts. The suspending bolts can also be attached to the crossbeams and the boiler hangers; but do not remove the blocking underneath the shell until the columns are braced firmly in every direction.

Carry brickwork as high as possible before removing the braces and blocking, and take great care when removing the blocking to prevent any rolling or swinging of the shell that would throw a sudden load or side thrust on columns and braces. When the bottom blocking has been removed, carry the walls up to full height and finally discard the column bracing.

Practically all multidrum water-tube boilers are supported by steelwork erected before the drums are set in place and the setting built. Observe the same precautions in securely bracing columns and beams, so there is no possibility of movement during the assembling of boiler parts and the building of the enclosing setting.

BOILER SETTINGS, REFRACTORIES

Q Describe the solid-wall boiler setting.
A This is the commonest form of setting wall construction in older small and medium-sized boiler plants. The interior of the combustion chamber and furnace is lined with first-grade firebrick or a plastic refractory material made especially for this purpose. The outer wall is constructed of second-grade firebrick or a good quality of common red brick. The firebricks forming the furnace lining are not laid in mortar but are merely dipped into a thin mixture of fire clay and water and rubbed into place so as to leave as thin a joint as possible. Common practice is to make the firebrick lining one brick thick, with a row of headers (bricks laid endwise) every fifth or sixth row to bond the lining to the outer wall.

Bricks in the outer part of the wall are usually laid in a mortar of lime and sand, or of cement, lime, and sand in the approximate proportions 1:1:5 by volume. Thickness of the setting walls depends on size of boiler and weight (if any) to be carried by the walls.

Q How are plastic materials used in the construction of boiler settings?
A Plastic refractory materials for lining boiler settings are usually supplied by the manufacturer in the form of moist slabs packed in airtight cartons or containers to prevent drying out before use. Metal anchors of various shapes are used to tie the plastic material to the outer wall, and the lining is built by placing the plastic slabs against the furnace wall and pounding each successive row firmly into place with a heavy hand hammer or air-driven ram. When finished, the lining is trimmed off so as to present a smooth surface. Being moist and soft to begin with, the plastic material

forms a homogeneous, monolithic mass with no cracks or joints. After installation, a plastic lining should be allowed to set for a few days, and then should be baked out with a slow fire to complete the hardening process.

Plastic refractories have several advantages over solid brick or tile. There are no joints to offer an opening for furnace heat to crumble or damage the lining. Repairs can be made to damaged plastic or brick lining without having to renew the entire lining. Odd shapes such as arches and baffle walls that would require special brick or tile and would be difficult to build are easily molded in place from plastic refractories.

Refractory material of similar composition to the plastic slabs can be procured in dry form. It is mixed with water to any required consistency and used in places where the material has to be cast or poured like concrete.

Q What are super refractories?
A Super refractories are used in the fuel-bed area of furnace walls. They are an 85 percent silicon-carbide brick that has excellent abrasion resistance. Being very dense and not subject to corrosion by most coal ash, they resist clinker adhesion and erosion far better than fireclay.

For coals with ashes that fuse easily, silicon carbide works best when placed to fuel-bed depth because it resists abuse and wear. In higher-temperature regions above the fuel bed, harmful chemical conditions caused by floating ash particles containing much iron may cause slagging of silicon-carbide bricks. This doesn't happen in the cooler zone lower down.

For upper walls, the best refractories are made of mullite, which is a composition of aluminum and silicon oxides. This refractory is ideal because it forms inert crystals of mullite that prevent fluxing with ash. To ensure the correct composition, mullite is produced artificially from electric-furnace melts.

Bonded mullite and bonded silicon-carbide brick both stand temperatures well above 3000°F. Mullite has abrasion resistance much greater than firebrick. Its resistance to clinker attack exceeds that of silicon carbide a little, but firebrick by a wide margin.

Some shapes are laid, hung, wedged, or bolted in place. They should be laid in cements of their own composition with thin joints and carefully bonded into the firebrick backing. In most cases, they are installed as standard brick shapes and laid with standard bricklaying methods. When chosen and applied correctly, their life is several times that of common refractories.

WATER WALLS

Q What are water walls?
A They are rows of vertical or inclined water tubes lining the inner walls

of boiler furnaces, and connected top and bottom to the other parts of the boiler so that there is continuous rapid circulation of water through the water walls. They provide a large additional area of boiler heating surface and also protect the refractory from the intense furnace heat. Water-wall tubes may be:

1. Plain tubes set close together to present a metal wall almost solid, or spaced a few inches apart, and backed in both cases by a refractory wall. Figure 19-12 shows the latter construction.

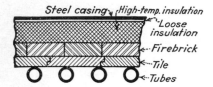

FIG. 19-12. Water-wall boiler tubes reduce heat in insulation.

2. Tubes having metal fins welded to the sides and installed so that the fins touch and thus expose a complete metal surface to the furnace heat.

3. Tubes that have a great number of short metal studs welded on their sides and that are backed by and covered with a refractory material so that only the stud ends are exposed to the heat.

In stoker-fired boilers, water-wall tubes at the sides of the grates, and for a few feet above, are often protected by cast-iron blocks bolted to the tubes.

Q What are buck stays?
A Used to strengthen and support walls of brick boiler settings, buck stays are long, narrow steel or iron castings, flat on one side with a heavy stiffening rib on the other. They are placed with the flat side against the walls on both sides of the setting and held together by long bolts passing under and across the setting. Two or more pairs of stays support the side walls. Sometimes stays are also placed against the rear walls, with bolts extending the entire length of the setting to the boiler-front plate castings. Figure 19-13 shows side stays.

Q What are the disadvantages of the solid-wall boiler setting, and what other forms of construction overcome them?
A The solid-wall setting has two serious disadvantages: the heat loss through the wall is considerable, and the extreme difference in temperature between the wall's outer and inner sides has a strong disintegrating effect on the brickwork. Some wall constructions (Fig. 19-14) designed to offset these disadvantages without sacrificing too much strength are:

1. Brick wall (Fig. 19-14*a*) having a firebrick lining and an air space between the lining and an outer red brick or second-grade firebrick wall, with long bricks or tiles tying the two walls together at intervals.

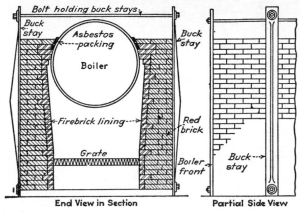

FIG. 19-13.   Solid-wall boiler setting with buck stays in place.

2. Brick wall (Fig. 19-14b) having a firebrick lining and an outer wall of red brick or second-grade firebrick, with a layer of some kind of special insulating brick between lining and outer wall.

3. An inner firebrick wall (Fig. 19-14c) carried on steel bearers, separated from an outer insulated steel casing by an air space through which air circulates constantly.

Wall (a) is much weaker than the solid wall, and the reduction in heat loss is hardly great enough to justify its use. Construction (b) cuts down heat loss and does not weaken the wall to the same extent as the dead air space. The

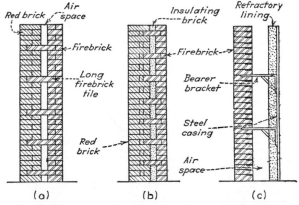

FIG. 19-14.   Three older types of furnace-wall construction.

air-cooled setting in (*c*) is a more efficient construction than either of the first two. It also serves as a preheater for combustion air.

HRT BOILER SUPPORTS

Q How are hrt boilers supported?
A Up to and including 54 in. diameter and 14 ft length, hrt boilers may be suspended from steel crossbeams resting on steel columns set outside the setting walls, or may be supported by four steel or cast-iron brackets riveted or welded to the shell and resting on the side walls of the setting. Boilers from 54 to 72 in. diameter may be supported by steel columns, beams, and hangers or by eight steel or cast-iron brackets set in pairs, four on each side and resting on the side walls. Boilers over 72 in. must be supported by the outside suspension method.

With bracket supports, front brackets rest on steel plates set on top of the brickwork, but rollers are placed between rear brackets and plates to confine movement due to expansion and contraction to the rear end of the boiler.

The bracket support is shown in Fig. 19-15 and the outside suspension method in Fig. 19-16.

Q Sketch in detail one method of attaching suspension bolts to the boiler shell.
A Figure 19-17 shows a suspension bolt attached to a steel lug welded to the shell.

Q Sketch the method of attaching suspension bolts to supporting crossbeams.
A Figure 19-18 shows a detailed sketch.

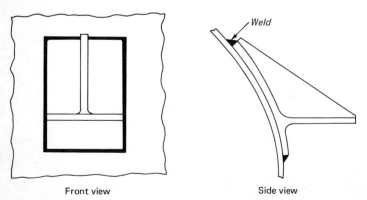

Front view   Side view
FIG. 19-15.   Supporting bracket used in older-type hrt boilers.

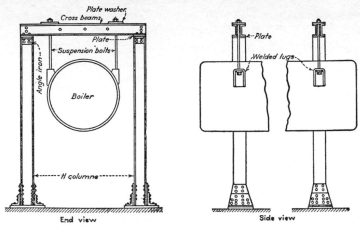

FIG. 19-16. Supporting steelwork of hrt boiler is critical because it upholds great weight of water filled drum above the fire.

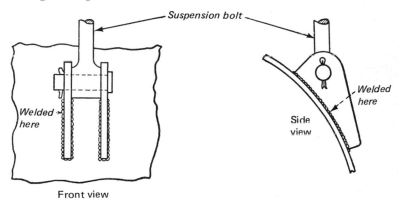

Front view

FIG. 19 17. Supporting lug for bolt is welded to the shell of boiler.

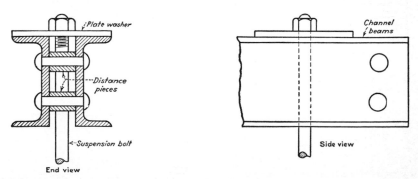

FIG. 19-18. Attaching suspension bolts to crossbeams must be done as shown here.

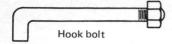

Hook bolt

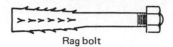

Rag bolt

Wedge bolt

FIG. 19-19.   Foundation bolt types.

Q Sketch typical foundation bolts for anchoring machinery.
A Three foundation bolts are shown in Fig. 19-19. The hook bolt is easily made but does not have the gripping power of the rag or wedge bolts. Hook and rag bolts are commonly used where the bolts can be placed in position before the concrete is run. Use the wedge bolt when the holes for foundation bolts have to be drilled in solid concrete. When a wedge bolt is placed in a drilled hole, drive it down on the wedge, taking care not to damage the bolt threads. This expands the split head so that it grips the sides of the hole.

MISCELLANEOUS REPAIRS

Q Is anyone allowed to make welding repairs on a boiler?
A *No.* Only qualified welders, and such repairs must be approved by an authorized inspector.

Q How do you remove and replace a defective stay bolt?
A Cut off the bolt heads. Drill a hole in each end slightly deeper than the thickness of the plate, using a drill that leaves only a thin shell of metal in the sheet. Close in this metal shell with a round-nosed chisel, taking care not to damage the threads in the plate. The piece of stay bolt in the boiler

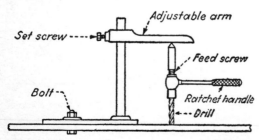

FIG. 19-20.   Setting up hand ratchet for drilling.

then drops down into the water leg, and can be fished out through a hand-hole.

If the threads have not been damaged, put in a stay bolt of the same size as the old one, but if the threads are spoiled it may be necessary to run an oversized tap through the holes and put in a larger-diameter stay bolt. In either case, the new bolt is screwed in, cut off on each end, leaving a few threads for riveting, and riveted on each end to form a head. When one end is being riveted, a heavy hammer is held against the other end. It is important that the new bolt be a tight fit in the threads. If it is loose to begin with, it will leak in a short time and the threads may possibly corrode in the plate.

A skillful welder can burn out the center of an old stay bolt with a torch in much less time than it takes to drill it out, but he must exercise great care not to burn or melt the plate. Drilling, though slower, is probably safer.

Q How is a hand ratchet set up for drilling holes?

A If there is a convenient place for fastening it to a bolt or stud, a boring post is set up, as in Fig. 19-20. The base is bolted rigidly to the job and the adjustable arm clamped at the right height for the ratchet. The drill is fed into the work by the feed screw on the opposite end of the ratchet. If a boring post cannot be set up in this fashion, place some heavy object against the ratchet to hold it, or set up and brace a plank from an adjacent wall or another boiler or machine. To apply pressure to the drill, use the same methods as for electric or air-driven drills.

NONDESTRUCTIVE TESTING

Q What is a nondestructive test?

A As opposed to bending, pulling or twisting a piece of material until it breaks, nondestructive testing checks the soundness of the material without affecting it physically or chemically. Today these tests include radiography (x-rays), dye checks, and ultrasonic and hydrostatic testing. In power boiler testing, in addition to visual examination of the weld, x-rays and gamma rays are used and required. See the suggested reading at end of this chapter for more detailed information.

SUMMER PROTECTION FOR HEATING BOILERS

Q Steam-heating boilers may deteriorate more during the summer shutdown than in an entire heating season unless properly prepared. List preventive maintenance at time of lay-up to retard corrosion and help assure trouble-free operation during the next heating season.

A Take these steps, where applicable, as soon as possible after the end of the heating season:

1. Remove all fuses from the burner circuit.
2. Remove soot and ash from the furnace, tubes, and flue surfaces.
3. Drain the boiler completely.
4. Flush the boiler to remove all sludge and loose scale particles.
5. Repair or replace leaking tubes, nipples, stay bolts, packing, and insulation.
6. Clean and overhaul all boiler appurtenances such as safety valves, gage glasses, and firing equipment. Special attention should be given to low-water cutoffs and feedwater regulators to ascertain that float (or electrode) chambers and connections are free of deposits.
7. Check the condensate return system for tightness of components.

Q What precautions should be taken with steel boilers?

A Steel boilers should be left open and dry. A sign or tag must be placed on the unit to warn that it is empty and *not* be fired. Although dry lay-up is preferred for steel boilers (see *Standard Plant Operators' Manual* for detailed sketches, showing each operation), they may have to be kept ready for operation on short notice in some plants. The water should contain a suitable corrosion-inhibiting chemical and should fill the boiler completely. To be sure, consult your boiler-water-treatment specialist.

Q What precautions do cast-iron boilers need?

A Cast-iron (c-i) sectional boilers should be filled to the top with water. A sign or tag should be attached cautioning that the boiler must be drained to normal water level before firing. Internal corrosion in c-i boilers may not be a serious problem. But if the water used is highly corrosive, follow the lay-up procedure for steel boilers.

If the c-i boiler is exposed to humid atmospheres and rapid temperature fluctuations, dry lay-up is best to prevent sweating and thus corrosion of external surfaces.

Q How should the boiler be returned to service after the summer lay-up season?

A Fill with water to normal operating level. Replace fuses. Immediately after firing the boiler, test all automatic controls including feedwater regulator, low-water fuel cutoffs and alarm, safety valves, and combustion safeguards.

> CAUTION: *Do not* leave an automatically fired boiler unattended after the initial start-up until it has gone through several firing cycles and you are sure that all controls are functioning properly.

Q How should hot-water-heating boilers be protected for the summer season?

A   Most of the steps recommended for laying up steam-heating boilers also apply to hot-water-heating boilers. An exception is that complete draining and flushing are usually *not* necessary. The recommended procedure is to drain from the bottom of the boiler while it is still hot (180 to 200°F) until the water runs clean, then refill to normal water-fill pressure. If water treatment is used in the system, sufficient treatment compound should be added to condition the added water.

SUGGESTED READING

Elonka, Stephen M.: *Standard Plant Operators' Manual,* 3d ed., McGraw-Hill Book Company, New York, 1980.

20

BOILER-ROOM MANAGEMENT

Firemen or operating engineers are not worth their salt if they are not good housekeepers. That means maintaining equipment in proper repair and keeping log sheets and other records. Some machinery insurance companies furnish log sheets for boilers, refrigeration equipment, etc. Here we show you a log sheet for low-pressure steam heating boilers, requiring weekly readings by the operator, and we also show you how to calculate the cost of steam generated per 1000 lb.

RECORDS

Q In addition to the daily or weekly logs (Fig. 20-1) kept for a boiler room, what monthly checks must be made and recorded?

A Check the safety valve by pulling the try lever to open position with the steam at full pressure. Release to allow the valve to snap closed.

Check the low-water fuel cutoff at least monthly by actually lowering the water level in the boiler slowly to simulate a developing low-water condition.

> CAUTION: Do *not* lower the water below the bottom of the water gage glass. Should the cutoff not function properly under this test, it must be immediately overhauled and placed in operating condition.

Dismantle the low-water fuel cutoff for complete overhaul at regular intervals. All the internal and external mechanism (including linkage, contacts, mercury, bulbs, floats, and wiring) should be carefully checked for defects. Always refer to manufacturer's recommendations. Check boiler and entire system carefully and completely for leakage or other defects at regular intervals at pipe connections, flanges, traps, and valves. In the interest of good housekeeping, any unsatisfactory condition should be noted and corrected.

Service oil or gas burner controls—and thoroughly check all operating and protective burner controls—at least monthly. In general, this service should be obtained from a reliable outside service organization. If serviced by the operator, a complete record of work done should be entered in the log. First, check the manufacturer's instructions and recommendations.

Clean the fire side of the boiler at least monthly. At the time of this cleaning, all brickwork and refractory should be checked and repaired as needed.

Wash out the boiler. The internal surfaces of the boiler should be checked at regular intervals to determine if scale or corrosion is present. Contact your boiler inspector if any unusual condition is noted. Length of time between washouts should be varied in accordance with conditions noted at the time of washout.

Q What other records should be kept?
A The *water analysis report* (Fig. 20-2) is an example. Whether tests are made by operator or service organization, water analysis reports must be kept along with those of major repairs, such as new tubes rolled into a boiler, new brickwork, and new boiler fittings installed.

Records of inventory should also be kept on such items as gage glasses, boiler tubes, oil burner plates, packings, and gaskets (manhole, handhole plate) so that the operator isn't "caught short" when an emergency repair must be made. Waiting for critical components to be delivered causes costly downtime.

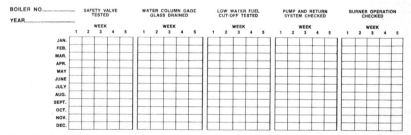

FIG. 20-1. Log sheet for low-pressure steam-heating boiler requiring weekly readings can be obtained from Hartford Steam Boiler Inspection and Insurance Company, Hartford, Conn. 06102. They also have log sheets for other power-plant machinery.

When ordering such components and materials, always give the manufacturer complete *nameplate* data, being careful to give the *serial number* accurately. Give an accurate *description* of the part needed, including *part number* and *drawing number* from which the part number is obtained. If the nameplate is lost, give the manufacturer as much information as possible, including the name of the unit and the date of purchase.

A *repair record* book should be kept to show *all* major repairs in the life of a boiler, recording each repair, by whom made, and the date. Such a report "history book" is as important as your medical history is to your doctor.

The operator in charge of the boiler room should keep a file drawer containing *blueprints* of all major equipment in the boiler room, *manufacturer's literature* of all equipment, with instruction sheets for safe operation, *fuel deliveries,* and filled-in charts from recording instruments, etc.

Q   How should records of repairs be kept?

A   Some operators keep records of repairs in full detail. For example, on a line drawing of the front of a typical sinuous-header water-tube boiler, all the tubes may be identified by giving each a number. Then a card record

WATER ANALYSIS REPORT			Collected _4/26/75_
Sample No. _605_			Analyzed _4/27/75_
For _ABC Co._			Reported _5/4/75_

	Ion		epm	ppm as $CaCO_3$
Cations	Calcium as Ca	_62_ ppm	3.10	155
	Magnesium as Mg	_31_ ppm	2.54	127
	Sodium and potassium as Na	_38_ ppm	1.64	83
	Total cations		7.28	365
Anions	Bicarbonate as HCO_3	_250_ ppm	4.10	205
	Carbonate as CO_3	_0_ ppm	0	0
	Hydroxide as OH	_0_ ppm	0	0
	Chloride as Cl	_11_ ppm	0.31	15
	Sulfate as SO_4	_138_ ppm	2.87	145
	Nitrate as NO_3	_____ ppm		
	Total anions		7.28	365

Silica as SiO_2	_5_ ppm	Total hardness	_282_ ppm $CaCO_3$
Iron as Fe_2O_3	_1.2_ ppm	Methyl orange	
Total dissolved solids	_536_ ppm	alkalinity	_205_ ppm $CaCO_3$
Suspended solids (weight)	_5_ ppm	Phenolphthalein alkalinity	_0_ ppm $CaCO_3$
Chloroform-extractable matter	_____ ppm		
Turbidity (after shaking)	_5_ ppm	pH _7.7_ Color_____	
Carbon dioxide as CO_2	_10_ ppm	Sp conductance_____ μmhos	

FIG. 20-2.   Water analysis report is important record.

can be made of anything that happens to each one. These data become very valuable when a leak occurs at the rolled joint of a tube end. It won't take long to decide whether to attempt rerolling the tube or to go ahead with its replacement. The card records will show if the tubes were installed 15 or 20 years ago. If they were, it is very likely they will not stand much rerolling.

A card file is a valuable system of recording performance and maintenance of a boiler, pump, feedwater heater, or other machinery. The cards should be filed in alphabetical order in a box of convenient size or in a metal filing cabinet. Typical equipment cards should carry all notations of value in a well-designed maintenance program.

Q What entries should be made in a daily log book?

A At the beginning of each shift the boiler operator enters (1) blew down water column, (2) blew down gage glass, (3) operated try cocks, (4) lifted safety valve, except for boilers in the 1000-psi range, (5) blew down boilers, unless on a specially controlled schedule, and (6) examined boilers generally for leaks or unusual conditions. If some unusual condition arises, or if an accident occurs, the first thing a boiler inspector wants to see, if available, is a record of routine observations of boiler operators.

SHIFT SCHEDULES

Q Design a practical shift schedule for four firemen in a typical large boiler heating-plant.

A Figure 20-3 is for a plant of four firemen that has a chief and an assistant. If one fireman is sick or off work, then either the chief or his assistant must take the shift. The schedule gives each man a complete cycle of shifts

Shift			Day of week																		
	S	M	T	W	T	F	S	S	M	T	W	T	F	S	S	M	T	W	T	F	S
First	B	B	B	B	C	C	C	C	C	C	C	A	A	A	A	A	A	A	D	D	D
Second	C	C	A	A	A	A	A	A	A	D	D	D	D	D	D	D	B	B	B	B	B
Third	D	D	D	D	D	B	B	B	B	B	B	B	B	C	C	C	C	C	C	C	A	A
First	D	D	D	D	B	B	B	B	B	B	B	C	C	C	C	C	C	C	A	A	A
Second	B	B	C	C	C	C	C	C	C	A	A	A	A	A	A	A	D	D	D	D	D
Third	A	A	A	A	A	D	D	D	D	D	D	B	B	B	B	B	B	B	C	C	C
First	A	A	A	A	D	D	D	D	D	D	D	B	B	B	B	B	B	B	C	C	C
Second	D	D	B	B	B	B	B	B	B	C	C	C	C	C	C	C	A	A	A	A	A
Third	C	C	C	C	C	A	A	A	A	A	A	A	D	D	D	D	D	D	D	B	B

FIG. 20-3. Shift schedule for four firemen during 4-week period is fair to all.

and hours in a four-week period (underlines on chart). The man who works the 12-to-8 A.M. shift on Sunday has the sixth shift for that week, which means overtime for him.

This arrangement is more popular than a schedule of 8 hours on, 24 hours off. First shift is from midnight to 8 A.M. second shift from 8 A.M. to 4 P.M., and the third shift from 4 P.M. to midnight. (For 12 additional shift schedules, see Chap. 18 in *Standard Boiler Operators' Questions and Answers.*)

CALCULATING ENERGY COSTS

Q Why is keeping tabs on the cost of energy generated important to the engineer in charge of a power plant?
A Month to month or year to year records of costs for the generation of steam, electricity, and heat for process and buildings indicate at a glance if the plant is operated efficiently or inefficiently, improving or deteriorating. Cost of energy generated also indicates in dollars and cents the worth of new improvements. It also tells management if buying energy from the public utility is cheaper than generating it.

Q Give a simple method of calculating steam costs.
A Add up dollars spent for fuel, labor, makeup water, repairs, and all expenses. Then divide the total costs by pounds of steam produced per month (or per year). Example:

Boiler-Room Costs
(For year 1981)

What Went In

Fuel (2,000,000 gal at $1.00/gal)	$2,000,000
Labor (total payroll for boiler room)	74,000
Makeup water (3,000,000 ft³ at $4/1000 ft³)	12,000
Expenses (lube oil, feedwater treatment, etc.)	9,000
Repairs (tubes, refractory, machinery components, etc.)	10,000
Total cost of steam	$2,105,000

What Came Out

Total steam generated	230,000,000 lb
Steam to boiler-room auxiliaries and waste	30,000,000 lb
Total steam delivered to engine room	200,000,000 lb

What It Cost

Steam cost (total cost $2,105,000/steam delivered 200,000) = $10.00 per 1000 lb. *Ans.*

NOTE: Meters needed for these calculations are make-up water meter, steam-flow meter, and fuel-oil meter (or dip stick).

BOILER AND MACHINERY INSURANCE

Q What is boiler and machinery insurance?
A In any power plant are certain hazards that could at some future time cause a serious accident resulting in property damage and loss of money (see Table 12-1 "Accidents Reported . . ."). The damage could put your company out of business. Rather than run this risk himself, the owner takes out insurance on his boilers or power machinery. He pays money to the insurance company to take the risk off his back. In return for these premiums the insurance company stands ready to reimburse the owner for damage losses from pressure vessels and machinery accidents.

At the same time, to prevent accidents, the insurance company provides inspectors who keep an eye on the insured machinery and the way it is operated. If not satisfied with sloppy housekeeping or operation, the insurance company may request that the operators be replaced. The agreement between owner and insurance company is called a policy. It states how much the owner pays for a certain amount of protection.

Q Who else cares about boiler safety?
A There are several organizations: (1) the American Society of Mechanical Engineers (ASME), (2) the inspectors (state, province, or city), (3) the National Board of Boiler and Pressure Vessel Inspectors, and (4) the Uniform Boiler and Pressure Vessel Laws Society. Plant operators should be familiar with the service these organizations offer all plant owners.

SUGGESTED READING

Elonka, Stephen M.: *Marmaduke Surfaceblow's Salty Technical Romances*, Krieger Publishing Company, Inc., Melbourne, Fla., 1979 (has ingenious solutions to 122 baffling plant problems, written as technical fiction, with complete technical index for ready reference in case of emergencies).

National Authorities on Boilers

American Society of Mechanical Engineers, 345 E. 47th St., New York, N.Y. 10017.

American Boiler Manufacturers Association, 1500 Wilson Blvd., Arlington, Va. 22209.

Edison Electric Institute, 750 Third Ave., New York, N.Y. 10017.

National Board of Boiler and Pressure Vessel Inspectors, 1055 Crupper Ave., Columbus, Ohio 43229.

National Fire Protection Association, 60 Batterymarch St., Boston, Mass. 02100.

Uniform Boiler and Pressure Vessel Laws Society, 57 Pratt St., Hartford, Conn. 06103.

APPENDIX

MEASUREMENTS

Length

12 inches = 1 foot
3 feet = 1 yard
5280 feet = 1 statute mile
6080 feet = 1 nautical mile
1 mil = 0.001 inch

Area

144 square inches = 1 square foot
9 square feet = 1 square yard
Cross-sectional area in circular mils = square of diameter in mils

Volume

1728 cubic inches = 1 cubic foot
27 cubic feet = 1 cubic yard

Weight

16 ounces = 1 pound
2000 pounds = 1 short ton
2240 pounds = 1 long ton

Liquid

2 pints = 1 quart
4 quarts = 1 gallon = 231 cubic inches
1 gallon (U.S.) = 0.83267 British Imperial gallon
1 gallon (British Imperial) = 1.20095 U.S. gallons
42 gallons = 1 barrel (of oil)

Circular Measure

60 seconds = 1 minute
60 minutes = 1 degree
360 degrees = 1 circle
90 degrees = 1 right angle
$11^1/_4$ degrees = 1 point on the compass

Time

60 seconds = 1 minute
60 minutes = 1 hour
24 hours = 1 day
365 days = 1 year

CONVERSION FACTORS

Atmosphere (standard) = 29.92 inches of mercury
Atmosphere (standard) = 14.7 pounds per square inch
1 horsepower = 746 watts
1 horsepower = 33,000 foot-pounds of work per minute
1 British thermal unit = 778 foot-pounds
1 cubic foot = 7.48 gallons
1 gallon = 231 cubic inches
1 cubic foot of fresh water = 62.5 pounds
1 cubic foot of salt water = 64 pounds
1 foot of head of water = 0.434 pound per square inch
1 inch of head of mercury = 0.491 pound per square inch
1 gallon of fresh water = 8.33 pounds
1 barrel (oil) = 42 gallons
1 long ton of fresh water = 36 cubic feet
1 long ton of saltwater = 35 cubic feet
1 ounce (avoirdupois) = 437.5 grains

THERM-HOUR CONVERSION FACTORS

1 therm-hour = 100,000 Btu per hour
1 brake horsepower = 2544 Btu per hour

$$1 \text{ brake horsepower} = \frac{2544}{100,000} = 0.02544 \text{ therm-hour}$$

$$1 \text{ therm-hour} = \frac{100,000}{2544} = \frac{39.3082 \text{ brake horsepower}}{(40 \text{ hp is close enough})}$$

$$1 \text{ therm-hour} = \frac{100,000}{33,475} = \frac{2.9873 \text{ boiler horsepower}}{(3 \text{ hp is close enough})}$$

EXAMPLE: How many therm-hours in a 100-hp engine?

ANSWER: $100 \times 0.02544 = 2.544$ therm-hours

BOILER HORSEPOWER

1 boiler hp = 33,475 Btu/hr
           = 34.5 lb steam per hour at 212°F
           = 139 ft² EDR (equivalent direct radiation)
1 EDR = 240 Btu/hr
1 kW = 3413 Btu/hr

WEIGHT OF WATER AT 62°F

1 in.^{3} = 0.0361 lb (of water)
1 ft^3 = 62.355 lb
1 gal = 8.3391 lb ($8^1/_3$ close enough)

METALS

	Weight per ft³; lb	Melting point, °F
Aluminum	168.5	1221
Bronze	543–554	1535–1832
Copper	550	1981.4
Iron, cast	450	2795
Mercury	885	2300
Nickel	555	2651
Steel, rolled	474–486	2714–2768
Zinc	445	787.1

COMMON SUBSTANCES

Weight, lb/ft³	
Air (60°F, 29.92 in. Hg)	0.0763
Coal, bituminous	47–56
Ice	58.7
Lignum vitae, dry	83
Oak, dry	52
Petroleum	55
Sulfur	125

SPECIFIC HEAT, VARIOUS SUBSTANCES

NOTE: Specific heat of any substance is heat (Btu) required to raise 1 lb of it 1°F

Air (constant pressure)	0.24
Dry flue gases (constant pressure)	0.24
Water vapor (atmospheric pressure)	0.48
Ice (0 to 32°F)	0.50
Steel	0.117

MISCELLANEOUS DATA

3413 British thermal units (Btu) = 1 kilowatt hour (kWh)
1000 watts = 1 kilowatt (kW)
1.341 horsepower (hp) = 1 kilowatt
2545 Btu = 1 horsepower-hour (hp-hr)
0.746 kilowatt = 1 hp
1 micrometer (formerly micron) = one millionth of a meter (unit of
length)

METRIC CONVERSION

NOTE: With imports of foreign machinery skyrocketing, the plant operator often must repair and replace components made to the metric system. Remember these three principal units: *meter,* the unit of length; *liter,* the unit of capacity; *gram,* the unit of weight.

Standard Basic Math and Applied Plant Calculations has 23 pages of metric conversions. Here are a few often used by operating engineers.

To convert	Into	Multiply by
Btu	kilocalories	0.252
bushels	liters	35.24
cubic feet	liters	28.32
cubic inches	liters	0.01639
cubic yards	liters	764.6
feet	centimeters	30.48
feet	meters	0.3048
gallons	liters	3.785
inches	centimeters	2.540
miles (statue)	meters	1609.
ounces (fluid)	liters	0.02957
pints (liquid)	liters	0.4732
pounds	kilograms	0.4536
pounds per square foot	kilograms per square meter	4.882
pounds per square inch	kilograms per square meter	703.1
quarts (liquid)	liters	0.9463
square feet	square meters	0.09290
square inches	square centimeters	6.452
square miles	square kilometers	2.590
square yards	square meters	0.8361
tons (long)	kilograms	1016.
tons (short)	tons (metric)	0.9078
watt-hours	kilowatt-hours	0.001
yards	meters	0.9144

INDEX

Absolute humidity, 80
Absolute pressure, 80, 210
Absolute temperature, 80, 153
Absolute zero, 80
Accidents, boiler, 249–251
Adamson ring, 268, 269
Addition (arithmetic), 61
Adiabatic compression, 160
Adiabatic expansion, 160
Aerosol, 149
Air:
 definition of, 85
 effects of, in feedwater, 381
 excess, 94, 131–132
  percentages of, table, 131
 overfire, 96
 preheated, 96, 202
 primary, 96
 required for combustion, 133–134
 secondary, 123
 standard, 82
 tertiary, 96
Air chamber for pumps, 304, 309–310
Air-fuel ratio, 95

Air heaters, 199, 200
 rotary regenerative, 203, 204
 tubular, 203, 204
Air lift, 319–320
Air-pressure switch, 120
Air pump, 320–321
Alarms, high- and low-water, 214–216
Alkalinity, 80
Alleviator, 310–311
Ambient temperature, 80
American National Standards Institute, 59
American Petroleum Institute (API)
 formula for gravity, 121
American Society of Mechanical Engineers (ASME), 58, 59
American Society for Testing and Materials (ASTM), 59
Analyzers:
 in-stack continuous gas, 149
 Orsat flue-gas, 124, 126–130
Annealing metals, 284, 289
Anthracite coal, 108
Arithmetic (see Mathematics)

Atmospheric pressure, measuring, 324
Atom, definition of, 83
Atomic weight, 83–85
    table, 85
Atomizing oil burners, 114–119
    combination coal-oil-gas burner, 118
    mechanical-type, 116
    steam-type, 115
Attemperators, 204
Automatic combustion control, 244–249
    basic types of, 247–248
        metering-type control, 247, 248
        on-off system, 247, 248
        positioning system, 247, 248
    programming sequence for, 249
Automatic draft control, 141–143
Automatic feedwater regulator, 237–238
Automatically fired boilers, caution for, 410

Bagasse, 120
Barometers, normal, 81
Barometric pressure, 80
Bituminous coal, 108
Blast-furnace gas, 120
Blowdown tank, 229–230
Blowoff valves, 227–229
Blowtorch effect, 121
Boiler accidents, 249–251
Boiler failures:
    causes of, 249–253
    preventing, 253
Boiler horsepower, 420
Boiler inspection (see Inspection of boilers)
Boiler inspector, assisting the, 388
Boiler records (see Records)
Boiler regulations, 58–59, 383–384
Boiler repairs (see Repairs)
Boiler-room management, 412–417
    calculating costs, 416
    insurance, 417
    record-keeping (see Records)
    safety organizations, 417
    shift schedules for firemen, 415–416

Boiler trim, 206–224
    fusible plugs, 223–224
    low-water cutoffs (see Low-water cutoffs)
    pressure gages, 206–210
    safety valves (see Safety valves)
    water alarms, 214–216
    water gages and columns, 210–214
Boilers (steam generators), 177–197
    accidents involving, 249–251
    automatically fired, caution for, 410
    cast-iron, 195–196
    classification of, 179
    coil-type, 194
    construction of, 178–179
        rules for, 58–59
        (See also Construction materials used in steam boilers)
    cooling down, 393
    dangerous defects in, 384, 389
    defective tubes, replacement of, 395–398
    definition of, 177
    efficiency of, 169
        testing, 171
    electric, 195–196
    erection of, 401
    failures: causes of, 249–253
        preventing, 253
    fire-tube, 179–186
        classification of, 179
        dry-back marine type of, 183, 186
        horizontal-return tubular (hrt), 180–181, 406–407
            weight of water in, 406
        packaged, 183–185
        pressure limitation of, 179
        three-pass firebox design, 183, 185
        vertical, advantages and disadvantages of, 183
        vertical dry-top, 181–182
        vertical wet-top, 182
    generator tubes (boiler bank), 199
    grate surface of, 178
    heating surface of (see Heating surface)
    horsepower calculation, 171

Boilers (steam generators) (*Cont.*):
  hot-water heating, circulator for,
    242–243
  new, firing, 387–389
  nozzle for openings, 275
  once-through, 194, 195
  operating problems, 377–380
  operation of, 383–389
  rating of, 169
  safety organizations for, 417
  scotch marine dry-back, 186
  scotch marine wet-back, 183
  settings for, 388, 402–403
  special types of, 192–197
  standards for, 58–59
  summer protection for, 409–411
  temperature control of, 199–205
  water-tube, 187–197
    flexible-tube, 196–197
    industrial, 190, 192, 193
    longitudinal-drum, 187–188
    multiple-coil, 194
    packaged, 191–193
    Stirling, 188–189
    vertical, 189–190
Bolts, 76
  stay, 256–258
    replacement of, 408–409
  suspension, 406, 407
Boyle's law, 159–161
Brazing, 285
Breeching, 149
British thermal unit (Btu), 92
Buck stays, 404
Bunsen burner, 89
Burners:
  functions of, 91
  gas (*see* Gas burners)
  oil (*see* Atomizing oil burners)
  pulverized-coal, 106–109
  (*See also* Automatic combustion con-
    trol)
Burning coal, oil, and gas (*see* Coal;
  Fuel oil; Fuels; Gas)
Butt joint, 259–260
  double-riveted, 259
  triple-riveted, 259
Butt welding, 262, 337

Caking coals, 121
Calculations:
  combustion, 132–134
  cost of energy, 416
  decimal places, 291–292
  draft, 144–146
  of equivalent evaporation, 170
  for gases, 160–161
  of heating surface, 290–293
  pipe, 358–362
  pumping, 325–334
  for riveted joints, 263–266
  for specific gravity, 112
  surface areas, 292–293
  for through stays, 267
  working pressure, 268–271
  (*See also* Mathematics)
Calking of joints, 261
Calorie, 80
Calorific value of fuels, 121
Calorimeters:
  separating, 174–175
  throttling, 173–174
Canadian license requirements (by
  province), 45–54
Carbon dioxide (CO_2) in flue gas, 130
  132
Carbon-steel tools, 284
Case hardening, 284
Caustic embrittlement, 378–379
Celsius degrees, 154
Centrifugal pumps (*see* Pumps, centrif-
  ugal)
Certificates for power plant, 383
Certification of engineers, 58, 383
Change of state, 80
Charles's law, 159
Chemical combination, 84
Chemical symbols, 85
Chemistry of matter, 83–85
Chimney effect, 149
Chimneys, 143–149
  draft calculations, 144–146
  linings for, 144
  maintaining a clean stack, 146–
    149
  types of, 143–144
Cleaning tubes, 390–393

Coal, 92–95
  analyses of, 95, 132
  anthracite, 91, 108
  bituminous, 92, 108
  burning: hand firing, 97–100
    mechanical stokers, 100–106
    pulverized coal, 93, 106–109
  caking, 121
  classes of, 92–93
  lignite, 108, 122
  meta-anthracite, 122–123
  pulverized, 93, 106–109
Coal bunkers, 79
Coal gas, 121
Combustible elements in fuels, 85–90,
    94
Combustion, 85–90
  automatic control of (see Automatic
    combustion control)
  complete, 133
  preheated air for, 96, 202
  products of, heating surface and, 178
  secondary, 96
  (See also Fuels)
Combustion calculations, 132–134
Compression, 76
  adiabatic, 160
  isothermal, 160
Concrete foundations, 398–400
Conduction, 157
Construction materials used in steam
    boilers, 276–285
  heat treatment of, 284–285
  physical properties of, 278–280
  strength of, 280–284
  tests for, 283
Convection, 157
Conversion factors, 419
  metric, table, 421–422
  therm-hour, 420
Cooling down boilers, 393
Corrosion, 389
  prevention of, 390
Cracking, definition of, 89

Deaerator, 381–382, 390
Desuperheaters, 204

Division (arithmetic), 63
Draft, 138–142
  automatic control of, 141–143
  balanced, 138, 141
  chimney-produced, 145, 146
  definition of, 134
  forced, 138–140, 142
  induced, 138, 140–142
  measurement of, 134
    (See also Draft gages)
  mechanical, 138, 142–143
  natural, 138, 142, 143
Draft calculations, 144–146
Draft control, automatic, 141–143
Draft fans, 140
Draft gages, 134–137
  diaphragm-operated, 136–137
  inclined-tube, 136–137
  U-tube, 134–136
Dry saturated steam, 162
Dry steam, 176
Dulong's formula, 133
Dynamic suction lift, 323

Economizers, 199–203
Efficiency of machines, examples of,
    68–70
Ejectors, 299
Elastic limit, 281
Electric boilers, 195–196
Electric welding, 287–288
Embrittlement, 389
  caustic, 378–379
Emission detector, continuous, 148
Energy cost calculations, 416
Engineering standards, 59
Engineers:
  certification of, 58, 383
  duties of, 55–56, 384–385
    on discovering defects, 389–390
  educational background of, 58
  operating experience of, 56–57
  penalties incurred by, 56
Enthalpy, 80
Equivalent direct radiation (EDR), 81
Equivalent evaporation, 165, 169,
    170

Evaporation:
  equivalent, 165, 169, 170
  factors of, 170–171
    table, 170
Evaporators, bent-tube, 376–377
Examinations, license, hints for, 55–60
Excess air, 94, 131
Expansion:
  adiabatic, 160
  isothermal, 160
  linear, coefficients of, table, 158

Failures, boiler:
  causes of, 249–253
  preventing, 253
Fans, forced-draft, 140
Feed systems, 225–227
Feedwater:
  analysis of, 366–368
  ASME code requirements for sup-
    plying, 226
  impure, effects of, 364–365
  impurities in:
    removing, 380–382
    table, 365
  oil in, 380–381
  pH value of, 379–380
  purifying, 368–374
  routine water tests, 366–368
  treatment of (see Feedwater treat-
    ment)
  zeolite process of softening water,
    371–374
Feedwater connections, 225–227
Feedwater-control systems, 237–243
Feedwater heaters, 374–377
  closed, 375, 376
  open, 374–376
Feedwater regulator, automatic, 237–
  243
Feedwater treatment, 364–382
  chemical treatment, 364, 366, 368,
    370
  coagulation, 370
  color indicator for pH, 366
  hot-process, 371
  lime-soda treatment, 371

Feedwater treatment (Cont.):
  operating problems, 377–380
    caustic embrittlement, 378–379
    corrosion, 378
    foaming, 377–378
    priming, 378
  thermal methods of, 374
  water hardness, 365–368
    monitor for, 373
    tests for, 366–368
Fire line, 178
Fire safety switch, 253–254
Fire-tube boilers (see Boilers, fire-tube)
Firebox boiler, three-pass, 183, 185
Firing new boilers, 387–389
Fishtail gas burner, 88
Flame(s), 90
  definition of, 122
  yellow and blue, 88
Flame afterburners, 122
Flame detectors, 121
Flash point, 95
Flue gas, 124–132
  analysis of, 124–125, 131
  average readings, 130–131
Fly ash, 121
Foaming in boilers, 377–378
Foot valve, 312–313
Force multipliers, 70–72
Forced-draft fans, 140
Forge welding, 285, 286
Formula for gravity, API, 121
Fouling, 150
Foundations for boilers, 398–400
  forms for, 400
  gravel and sand for, 400–401
  ground-bearing capacity of, 399
Fractions (arithmetic), 64–66
Fuel-air ratio, 95
Fuel-management system, 244–245
Fuel oil:
  burners for, 121
    (See also Atomizing oil burners)
  burning of, 109–114
  commercial grades of, 121
  composition of, 93
  fire point of, 113
  flash point of, 95, 113

Fuel oil (*Cont.*):
  formation of, original, 109
  heat value of, 93, 112
  moisture and sediment in, 114
  pour point of, 114
  properties of, 109
  sulfur content of, 114
Fuel-oil atomizer (*see* Atomizing oil
    burners)
Fuels, 83–96
  air-fuel ratio, 95
  ash-free basis, 120
  combustible elements in, 85–90, 94
  commercial types of, 90
  harmful elements in, 94
  heat value of, 91–93
    Dulong's formula for, 133
  noncombustible elements in, 94
  principal constituents of, 91
  (*See also* Coal; Fuel oil; Gases)
Fume afterburners, 122
Furnace walls, 404–405
Furnaces:
  gas-burning: air-pressure switch for,
    120
    purging protection for, 120, 385
  safely lighting off, 385–387
  slag-tap, 123
Fusible plugs, 223–224
  placement of, 224
Fusion, 81
Fusion welding, 286–287

Gage cocks, 207, 212–214
Gage pressure, 210
Gages:
  draft, 134–137
  pressure, 206–210
  steam, 207–210
    siphon for, 208–209
    testing, 209–210
  water, 206, 211–218
Gas analyzer, in-stack continuous, 149
Gas burners, 88–89, 118–120
  air-pressure switch for, 120
  combination oil and gas burner, 117
  fishtail, 88
  low-and high-pressure, 118–119

Gas firing, precautions for, 120, 386
Gases, 92, 159–161
  burning, 86, 93, 119–120
  calculations for, 160–161
  carbureted water, 122
  characteristics of, used for power
    generation, table, 87
  coke-oven, 122
  expansion or compression of, 160
  liquefied petroleum gas (LPG), 122
  natural (*see* Natural gas)
Gate valves, 231–232, 235
Gay-Lussac's law, 159–161
Generator tubes (boiler bank), 199
Generators:
  steam (*see* Boilers)
  trim for (*see* Boiler trim)
Globe valves, 234, 235
Grate bars, 99
  air space in, 101
Grate surface of boiler, 178
Gravel and sand for boiler founda-
    tions, 400–401
Gravity, specific, 112
  API formula for, 121
Grindability, 122

Hand firing of coal, 97–100
Handholes, 273–274
Heat:
  definition of, 151
  latent, 162, 176
  measuring, 156
  mechanical equivalent of, 176
  regenerative, 203, 204
  sensible, 161–162, 176
  specific (*see* Specific heat)
  total, 162
  transfer of, 157
  (*See also* Steam; Temperature)
Heat balance, 95
Heat exchanger, 379
Heat stresses, 77
Heat-transmission coefficient, 176
Heat value of fuel, 91–93, 133
  higher, 122
Heaters:
  air, 199, 200, 204

Heaters (*Cont.*):
  deaerating, 390
  feedwater, 374–377
  hot-water, 175
Heating medium, 176
Heating surface:
  calculating, 290–293
    of flue and tubes, 291
  cleaning, 390–395
  definition of, 178
Higher heat value, 122
Hogged fuel, 122
Horsepower:
  of boiler, 171
  of pump, 326–327
Hot-water heater, 175
Humidity:
  absolute, 80
  relative, 82
Hydrocarbon, burning, 86
Hydrogen, effect of, in fuel, 126
Hydrometers, 112
Hydrostatic test, 388–389

Impellers in centrifugal pumps, 313–
  314
Injectors, 294–300
  double-tube, 296–297
  single-tube, 294–296
Inspection of boilers, 388–390
  assisting the boiler inspector, 388
  examining boiler for defects, 389
  preparing boiler for inspection, 388
Insulation, thermal (lagging), 362–363
Insurance for boilers and power ma-
  chinery, 417
Internal inspection of boilers, 389–
  390
Inventory records, 413–414
Isobaric, 81
Isothermal compression, 160
Isothermal expansion, 160
Isothermic, 176

Joints:
  pipe, 342–345
  riveted (*see* Riveted joints)

Kerosene, 122
Kingsbury thrust bearing, 316

Lagging (thermal insulation), 362–363
Lap joint, 259
Lap welding, 286, 337
Latent heat, 162, 176
License examinations, hints for, 55–60
License requirements for male and fe-
  male stationary engineers:
  in Canada (by province), 45–54
  Nuclear Regulatory Commission, 1
  in the United States (by state), 3–36
  U.S. Merchant Marine, 37–44
Lignite coal, 108, 122
Linear expansion, coefficients of, table,
  158
Liquid-ash-removal system, 122
Load factor, 81
Log sheets, 413
Long-flame burner, 122
Low-water cutoffs, 216–219
  electrode probe, 217–218
  float linkage, 217, 218
  float magnet, 216, 218
  installation of, 219
  testing of, 251

Management:
  boiler-room (*see* Boiler-room man-
    agement)
  fuel-management system, 244–245
Manhole cover plates, 271–274
Manhole gaskets, 273
Manholes, 271–273
Materials of construction (*see* Construc-
    tion materials used in steam
    boilers)
Mathematics, 61–68
  addition, 61–62
  division, 63
  fractions, 64–66
  multiplication, 63–64
  powers and roots, 66–68
  subtraction, 62–63
Matter, definition of, 83
Measurements, tables, 418–419

Mechanical draft, 138–142
Mechanical equivalent of heat, 176
Mechanical mixture, 84
Mechanical stokers (*see* Stokers)
Mercury column, 81
Meta-anthracite coal, 122–123
Metals, 420
   annealing, 284, 289
   expansion of, table, 158
   (*See also* Construction materials used
     in steam boilers)
Metric conversion, table, 421–422
Micrometers, 81
MicroRingelmann chart, 147–148
Molecular weight, 84
Molecule, definition of, 84
Multiplication tables, 63–64

Natural circulation, 81
Natural draft, 134, 138–142
Ntural gas, 122, 126, 132
   heat value of, 93
Nitrogen, 81
Noncumbustible elements in fuels,
   94
Nondestructive testing, 409
Normal barometer, 81
Nozzles, boiler, 275
Nuclear Regulatory Commission li-
   cense requirements, 1

Oil (*see* Fuel oil)
Oil burners (*see* Atomizing oil burners)
Open-cup test, 113
Orsat flue-gas analyzer, 124, 126–130
Overfeed stoker, 101, 104
Overfire air, 96
Oxyacetylene welding, 287
Ozygen, 81
   in flue gas, 131
Ozone, 81

Particle size, 150
Particulate-emission detector, continuous, 148

Particulate matter, controlling, 8
Peak load, 82
Petroleum, 109
pH value of feedwater, 379–380
Physics, basic, 68–80
   efficiency, 68–70
   force multipliers, 70–72
   tank capacities, 77–79
   tension, compression, and shear,
     75–77
   torque and speed, 72–75
   water pressure, 80
Pipe calculations, 358–362
Pipe coils, length of, 360
Pipe dimensions, welded and seamless,
   table, 338
Pipe fittings, 341–342
   symbols for, 355
Pipe flanges, 342–345
Pipe joints, 342–345
Pipe materials, 336–339
Pipe sizes, table of, 338
Pipe threads, 339–341
Pipe welding, 337, 342–344
Pipelines:
   expansion of, 347–349
   identification of, symbols for, 355–
     356
   supports for, 346–347
Pipes, 336–349
   bending, 346
   bracket supports for, 346
   butt-welded, 337
   cast-iron, 337, 345
   lap-welded, 337
   wrought-iron, 337
   wrought-steel, 337
Piping layouts, 354–358
   symbols for, 356
Piston speed of pumps, table, 328
Pitting, 390
Plastic refractories, 402–403
Plenum, 81
Plug-cock blowoff valve, 229
Powers and roots (mathematics), 66–
   68
Practical suction lift, 325
Preheated air, 96, 202

Pressure:
  absolute, 80, 210
  atmospheric, measuring, 324
  barometric, 80
  gage, 210
  supercritical, 195
  working, calculating, 268–271
Pressure gages:
  draft, 134–137
  steam, 206–210
    ASME test requirements for, 209–
    210
Pressure vessels, 81, 262
Primary air, 96
Priming:
  in boilers, 377–378
  of pumps, 311–312
Pulsation, 123
Pulverized fuel, burning, 106–108
Pump capacity, estimating, 330–334
Pump governor, 308–309
Pumping calculations, 325–334
  estimating pump capacity, 330–334
  piston speeds, table, 328
  principal constants used in, 325
Pumping head, 325
Pumping theory, 322–326
  measuring atmospheric pressure,
  324
  static suction lift, 322–325
Pumps, 300–321
  air, 320–321
    liquid-piston-type, 320
  air chamber for, 304, 309–310
  alleviator for, 310–311
  centrifugal, 313–317
    characteristic curves for, 334–335
    impellers for, 313–314
    operation principle of, 313
    single-stage volute-type, 315
    starting and stopping, 317
    valves for, 316–317
  classification of, 300
  governors for, 308–309
  reciprocating, 300–308
    controls for, 306–307
    duplex direct-acting steam, 300–
    302

Pumps, reciprocating (Cont.):
    inside-packed piston, 302
    liquid-end valves for, 311–313
    outside-packed plunger, 302
    power-driven, 300
    priming of, 311–312
    setting valves of, 307–308
    simplex direct-acting steam, 300,
    303
    vertical simplex, 305, 306
    vertical triplex, 301
  rotary, 317–321
  vacuum, 320–321
Purge, 123
Purging protection for gas-burning
  furnace, 120, 385
Purifying feedwater, 368–374
Pyrometers, 155

Qualified welders, 286, 408
Quality of steam, 162
Quantity of heat, measuring,
  156

Radiation, 81, 156–157
Reaction, 82
Reciprocating pumps (see Pumps, reciprocating)
Records, 412–415
  daily log book, 415
  log sheets, 413
  records of inventory, 413–414
  repair record book, 414–415
  water analysis report, 413, 414
Refractories:
  plastic, 402–403
  super, 403
Regenerative heat, 203, 204
Regulators, automatic feedwater, 237–
  238
Relative humidity, 82
Repairs, boiler, 285–290
  code, 288–290
    defined by the National Board In-
    spection Code, 286
  miscellaneous, 408–409

Repairs, boiler (*Cont.*):
  records of, 414–415
  (*See also* Welding)
Residual fuel, 96
Retarders, 150
Retort, 123
Ringelmann chart, micro-, 147–148
Rivet holes, 261
Riveted joints, 259–268
  calculations for, 263–266
  efficiency of, 262–268
Rotary pumps, 317–321
  gear-type, 318

Safe working strength, 282
Safety, factor, of, 282
Safety appurtenances (*see* Boiler trim)
Safety organizations, 417
Safety valves, 206, 218–222
  ASME code requirements for,
    221
  blowdown adjustment on, 220
  calculations for, 222
  spring-loaded pop, 219–220
  testing of, 222, 252
Saturated steam, 162, 164
Saturated-steam tables, 163, 164
Saturation, 82
Saybolt Universal viscosimeter, 109,
    110
Secondary air, 123
Secondary combustion, 96
Sensible heat, 176
  of steam, 161–162
Separators, steam, 349–350
Shearing strength, 76
Shell openings, 271–275
Shift (watch), taking over, 383–385
Shift schedules, 415–416
Siphon for steam gage, 208–209
Slack, coal, 123
Slag-tap furnace, 123
Smoke, 146–149, 387
  formation of, 89–90
  measuring density of, 147–149
  prevention of, 91, 387
Soldering, 284–285

Soot:
  formation of, 89–90
  prevention of, 91
  removal of, 386, 390–391
Soot blowers, 386
Specific gravity, 112
  calculations for, 112
Specific heat, 81, 113, 156–159
  of common substances, 156, 421
  of superheated steam, 172
  of water, 172
Specific volume, 82
Spectroscopic analysis, 82
Sprinkler stoker, 100–103
Stacks (*see* Chimneys)
Standard air, 82
Static suction lift, 322–324
  maximum theoretical, 324, 325
Stay bolts, 256–258
  replacement of, 408–409
Stays, 255–258
  buck, 404
  diagonal, 256, 257
    ASME code for finding area of,
      268
  through, 255
    calculations for, 267
Steam, 161–173
  critical temperature and pressure of,
    165
  definition of, 161
  dry, 176
  effect of pressure changes, 164
  generation of, 151, 161
  latent heat of, 162
  moisture in, 161, 163
  properties of, 161, 163, 166–168
  quality of, 162
  saturated, 162, 164
    dry, 162
  saturated-steam tables, 163, 164
  sensible heat of, 161–162
  superheated (*see* Superheated steam)
  total heat of, 162
Steam-atomizing oil burner, 115
Steam gages, 207–210
Steam generators (*see* Boiler trim;
    Boilers)

Steam separators, 349–350
Steam traps, 350–354
Stirling water-tube boiler, 188–189
Stokers, mechanical, 100–106
  bin-fed, 101
  chain-grate, 100–104
  classes of, 100–101
  compared with hand firing, 100
  multiple-retort, 105
  overfeed, 101, 104
    front-fed, 104
  single-retort, 104, 105
  for small boilers, 101
  sprinkler, 100–103
  traveling-grate, 101, 102
  underfeed, 101, 104–106
Subtraction (arithmetic), 62–63
Suction lift, 322–325
  dynamic, 323
  practical, 325
  static, 322–324
    maximum theoretical, 324, 325
Supercritical pressure, 195
Superheated steam, 162, 164, 172–173
  degrees of, 176
  properties of, tables, 163, 166–168
  specific heat of, 172
Superheaters, 199, 200, 204, 205
Suspension bolts, 406, 407
Switches:
  air-pressure, 120
  fire safety, 253–254

Tank capacity, 77–79
Tanks:
  blowdown, 229–230
  welded seams for, 262
Temperature:
  absolute, 80, 153
  ambient, 80
  control of, 199–205
  critical, 82
  definition of, 152
  dry-bulb, 82
  measuring, 152–156
Tempering steel, 284

Tension, compression, and shear, 75–77
Tertiary air, 96
Therm-hour conversion factors, 420
Thermal conductivity, 157
  table, 158
Thermal insulation (lagging), 362–363
Thermit welding, 287
Thermometers, 153–154
  Celsius, 154
Thrust bearings, 315–316
  Kingsbury, 316
Torch, 123
Torque and speed, 72–75
Trim for boilers (see Boiler trim)
Tubes:
  cleaning, 390–393
  defective, replacement of, 395–398
  expanding of, 397
  ferrules for, 398
Turboblower, 140
Tuyeres, 123

U-tube draft gages, 134–136
Underfeed stoker, 101, 104–106
United States license requirements (by state), 3–36
U.S. Merchant Marine license requirements, 37–44

Vacuum pumps, 320–321
Valves:
  automatic nonreturn, 232–233
  automatically controlled, 236
  blowoff, 227–229
    precautions for, 229
  foot, 312–313
  gate, 231–232, 235
    inside-screw, 232
    OS&Y, 231, 232
  globe, 234, 235
  metals used for, 237
  outside-screw-and-yoke (OS&Y), 231, 232
  pump, 307–308, 316–317
  reducing, 235–236

Valves (*Cont.*):
  relief spring-loaded, 243
  safety (*see* Safety valves)
Van Stone flanges, 342
Vapor, 176
Viscosimeters, 110
Viscosity, 110–112
  Saybolt Seconds Universal and Saybolt Second Furol, 109, 110
Viscosity-temperature chart, 111

Waste fuel, 96
Waste heat, 96
Water (H_2O):
  analysis of, 366–368
  behavior of, under expansion and contraction, 159
  boiling point of, 161
  specific heat of, 172
  weight of, at 62°F, 420
  (*See also* Feedwater; Steam)
Water alarms, 214–216
Water analysis report, 413, 414
Water columns, 134, 210–213
Water gages, 206, 211–218
  bicolor, 216
  remote, 217
Water gas, 123
Water-heating system, 175, 242–243
Water line, 178
Water pressure, 80

Water treatment (*see* Feedwater treatment)
Water-tube boilers (*see* Boilers, water-tube)
Water walls, 403–406
Weight:
  atomic, 83–85
    table, 85
  of common substances, 421
  of metals, 420
  molecular, 84
  of water at 62°F, 420
Welders, qualified, 286, 408
Welding, 285–290, 408
  butt, 262, 337
  electric, 287–288
  forge, 285, 286
  fusion, 286–287
  lap, 286, 337
  oxyacetylene, 287
  preheating components for, 289
  thermit, 287
Wind box, 123
Wood, heat value of, 93
Working pressure, calculating, 268–271

X-ray test, 409

Zeolite water treatment, 371–374
Zero, absolute, 80
Zinc slabs in boiler, 171

About the authors

Stephen M. Elonka is a contributing editor to *Power* magazine (he was senior associate editor before retiring) and is the creator of the famous Marmaduke stories (technical fiction) published monthly in *Power* since 1948. He is a Licensed Chief Marine Engineer, Oceans, Unlimited Horsepower. He has worked as a machinist building diesel and airplane engines and has sailed aboard 21 merchant ships (both steam and diesel) and two U.S. Navy vessels during World War II.

A licensed regular instructor of vocational high school, New York State, as the United States entered World War II in 1942, he was assigned to the U.S. Merchant Marine Academy at King's Point, N.Y. After a year as assistant training officer, he completed the famous Steam Sea Project (study assignment for cadet midshipmen at sea) and requested sea duty "because that's where the action is." An engineering officer (Lieutenant Commander, U.S. Naval Reserve) aboard the *USS Wheatland,* at war's end, he was commissioned as Lieutenant Commander, U.S. Maritime Service, by Rear Admiral Telfair Knight, and was assigned as national engineering training officer based in Washington, D.C., for organizing the peace-time training program.

Steve Elonka has written hundreds of engineering articles and is the author and coauthor of thirteen books by ten publishers around the world, some translated into foreign languages.

Alex Higgins was a consulting engineer with many years' experience in teaching and power plant construction and operation. He was Director of Extension Courses at the Provincial Institute of Technology and Art, Calgary, Alberta, and a member of the American Society of Mechanical Engineers, the Engineering Institute of Canada, and the Alberta Association of Professional Engineers and was a Licensed Chief Engineer.